INTRODUCTION TO
MAMMALIAN REPRODUCTION

INTRODUCTION TO MAMMALIAN REPRODUCTION

edited by

Daulat Tulsiani, Ph.D.
Vanderbilt University School of Medicine
Nashville, Tennessee

KLUWER ACADEMIC PUBLISHERS
Boston / Dordrecht / London

Distributors for North, Central and South America:
Kluwer Academic Publishers
101 Philip Drive
Assinippi Park
Norwell, Massachusetts 02061 USA
Telephone (781) 871-6600
Fax (781) 681-9045
E-Mail: kluwer@wkap.com

Distributors for all other countries:
Kluwer Academic Publishers Group
Post Office Box 322
3300 AH Dordrecht, THE NETHERLANDS
Telephone 31 786 576 000
Fax 31 786 576 254
E-Mail: services@wkap.nl

Electronic Services < http://www.wkap.nl>

Library of Congress Cataloging-in-Publication Data

A C.I.P. Catalogue record for this book is available
from the Library of Congress.

Introduction to Mammalian Reproduction edited by Daulat Tulsiani
ISBN 1-4020-7283-X

Printed on acid-free paper.

Printed in the United States of America.

The Publisher offers discounts on this book for course use and bulk purchases. For further information, send email to <joanne.tracy@wkap.com>.

TABLE OF CONTENTS

Part 1. Male Gamete

Part II. Female Gamete

CONTRIBUTORS

Aida Abou-Haila, D.Sc.
UFR Biomedicale
Universite Rene Descartes Paris V
45, rue des Saints-Peres
F-75270 Paris, Cedex 6
France
aida.abou-haila@biomedicale.univ-paris5.fr

Yoshihiko Araki, M.D., D.Med.
Sci
Department of Immunology and
Parasitology
Yamagata University School of
Medicine
2-2-2 Iida-Nishi
Yamagata City 990-9585
Japan
yaraki@med.id.yamagata-u.ac.jp

Gail A. Cornwall, Ph.D.
Departments of Cell Biology &
Biochemistry
Texas Tech. University Health
Science Center
3601 4th Street
Lubbock, TX 79430
USA
gail.cornwall@ttmc.ttuhsc.edu

Benjamin J. Danzo, Ph.D.
Department of Obstetrics &
Gynecology
Vanderbilt University School of
Medicine
Room B-1100 MCN
Nashville, TN 37232-2519
USA
ben.danzo@vanderbilt.edu

M. Deng, Ph.D.
Department of Animal Science
University of Connecticut
3636 Horsebarn Road Ext U-40
Storrs, CT 06269
USA

Luis Dettin, Ph.D.
Department of Cell Biology
Georgetown University Medical
Center
SE 216 Medical-Dental Building
3900 Reservoir Road, NW
Washington, DC 20007
USA

Alan B. Diekman, Ph.D.
Ctr. for Research in Contraceptive
and Reproductive Health
Department of Cell Biology
Univ. of Virginia Health System
Box 800732
Charlottesville, VA 22908-0732
USA

Bonnie S. Dunbar, Ph.D.
Department of Cell Biology
Baylor College of Medicine
One Baylor Plaza
Houston, TX 77030
USA
bdunbar@bcm.tmc.edu

Martin Dym, Ph.D.
Department of Cell Biology
Georgetown Univ. Medical Ctr.
3900 Reservoir Road NW
Washington, DC 20007
USA
dymm@gunet.georgetown.edu

Janice P. Evans, Ph.D.
Department of Biochemistry &
Molecular Biology
Division of Reproductive Biology
Johns Hopkins University
Bloomberg School of Public
Health
615 N. Wolfe St., Room 3606
Baltimore, MD 21205
USA
jpevans@jhsph.edu

Asgerally T. Fazleabas, Ph.D.
Department of Obstetrics &
Gynecology
University of Illinois at Chicago
820 South Wood Street
Chicago, IL 60612-7313
USA
asgi@uic.edu

Chhanda Gupta, Ph.D.
Department of Pediatrics
University of Pittsburgh
Pittsburgh, PA 15213
USA
gchhanda@hotmail.com

John C. Herr, Ph.D.
Ctr. for Research in Contraceptive
and Reproductive Health
Department of Cell Biology
Univ. of Virginia Health System
Charlottesville, VA 22908-0732
USA

Barry T. Hinton, Ph.D.
Department of Cell Biology
Univ. of Virginia Health System
School of Medicine
P.O. Box 800732
Charlottesville, VA 22908-0732
USA
bth7c@virginia.edu

Gautam Kaul, Ph.D.
Senior Scientist
Department of Biochemistry
National Dairy Research Institute
Karnal 132001
Haryana, India

Firyal S. Khan-Dawood, Ph.D.
Department of Pathology
Morehouse School of Medicine
720 West View Drive SE
Atlanta, GA 30310 USA
fkhan@msm.edu

Gary Killian, Ph.D.
Dairy Breeding Research Center
Penn State University
University Park, PA 16802
USA
lwj@psu.edu

Lin Liu, Ph.D.
Department of Animal Science
University of Connecticut
3636 Horsebarn Road Ext U-40
Storrs, CT 06269
USA
lliu@canr.uconn.edu

Christoph R. Loeser, M.D.
Center for Dermatology and
Andrology
Justus-Liebig-University
Gaffkystr. 14
35385 Giessen
Germany
christoph.loeser@derma.med.uni-
giessen.de

Ben M.J. Pereira, Ph.D.
Department of Biosciences &
Biotechnology
Indian Institute of Technology
Roorkee
Roorkee-247 667,
India
benmjfbs@rurkiu.ernet.in

Ramasare Prasad, Ph.D.
Department of Biosciences &
Biotechnology
Indian Institute of Technology
Roorkee
Roorkee-247 667
India

Sarvamangala V. Prasad, Ph.D.
Department of Cell Biology
Baylor College of Medicine
One Baylor Plaza
Houston, TX 77030
USA
prasad@bcm.tmc.edu

Parul Pruthi, Ph.D.
Department of Biosciences &
Biotechnology
Indian Institute of Technology
Roorkee
Roorkee-247 667
India

Neelakanta Ravidranath, Ph.D.
Department of Cell Biology
Georgetown Univ. Medical Center
SE216 Medical-Dental Building
3900 Reservoir Road, NW
Washington, DC 20007
USA
ravindrn@gunet.georgetown.edu

Carmen M. Rodriguez, Ph.D.
Department of Cell Biology
Univ. of Virginia Health System
School of Medicine
P. O. Box 800732
Charlottesville, VA 22908-0732
USA
cmr4z@virginia.edu

Wolf-Bernhard Schill, M.D.
Center for Dermatology and
Andrology
Justus-Liebig-University
Gaffkystr. 14
35385 Giessen
Germany

Hans-Christian Schuppe, M.D.
Center for Dermatology and
Andrology
Justus-Liebig-University
Gaffkystr. 14
35385 Giessen
Germany

Thomas Stalf, Ph.D.
Institute of Reproductive
Medicine
Frankfurter Str. 52
35392 Giessen
Germany

Susan S. Suarez, M.S., Ph.D.
Department of Biomedical
Sciences
College of Veterinary Medicine
Cornell University
T5-006 Veterinary Research
Tower
Ithaca, NY 14853
USA
sss7@cornell.edu

X. C. Tian, Ph.D.
Department of Animal Science
University of Connecticut
3636 Horsebarn Road Ext U-40
Storrs, CT 06269
USA

Kiyotaka Toshimori, M.D.
Miyazaki Medical College
Kihara 5200, Kiyotake
Miyazaki, 889-1692
Japan
ktoshi@post.miyazaki-med.ac.jp

Daulat R.P. Tulsiani
Departments of Obstetrics &
Gynecology
and Cell Biology
Vanderbilt University School of
Medicine
Room D-3243 MCN
Nashville, TN 37232-2633
daulat.tulsiani@vanderbilt.edu

Srinivasan Vijayaraghvan, Ph.D.
Department of Biological Sciences
Kent State University
Kent, OH 44242
USA
svijayar@kent.edu

Pablo E. Visconti, Ph.D.
Center for Research in
Contraceptive
and Reproductive Health
Department of Cell Biology
Univ. of Virginia Health System
Charlottesville, VA 22908-0732
USA
pv6j@virginia.edu

V. Anne Westbrook, Ph.D.
Center for Research in
Contraceptive
and Reproductive Health
Department of Cell Biology
Univ. of Virginia Health System
Box 800732
Charlottesville, VA 22908-0732
USA
aw2p@virginia.edu

Xiangzhong Yang, Ph.D.
Department of Animal Science
University of Connecticut
3636 Horsebarn Road Ext U-40
Storrs, CT 06269
USA
tyang@canr.uconn.edu

[1]Email address of corresponding
authors

PREFACE

One of the goals of reproductive (gamete) biologists is to understand the biochemical processes and molecular mechanisms that regulate the formation and maturation of male and female gametes, and their ultimate union to form a zygote, a cell with somatic chromosome numbers. Development of the zygote begins immediately after sperm and egg haploid pronuclei come together, pooling their chromosomes to form a single diploid nucleus with the parental genes. The major difference between the reproductive and non-reproductive processes is that many events including interaction of the opposite gametes are species specific, and the knowledge gained in a given species may be applicable only in a few closely related species. Thus, the progress in understanding many aspects of gamete biology have been painfully slow. Despite slow advancement, many fascinating discoveries have been made. Recent successes of *in vitro* fertilization (IVF), and intracytoplasmic sperm injection (ICSI) techniques are noteworthy, and have helped many couples experience joy of parenthood. The assisted reproductive procedures are now being routinely used to increase the numbers of farm animals and endangered species. Many of these advances, in conjunction with recent successes in the cloning of laboratory and farm animals, were some of the factors behind my decision to undertake the task of organizing this book on mammalian reproduction.

So far as I know, there is no other book that systematically describes the formation and maturation of male and female gametes, and factors that regulate their union during the fertilization process, activation and implantation of fertilized egg, manipulation of the gametes for assisted reproduction, and environmental toxicants. That such book was needed became apparent to me when teaching a course on reproduction to the graduate and medical students at the Vanderbilt School of Medicine. Every attempt has been made to include a wide spectrum of topics (chapters) on morphological and physiological aspects of male and female gametes. These chapters are contributed by investigators currently engaged in "cutting-edge" research in the area of reproductive biology. Needless to say, I am very grateful to all the contributors, whose expertise, willingness to contribute, and hard work have made this book possible. My sincere hope is that the book will succeed in giving pertinent information to most of its readers, which are likely to include undergraduate, graduate and medical students, and perhaps their mentors. If my attempts generate a reasonable interest and stimulate a few young minds to expect exciting possibilities in the area of gamete biology, this book has fulfilled its purpose.

Daulat R.P. Tulsiani, Ph.D.

ACKNOWLEDGEMENTS

I was born in Village Gucherow of the District of Karachi in former British India and moved with my family to the independent India during the partition in 1947. I must acknowledge the help of many kind souls who assisted my refugee family to get settled in India. I am grateful to all my teachers in India for their continuous help and encouragement.

My first encounter in the U.S.A. came when I joined the laboratory of Professor Raul Carubelli in the Oklahoma Medical Research Foundation, Oklahoma City, Oklahoma, in 1968 as a postdoctoral fellow. I greatly benefited from the advice and encouragement of my mentor and his colleagues during my young years in Oklahoma.

I joined the research team of Professor and Chairman Oscar Touster at Vanderbilt University, Nashville, in 1972. The university provided a rich academic environment for my professional growth. It is not possible to list the names of all the colleagues, research fellows, and students who have been a continuing inspiration during my stay at Vanderbilt. However, I must acknowledge the following colleagues for their collaborations and discussions throughout my tenure in the area of reproductive biology: Drs. Marjorie D. Skudlarek, Marie-Claire Orgebin-Crist, Benjamin J. Danzo, and Michael K. Holland. I am grateful to Professor Stephen S. Entman, Chairman of the Department of Obstetrics & Gynecology, and Vanderbilt University for providing me with the space and facilities for editing this book.

My sincere thanks to the contributors who graciously sent their assigned chapters in a timely manner. Many chapters needed very little editing; however, there were some that needed extensive editing and formatting. I learned more by editing these chapters than from any other text book or research article.

I am deeply indebted to Loreita Little and Lynne Black for editorial assistance, and to Lynne Black for final preparation of the chapters for the camera-ready format. Without this assistance, the publication of this book would have been considerably delayed. Finally, I am grateful to Joanne Tracy, Editor of Biosciences, at the Kluwer Academic Publishers, for her faith in this project. The research in my laboratory is supported in part by research grants HD25869 and HD34041 from the National Institute of Child Health and Human Development.

Chapter 1

MAMMALIAN TESTES: STRUCTURE AND FUNCTION

Neelakanta Ravindranath, Luis Dettin, and Martin Dym
Georgetown University School of Medicine, Washington, DC, USA

INTRODUCTION

The male reproductive system consists of the primary sex organs, the two testes and a set of accessory sexual structures. The adult mammalian testis performs two important functions, spermatogenesis and male sex hormone production. It is an organ structurally designed to produce the haploid male gametes from diploid postnatal germ-line stem cells, i.e. type A spermatogonia. The process of morphological and functional differentiation of type A spermatogonia into the haploid male gamete, the spermatozoon, is termed spermatogenesis. In addition, the testis elaborates a steroid hormone, testosterone, that is responsible for maintaining the spermatogenic process as well as the secondary male sexual characteristics. Furthermore, testosterone is important for several different functions in various organ systems including the maintainance of muscle mass and bone density. The process of testosterone formation from its precursor, cholesterol, is termed steroidogenesis. In this chapter, we will discuss how the structure and form of the testis contributes to the processes of spermatogenesis and steroidogenesis.

MORPHOLOGY OF THE ADULT TESTIS

Each testis is covered with a thick fibrous capsule, the tunica albuginea. The thick infolding of the tunica albuginea at the posterior margin of the testis forms the mediastinum of the testis. Connective tissue septae originate from the mediastinum and pass into the interior of the testis, and subdivide it into several lobules. Within these lobules lie the convoluted folds of the seminiferous tubule. The space surrounding the seminiferous folds is occupied by the interstitial tissue. The seminiferous tubules form coiled loops that terminate at both ends into the rete testes located within the mediastinum. Spermatozoa and testicular fluid produced within the seminiferous tubule pass through the rete testes into the ductuli efferentes and epididymis.

Histologically, the adult testis can be divided into two compartments, a seminiferous tubular compartment and an interstitial compartment (Fig. 1). The tubular compartment consists of an outer layer (s) of peritubular myoid cells and an inner layer of seminiferous epithelium separated by an intermediate layer of acellular matrix or basement membrane. The interstititial compartment consists of Leydig cells, immune cells (macrophages and lymphocytes), and fibroblasts. In addition, it also contains blood and lymph vessels, nerves, and loose connective tissue. The tubular and interstitial compartments of the testis perform the defined functions of spermatogenesis and steroidogenesis, respectively.

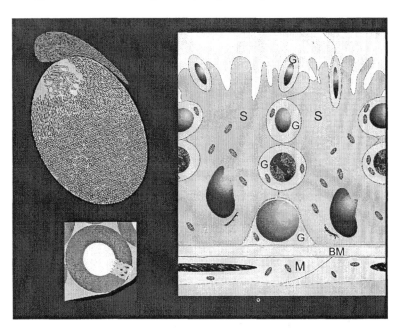

Figure 1. Schematic representation of a mammalian testis (top left), a cross section of a seminiferous tubule (bottom left), and the seminiferous epithelium (right) showing myoid cells (M), the basement membrane (BM), Sertoli cells (S), and germ cells (G).

SEMINIFEROUS EPITHELIUM AND SPERMATOGENESIS

The seminiferous epithelium rests on the acellular basement membrane and contains two types of cells, Sertoli cells and germ cells (Fig. 1). At the time of birth, the seminiferous epithelium consists of Sertoli cells and only one type of germ cell, i.e., the gonocyte, located in the central part of the seminiferous cord. Gonocytes migrate to the basement membrane during the early postnatal period and are now called type A spermatogonia. Type A spermatogonia could be called 'male germ-line stem cells' as they renew themselves and also differentiate into spermatozoa (1). During the process of differentiation into spermatozoa, the type A spermatogonia undergo several mitotic divisions to yield type B spermatogonia. Type B spermatogonia mitotically divide to yield primary spermatocytes. Primary spermatocyte through two successive meiotic divisions form haploid spermatids. The haploid spermatids morphologically differentiate into spermatozoa. Thus, in the adult testis, the seminiferous epithelium consists of various germ cell types with the stem cells resting on the basement membrane and more differentiated germ cell types arranged progressively towards the lumen. The germ cells at different stages of differentiation are in close anatomical and functional contact with the Sertoli cells. However, tight junctional complexes between adjoining Sertoli cells compartmentalize the seminiferous epithelium into a basal compartment and an adluminal compartment (2). The junctional complexes separate young germ cells, i.e. spermatogonia and the preleptotene spermatocytes, from later spermatocytes, spermatids, and spermatozoa. In addition, they form the morphological basis of blood-testis barrier. This barrier creates a unique microenvironment in the adluminal compartment. The germ cells in the basal compartment communicate with the neighbouring Sertoli cells, the basement membrane, the peritubular myoid cells, and the blood and lymphatic vessels. More advanced germ cells in the adluminal compartment derive substances in blood or lymph through the Sertoli cell (3). Thus, Sertoli cells interact with all types of germ cells via desmosomes and gap junctions (4, 5). In addition, Sertoli cells develop ectoplasmic specializations (actin-rich filaments sandwiched between plasma membrane and endoplasmic reticulum) and tubulobulbar complexes with spermatids (6). The development and degradation of these structural complexes between neighbouring Sertoli cells at the base of the seminiferous epithelium and between elongating spermatids and the Sertoli cell at the apical end of the seminiferous epithelium has been correlated with the movement of spermatocytes from basal to adluminal compartment and the release of sperm to the lumen, respectively (6).

Sertoli Cell

Generally, the Sertoli cells exhibit an infolded nuclear envelope with pores, a homogeneous nucleoplasm, and a single tripartite nucleolus (Fig. 2). Within the cytoplasm of Sertoli cells, a large Golgi apparatus, numerous mitochondria, lysosomes, multivesicular bodies, lipid droplets, and residual

4

bodies have been described. Sertoli cells present a profuse network of both rough and smooth endoplasmic reticulum suggesting their capability for both protein and steroid synthesis and secretion. Sertoli cells lack secretory granules, large vacuoles, and exocytotic vesicles (7, 8). Lack of these structures indicate that the synthesized proteins may be transferred to the plasma membrane where they are either secreted after cleavage or remain membrane-bound for interaction with the corresponding receptor on germ cell types. A classic example of the growth factor that is expressed by Sertoli cells in both membrane-bound and secretory form is stem cell factor (9). The corresponding receptor, c-kit, is expressed on the surface of type A spermatogonia (10, 11). This concept appears to be true as Sertoli cells have been shown to extend cytoplasmic processes (conical at the base, sheet-like in the middle, and tapered apical towards the lumen) that interact with the plasma membranes of spermatogonia, spermatocytes, spermatids, and spermatozoa (12). The shape of the Sertoli cells in 3-dimension

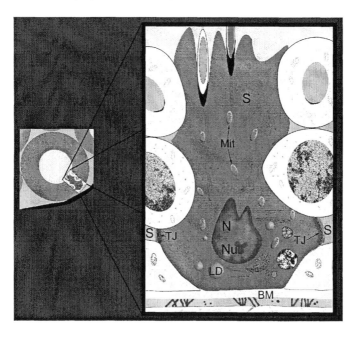

Figure 2. A schematic representation of a Sertoli cell. Morphological details of the Sertoli cell is shown in the magnified image of the portion of the seminiferous tubule from the drawing on the left. Note that the Sertoli cell (S) is placed perpendicular to the basement membrane (BM). It possesses an infolded nucleus (N) with cytoplasm containing numerous mitochondria (Mit) and lipid droplets (LD). A tripartite nucleolus (Nu) is apparent within the nucleus. A tight junction (TJ) between adjoining Sertoli cells is also shown.

changes continuously to accommodate the developing and differentiating germ cells and their mobilization from the base to the lumen (13). Apart from the above mentioned structures, Sertoli cells possess elaborate cytoskeleton consisting of microtubules and filaments that may be involved in transport of spermatids through the seminiferous epithelium (6).

Germ Cells

In the seminiferous epithelium of the adult testis where spermatogenesis is progressing actively, germ cell types begining with the most primitive germ cell, i.e., type A spermatogonia, to the most differentiated type, i.e., spermatozoa, are observed. The intermediary cell types during this differentiation pathway are type B spermatogonia, preleptotene spermatocytes, spermatocytes in different phases prior to meiotic division (leptotene, zygotene, and pachytene), secondary spermatocytes, and spermatids (round and elongating). A schematic representation of the stages of differentiation of type A spermatogonia into spermatozoa is shown in Fig. 3.

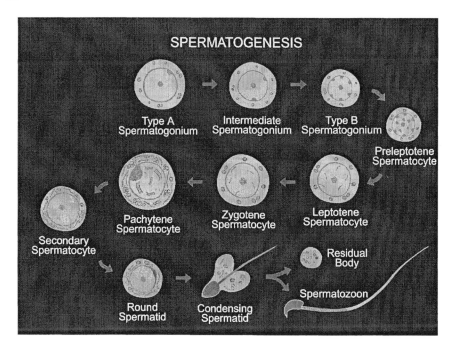

Figure 3. A schematic representation of the process of spermatogenesis. Type A spermatogonia that are present at the base of the seminiferous epithelium undergo a series of mitotic divisions to yield intermediate and type B spermatogonia. Further mitotic divisions of type B result in the formation of preleptotene spermatocytes. The preleptotene spermatocytes through leptotene, zygotene, and pachytene stages undergo the first meiotic division. The resultant secondary spermatocytes proceed through the second meiotic division to yield round spermatids. Round spermatids morphologically differentiate into spermatozoa.

Spermatogonia

Although the discovery of spermatogonia dates back to the second-half of the 19th century (14), it was Regaud (15) who classified spermatogonia of rats into 'dusty and crusty types of cells' based on the chromatin patterns in the nuclei. Later, dusty type and crusty type were renamed as type A and type B, respectively (16). The type A cells exhibit nuclei with fine, palely stained chromatin granulation (Fig. 4), and the type B cells, in contrast, exhibit coarse granules of heavily stained chromatin close to the nuclear membrane. An intermediate type of spermatogonia with fine plaques of chromatin close to the nuclear membrane has been recognized in rodents (17). In addition, type A spermatogonia have been further classified into five different subtypes A_0, A_1, A_2, A_3, and A_4 based on nuclear morphology (18, 19, 20). An alternate method of classification in rodents suggests the presence of A-single (A_s), A-paired (A_p), and A-aligned (A_{al}) spermatogonia before the formation of A_1 through A_4 (21, 22, 23). Unlike in the rodents, only two subtypes of type A spermatogonia have been observed in the human. These are termed dark and pale types (24). The characteristic feature of the dark type A spermatogonia is the central pale-stained area in the nucleus (nuclear vacuole) surrounded by densely staining chromatin. In contrast, pale type A spermatogonia lack the nuclear vacuole and exhibit palely-stained granular chromatin. In similarity with the human, dark and pale type A spermatogonia have been observed in monkeys (25, 26, 27). More recently, we have observed dark and pale type A spermatogonia with similar characteristic features in 6-day old immature mice testis (Dettin et al, unpublished results).

Several review articles (28, 29) have summarized the work on male germ line stem cells in the testis. The spermatogonial renewal and differentiation schemes described are based on a whole mount analysis of adult seminiferous tubules and depend upon morphology and position in the cycle of the seminiferous epithelium. More recently, using plastic sections from adult mice testes, Russell and colleagues defined distinct morphological characteristics of the A_s, A_{al}, and A_p. (30). In the 6-day-old mouse testis, there are no stages since germ cell development is barely initiated, thus it is difficult to identify the A_1 to A_4. All the type A cells appear homogeneous in section, although it is believed that among the type A there must be a subpopulation of cells that are the real stem cells – possibly the A_s. More recent studies indicate that spermatogonial stem population lacks *c-kit* while the differentiating spermatogonia express *c-kit* on their surface (31). What "keeps a stem a stem" is one of the most important unanswered questions in biology today (32). Based on the report that stem cells in skin possess integrin receptors and that their loss is associated with differentiation, Brinster and colleagues have reported that β1- and α6-integrins could be used as markers for spermatogonial stem cells (33), although Sertoli cells and other spermatogonia (type A, In, and B) also have integrin receptors. Using transgenic mice either over expressing GDNF (glial cell-derived neurotrophic factor) or lacking a GDNF allele, it has been shown that spermatogonial stem cells require GDNF for their survival and renewal (34). Thus, GDNF receptor could serve as a marker for the identification of spermatogonial stem cells.

In all mammalian species, spermatogonia are linked together by intercellular bridges (35). These bridges possess a thin rim of poorly stained cytoplasm with very few organelles. Although attempts have been made to correlate the structural features of various types of A spermatogonia (30, 36) with function, they are hampered by lack of biochemical markers to identify and separate these individual types of type A spermatogonia. Type B

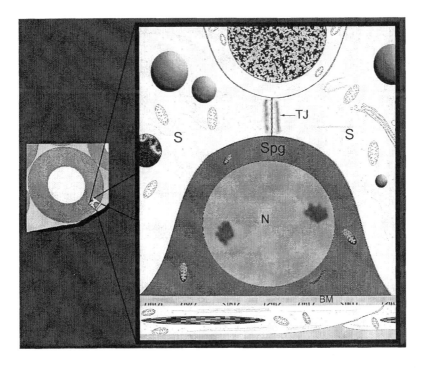

Figure 4. A schematic representation of a type A spermatogonium (right) within a cross section of a seminiferous tubule (left). Note that the spermatogonium (Spg) is located within the basal compartment of the seminiferous epithelium. Two adjoining Sertoli cells (S) along with the basement membrane (BM) create the basal compartment. A Sertoli-Sertoli tight junction (TJ) is also shown above the spermatogonium. Spermatogonia possess large nuclei (N) with a thin rim of cytoplasm and scarce cytoplasmic organelles.

spermatogonia are smaller in size compared to type A spermatogonia. The chromatin material is arranged towards the periphery of the nucleus. A prominent nucleolus is located centrally in the nucleus. These cells appear to have least contact with the basement membrane (37) suggesting that they are about to detach and move towards the adluminal compartment in the pathway of differentiation into primary spermatocytes.

Spermatocytes

Type B spermatogonia undergo mitotic division to yield preleptotene spermatocytes. These are the primary spermatocytes which prepare themselves for undergoing the first meiotic division. The primary spermatocytes are smaller in size than type B spermatogonia and possess less chromatin along the nuclear envelope (37, 38). They are still in contact with the basement membrane. During transition into the leptotene stage of meiotic prophase, they lose the contact with the basement membrane and become more rounded. At this stage, the peripheral chromatin is lost completely and fine condensed filamentous chromatin threads appear in the nucleus. This suggests active synthesis of DNA leading to chromatid duplication and eventual condensation of the chromatid DNA. The chromosomes remain unpaired. In the next stage, i.e., zygotene spermatocytes, the homologous chromosomes pair with each other forming a synaptonemal complex. Following the formation of the synaptonemal complex, the primary spermatocyte enter the pachytene stage in which there is further thickening and shortening of the paired chromosomes linked together at regions known as chiasmata. Functionally, genetic recombination between maternal and paternal homologous chromosomes occur by 'crossing over' in this stage. In addition, there is enormous growth of the cell with increased volumes of both the nucleus and the cytoplasm. The diplotene stage of prophase that follows the pachytene stage is very short, however, the paired chromosomes segregate except at chiasmatic regions of association. At the end of the diplotene stage, cells possess two pairs of each chromosome and enter metaphase, anaphase, and telophase in rapid succession completing the first meiotic division leading to the formation of secondary spermatocytes with a single pair of each chromosome. Secondary spermatocytes have a very short life and lack special distinguishing features (38). These cells are smaller than pachytene spermatocytes but larger than round spermatids. The nuclei present evenly distributed chromatin with a centrally located nucleoli. The secondary spermatocytes undergo the second meiotic division resulting in the formation of round spermatids that possess haploid number of chromosomes.

Spermatids and Spermatozoa

Spermatids are the haploid cells derived from the secondary spermatocytes. These spermatids undergo a unique morphological transformation to yield spermatozoa. This process of transformation is defined as spermiogenesis. Spermiogenesis involves formation of an acrosome, changes in nuclear morphology and position, and formation of a tail. Nineteen steps have been identified in the transformation of immature round spermatids into spermatozoa in the rat (39) and these steps could be subgrouped into a Golgi phase, cap phase, acrosomal phase and maturation phase. Very recently formed spermatids are spherical in shape with a centrally placed spherical nucleus, a well developed perinuclear Golgi apparatus, and adjacent centrioles. A chromatid body, an electron dense mass, is also visible in the perinuclear region (37). These cells do not have an acrosome. They are classified as step 1 spermatids. Through step 2 to step 7 in the rat, the Golgi apparatus deposits proacrosomal vesicles (containing acrosomal granules) which form an acrosomal cap at one end of the nucleus. This process has been described in much detail in Chapter 2. Until step 7, the spermatids are round in shape with a centrally located nucleus. Beyond step 7, the nucleus becomes eccentric with the acrosomal end coming in contact with the cell membrane, and the process of elongation of spermatids begin. The acrosomal end is the anterior pole of the sperm head and is oriented towards the base of the seminiferous epithelium. Progressively through each step, nuclear chromatin condenses to eventually form an electron-dense homogeneous mass. In addition, the nuclear volume decreases, and the shape of the nucleus changes in a species-specific manner. The sperm tail formation begins at the immature spermatid stage in step 1. An axial filament arises from one of the centrioles and pushes the cell membrane outwards. This filament known as the axoneme forms the core of the sperm tail and consists of nine peripheral doublets with a central pair of microtubules. The axoneme is located at the opposite end of the acrosomal pole. During the cap phase of development, the chromatid body surrounds the origin of the flagellum. In the acrosomal phase, modifications to the centriolar apparatus results in a neck piece. As the flagellum consisting of the axoneme surrounded by the cell membrane elongates, the central pair of microtubules are symmetrically enveloped by nine columns of coarse fibers, and are united to nine segmented columns in the neck piece. In the developing flagellum extending from the neck, a middle piece, an annulus, a principal piece, and an end piece can be identified. In the middle piece which is separated from the principal piece by the fibrous ring (annulus), mitochondria align along the coarse fibers. The principal piece is devoid of mitochondria and the nine columns of coarse fibers are enveloped by a dense fibrous sheath. The end piece consists of the central and outer pairs of microtubules covered by the cell membrane. The functional significance of

10

the ultrastructures in spermatids and spermatozoa are described in chapters 2 and 5. In the maturation phase, the cytoplasm is extruded as residual bodies and is phagocytosed by the Sertoli cells.

Cell Associations and the Wave of the Seminiferous Epithelium

The structure-function relationship in the tubular compartment is best defined by stages, the cycle, and the wave of the seminiferous epithelium. Although there are variations among species, characteristic cell-cell associations have been well defined for rat, mice, and the human (38, 40-42). In the rat, there are 14 stages or cell associations, and the germ cells seen at each stage is shown in Fig. 5 (20).

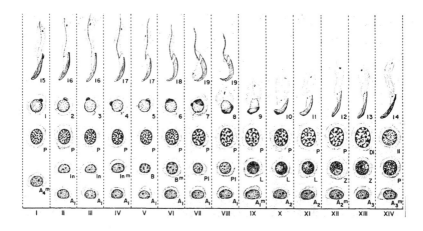

Figure 5. Cell associations or stages of the cycle of the seminiferous epithelium in the rat. Each stage of the cycle is represented by a Roman numeral. The column above the numeral indicates the composition of germ cell types during that particular stage. For example, in column V, type A1 spermatogonia, type B spermatogonia, pachytene spermatocytes, round spermatids, and elongating spermatids in step 17 of spermiogenesis are seen. This segment of the seminiferous epithelium is in stage V of the cycle. As time progresses, the same segment will progress to stage VI with type A_1 and B spermatogonia, pachytene spermatocytes, round spermatids in cap phase, and elongating spermatids in step 18 of spermiogenesis.

During the process of development and differentiation of spermatogonia in the basal compartment to spermatozoa in the adluminal compartment, a specific cell association, e.g., VII, exists in a small segment of the tubule. In that segment of tubule, the cell association will mature to successive cell associations until there is a return to the starting cell association, i.e., VII. This is referred to as the cycle of the seminiferous epithelium. A particular cell association pattern within the cycle is called a stage of the cycle. Along the length of the tubule, a continuous series of successive cell associations are present. For example, a particular segment of tubule is in stage VII of the cycle. The next segment along the longitudinal axis of the tubule will exhibit stage VIII of the cycle, and the following segment will be in stage IX of the cycle. A progressive change in stages of the cycle is observed until stage XIV in rodents after which stage I begins. Beginning with a particular stage of the cycle and ending at the same stage of the cycle along the longitudinal axis of the seminiferous tubule is known as the wave of the seminiferous epithelium.

PERITUBULAR MYOID CELL

Originally described by Regaud (15) as connective tissue cells, peritubular myoid cells were renamed based on their smooth muscle cell-like ultrastructure and their contractile nature (43-45). While only a single layer of myoid cells has been observed in rodents (44, 46), larger animals and humans have multiple layers of myoid cells (47, 48). Ultrastructurally, the myoid cells contain central elliptical nuclei, sparse cytoplasmic organelles, and abundant cytoplasmic filaments made up of α-smooth muscle isoactin (49). The actin-containing filaments form bundles that run in different directions in a species-specific manner with the most complicated arrangement seen in the human seminiferous tubules (50). The arrangement of actin filaments may contribute to the direction and movement of the cells. Apart from actin, myoid cells also contain myosin, another cytoskeletal protein, that may be involved in the contractile process (51). Other cytoskeletal proteins such as desmin, vimentin, alpha-actinin, and vinculin have been demonstrated in the cytoplasm of myoid cells in a species-specific manner (50). Based on the structural features, contractility of seminiferous tubules (52, 53) has been attributed to the myoid cells. Efforts have also been made to associate the contractility of seminiferous tubules to the ongoing process of spermatogenesis and spermiogenesis within the tubules (54) and the luminal content of spermatozoa and testicular fluid (55). In addition to providing the contractile function to the seminiferous tubule, the myoid cells contribute to the blood-testis barrier partially by preventing the entry of high molecular weight substances (2, 56). With their close proximity to the Sertoli cells and germ cells (basal compartment), myoid cell-secretory products may modulate spermatogenesis (Fig. 5). PModS, a protein secreted by myoid cells, has been shown to affect the function of Sertoli cells (57). Recently, it has been shown that leukemia inhibitory factor (LIF) elaborated by myoid cells may modulate the function of both Sertoli and spermatogonial cells that express the receptors for LIF (58). Although the secretion of several other factors by myoid cells has been demonstrated, their functional significance is not known.

12

BASEMENT MEMBRANE

The seminiferous epithelium rests on a thin zone of extracellular matrix known as the basement membrane (59). In between the basement membrane and the peritubular myoid cells, lie a zone of collagen fibrils arranged in varying orientation (60). The basement membrane in the rat is composed of laminin, type IV collagen, heparan sulfate proteoglycan, and nidogen/entactin (61) (Fig. 6).

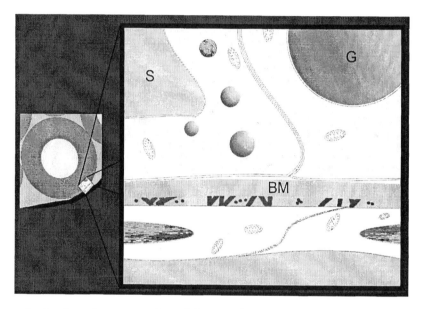

Figure 6. A schematic representation of the acellular basement membrane separating the seminiferous tubule from the interstititial compartment. Note that the basement membrane (BM) underlies the plasma membranes of both Sertoli cells (S) and early germ cells (G) in the basal compartment of the seminiferous tubules.

The individual components of the basement membrane have been shown to induce structural and morphological changes in the cells of the basal compartment of the seminiferous epithelium, specially in the Sertoli cells (62-64). Sertoli cells acquired the characteristic *in vivo* appearance with a columnar shape and polarity. When cultured on Matrigel (a reconstituted basement membrane derived from the Engelbreth-Holm-Swarm tumor), isolated Sertoli cells and germ cells were able to differentiate and form testicular cords in the presence of a few peritubular myoid cells (65). Not only does basement membrane induce changes in the structure of cells at the base of the seminiferous epithelium, it also induces functional changes in them. Most pronounced effects have been observed in terms of responsiveness of Sertoli cells to follicle stimulating hormone. These include enhanced stimulation of the expression of the c-fos gene (66), and the enhanced secretion of androgen binding protein and transferrin (64, 65). The enhancement in the function of Sertoli cells in response to FSH in the presence of basement membrane has also been correlated with the modulation of the signal transducers involved, i.e., cAMP and intracellular calcium (67, 68). Despite an important role for FSH in Sertoli cell function, the survival of Sertoli cells is exclusively dependent on the presence of underlying basement membrane (69). Basement membrane may also play a specific role in the survival of spermatogonia in the basal compartment, and detachment of spermatogonia from the basement membrane may signal either differentiation or apoptosis.

LEYDIG CELLS AND STEROIDOGENESIS

Leydig cells were first identified as masses of cells with fatty vacuoles and pigment inclusions in the intertubular space (70). They are frequently associated with blood vessels in the intertubular spaces. Leydig cells are either spherical or irregularly polyhedral in shape (39). The nuclei are round or oval and generally are eccentrically placed. Ultrastructurally, the nuclear envelope may present numerous indentations (37). Clumps of heterochromatin associate with the inner surface of the nuclear envelope. The nucleolus exhibits dense granular and amorphous areas. The characteristic feature of the cytoplasm of the mammalian Leydig cells is the abundance of smooth endoplasmic reticulum (71). It consists of a random network of interconnected tubules packed in sheets. They may also appear as arrays of fenestrated cisternae or concentric whorls of smooth membranes. There are species-specific variations in the appearance and abundance of smooth endoplasmic reticulum. Efforts have been made to correlate the relative abundance of smooth endoplasmic reticulum with the steroidogenic ability of Leydig cells (72, 73). There exists a strong correlation between androgen secretion and the amount of smooth endoplasmic reticulum and Golgi membranes within the cytoplasm of Leydig cells (74). The smooth endoplasmic reticulum associates with the large lipid droplets found in the cytoplasm of Leydig cells. The lipid droplets contain cholesterol in the esterized form. Most of this cholesterol is synthesized by the smooth endoplasmic reticulum of the Leydig cells (75). The first step in steroidogenesis is the movement of cholesterol present in the lipid droplets onto the surface of mitochondria. Structurally, the movement may involve the actin and the vimentin cytoskeletal filaments found in the cytoplasm of Leydig cells (76). The size and number of mitochondria present in Leydig cells vary depending upon the species of animals. The mitochondria are either ovoid or rod-shaped, and consist of an outer membrane and inner membrane including the cristae. Cholesterol present on the outer membrane is transported to the inner membrane where it is cleaved by the side chain cleavage enzyme (P450$_{SCC}$) and converted into pregnenolone. Recent evidence suggests that the movement of cholesterol from outer to inner membrane of mitochondria is facilitated by steroiogenic acute regulatory protein (StAR) and/or by peripheral benzodiazepine receptor (77) (78). Pregnenolone in the presence of 17α-hydroxylase, 17-20 lyase, and 3β-hydroxysteroid dehydrogenase is converted into testosterone in the smooth endoplasmic reticulum. Testosterone could also be converted further into estradiol in the presence of aromatase enzyme in the testis. Thus, the structural features of Leydig cells, i.e., lipid droplets, abundant smooth endoplasmic reticulum, and mitochondria contribute to the process of steroidogenesis. The current model of Leydig cell steroidogenesis is depicted in Fig. 7.

CONCLUSIONS

This chapter focuses exclusively on histomorphological features of the mammalian testis and their relationship to the function of the testis. The testis can be subdivided into a tubular compartment and an interstitial compartment based on structure and function. The tubular compartment is the site of spermatogenesis and consists of seminiferous epithelium on the inside and a peritubular myoid cell layer on the outside. Sertoli cells, the somatic cell component of the epithelium, provides structural and nutritional support to the dormant and developing germ cells. In addition, it may also regulate the process of spermatogenesis through the release of several growth factors and cytokines. Spermatogenesis begins with the diploid type A spermatogonial stem cells and proceeds through type B spermatogonia, primary spermatocytes, and secondary spermatocytes to yield haploid round spermatids. Haploid round spermatids morphologically differentiate into spermatozoa. The tubular compartment on the outside is surrounded by one or multiple layers of myoid cells. Peritubular myoid cells provide contractility to the seminiferous tubule and may elaborate secretory products that regulate the functions of Sertoli, Leydig, and germ cells. These cells are separated from the seminiferous epithelium by an acellular layer of basement membrane.

Basement membrane affords structural integrity to Sertoli cells and probably to the early spermatogenic cells. It also acts as a survival factor for these cells. The interstitial compartment with a high vascular and lymphatic supply consists predominantly of Leydig cells, macrophages, and fibroblasts. Leydig cells possess the machinery for steroid biosynthesis and secretion. Structurally, the cell is endowed with lipid droplets containing cholesterol (the precursor for steroid synthesis), and an abundant smooth endoplasmic reticulum and mitochondria to convert cholesterol into testosterone.

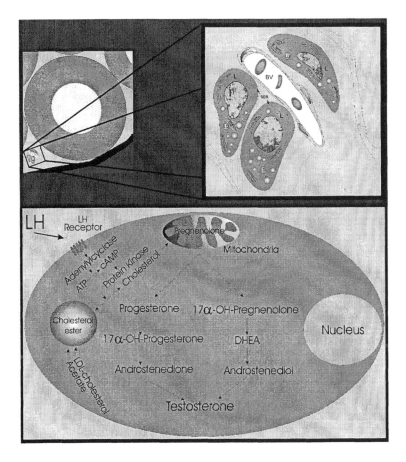

Figure 7. Morphological details of the Leydig cell are shown in the magnified image of the portion of the seminiferous tubule from drawing on the left. Note that the Leydig cells (L) are large in size and are located in close proximity to the blood vessel (BV) (top right.) They possess numerous lipid droplets (LD) and a rich network of smooth endoplasmic reticulum (SER) for steroid biosynthesis (bottom).

ACKNOWLEDGEMENTS

Aspects of the work by the authors presented in this report were supported in part by grants HD16260 and HD33728 from the National Institute of Child Health & Human Development.

REFERENCES

1.   Dym, M., Spermatogonial stem cells of the testis. Proc Natl Acad Sci U S A, 1994. 91: 11287-9.

16

2. Dym, M. and D.W. Fawcett, The blood-testis barrier in the rat and the physiological compartmentation of the seminiferous epithelium. Biol Reprod, 1970. 3: 308-26.
3. Setchell, B.P., Maddocks, S., and Brooks, D.E., Anatomy, vasculature and fluids of male reproductive tract., in Physiology of Reproduction, E. Knobil, Neill, J.D., Editor. 1994, Plenum: New York.
4. Byers, S.W., Pelletier, R.-M., and Suarez-Quian, C., Sertoli-Sertoli cell junctions and the seminiferous epithelium barrier., in The Sertoli Cell, L.D. Russell, Griswold, M.D., Editor. 1993, Cache River Press: Clearwater, FL.
5. Russell, L.D., Morphological and functional evidence for Sertoli-germ cell relationships., in The Sertoli Cell, L.D. Russell, Griswold, M.D., Editor. 1993, Cache River Press: Clearwater, FL.
6. Vogl, A.W., Pfeiffer, D.C., Redenbach, D.M., and Grove, B.D., Sertoli cell cytoskeleton, in The Sertoli Cell, L.D. Russell, Griswold, M.D., Editor. 1993, Cache River Press: Clearwater, FL.
7. de Kretser, D.M., Kerr, J.B., The Cytology of the Testis, in The Physiology of Reproduction, E. Knobil, Neill, J.D., Editor. 1994, Plenum Press: New York.
8. Bardin, C.W., Cheng, C.Y., Musto, N.A., Gunsalus, G.L., The Sertoli Cell, in Physiology of Reproduction, E. Knobil, Neill, J.D., Editor. 1994, Plenum Press: New York.
9. Rossi, P., et al., Expression of the mRNA for the ligand of c-kit in mouse Sertoli cells. Biochem Biophys Res Commun, 1991. 176: 910-4.
10. Yoshinaga, K., et al., Role of c-kit in mouse spermatogenesis: identification of spermatogonia as a specific site of c-kit expression and function. Development, 1991. 113: 689-99.
11. Dym, M., et al., Expression of c-kit receptor and its autophosphorylation in immature rat type A spermatogonia. Biol Reprod, 1995. 52: 8-19.
12. Russell, L.D., Form, dimensions, and cytology of mammalian Sertoli cells., in The Sertoli Cell, L.D. Russell, Griswold, M.D., Editor. 1993, Cache River Press: Clearwater, FL.
13. Morales, C., Clermont, Y., Structural changes of the Sertoli cell during the cycle of the seminiferous epithelium., in The Sertoli Cell, L.D. Russell, Griswold, M.D., Editor. 1993, Cache River Press: Clearwater, FL.
14. von Ebner, V., Untersuchungen uber den Bau der Samenkanalchen und die Entwicklung der Spermatozoiden bei den Sangentieren und beim Menschen., in Rollet's Unterschunger aus dem Institut fur Physiologie und Histologie. 1871: Graz, Leipzig.
15. Regaud, C., Etudes sur la structure des tubes seminiferes et sur la spermatogenese chez les mammiferes. Arch Anat Microsc, 1901. 4: 101-156.
16. Allen, E., Studies on cell division in the albino rat. J Morphol, 1918. 31: 133-185.
17. Clermont, Y., Le Blond, C.P., Renewal of spermatogonia in the rat. Am J Anat, 1953. 93: 475-502.
18. Clermont, Y., Quantitative analysis of spermatogenesis of the rat: a revised model for the renewal of spermatogonia. Am J Anat, 1962. 111: 111-129.
19. Clermont, Y. and E. Bustos-Obregon, Re-examination of spermatogonial renewal in the rat by means of seminiferous tubules mounted "in toto". Am J Anat, 1968. 122: 237-47.
20. Dym, M. and Y. Clermont, Role of spermatogonia in the repair of the seminiferous epithelium following x-irradiation of the rat testis. Am J Anat, 1970. 128: 265-82.
21. Huckins, C., The spermatogonial stem cell population in adult rats. I. Their morphology, proliferation and maturation. Anat Rec, 1971. 169: 533-57.
22. Oakberg, E.F., Spermatogonial stem-cell renewal in the mouse. Anat Rec, 1971. 169: 515-31.
23. de Rooij, D.G., Proliferation and differentiation of spermatogonial stem cells. Reproduction, 2001. 121: 347-354.
24. Clermont, Y., Spermatogenesis in man. A study of the spermatogonial population. Fertil Steril, 1966. 17: 705-21.
25. Clermont, Y. and M. Antar, Duration of the cycle of the seminiferous epithelium and the spermatogonial renewal in the monkey Macaca arctoides. Am J Anat, 1973. 136: 153-65.

26. Marshall, G.R., D.S. Zorub, and T.M. Plant, Follicle-stimulating hormone amplifies the population of differentiated spermatogonia in the hypophysectomized testosterone-replaced adult rhesus monkey (Macaca mulatta). Endocrinology, 1995. 136: 3504-11.

27. Weinbauer, G.F., et al., Quantitative analysis of spermatogenesis and apoptosis in the common marmoset (Callithrix jacchus) reveals high rates of spermatogonial turnover and high spermatogenic efficiency. Biol Reprod, 2001. 64: 120-6.

28. Meistrich, M.L., Van Beek, M.E.A.B., Spermatogonial stem cells, in Cell and Molecular Biology of the Testis, C. Desjardins, Ewing, L.L., Editor. 1993, Oxford University Press: New York, NY.

29. de Rooij, D.G., Stem cells in the testis. Int J Exp Pathol, 1998. 79: 67-80.

30. Chiarini-Garcia, H. and L.D. Russell, High-resolution light microscopic characterization of mouse spermatogonia. Biol Reprod, 2001. 65: 1170-8.

31. Ohta, H., et al., Real-time observation of transplanted 'green germ cells': proliferation and differentiation of stem cells. Dev Growth Differ, 2000. 42: 105-12.

32. Watt, F.M. and B.L. Hogan, Out of Eden: stem cells and their niches. Science, 2000. 287: 1427-30.

33. Shinohara, T., M.R. Avarbock, and R.L. Brinster, beta1- and alpha6-integrin are surface markers on mouse spermatogonial stem cells. Proc Natl Acad Sci U S A, 1999. 96: 5504-9.

34. Meng, X., Regulation of cell fate decision of undifferentiated spermatogonia by GDNF. EMBO Journal, 2000. 19: 453-62.

35. Dym, M. and D.W. Fawcett, Further observations on the numbers of spermatogonia, spermatocytes, and spermatids connected by intercellular bridges in the mammalian testis. Biol Reprod, 1971. 4: 195-215.

36. Chiarini-Garcia, H., et al., Distribution of type A spermatogonia in the mouse is not random. Biol Reprod, 2001. 65: 1179-85.

37. de Kretser, D.M., Kerr, J.B., The Cytology of the Testis, in The Physiology of Reproduction, E. Knobil, Neill, J.D., Editor. 1988, Raven Press: New York, NY.

38. Russell, L.D., Ettlin, R.A., Sinha Hikim, A.P., Clegg, E.D., Histological and Histopathological Evaluation of the Testis. 1 ed. 1990, Clearwater, FL: Cache River Press.

39. Dym, M., The Male Reproductive System, in Histology, L. Weiss, Editor. 1983, Elsevier Biomedical: New York, USA.

40. Perey, B., Clermont, Y., Leblond, C.P., The wave of the seminiferous epithelium in the rat. Am J Anat, 1961. 108: 47-75.

41. Clermont, Y., Kinetics of spermatogenesis in mammals: seminiferous epithelium cycle and spermatogonial renewal. Physiol Rev, 1972. 52: 198-236.

42. Clermont, Y., The cycle of the seminiferous epithelium in man. Am J Anat, 1963. 112: 35-51.

43. Clermont, Y., Contractile elements in the limiting membrane of the seminiferous tubules of the rat. Exp Cell Res, 1958. 15: 438-440.

44. Ross, M.H., The fine structure and development of the peritubular contractile cell component in the seminiferous tubules of the mouse. Am J Anat, 1967. 121: 523-57.

45. Fawcett, D.W., P.M. Heidger, and L.V. Leak, Lymph vascular system of the interstitial tissue of the testis as revealed by electron microscopy. J Reprod Fertil, 1969. 19: 109-19.

46. Gardner, P.J.a.H., E.A., Fine structure of the seminiferous tubule of the Swiss mouse. I. The limiting membrane, Sertoli cell, spermatogonia and spermatocytes. Anat Rec, 1964. 150: 391-404.

47. Wrobel, K.H., R. Mademann, and F. Sinowatz, The lamina propria of the bovine seminiferous tubule. Cell Tissue Res, 1979. 202: 357-77.

48. Ross, M.H. and I.R. Long, Contractile cells in human seminiferous tubules. Science, 1966. 153: 1271-3.

18

49. Tung, P.S. and I.B. Fritz, Characterization of rat testicular peritubular myoid cells in culture: alpha-smooth muscle isoactin is a specific differentiation marker. Biol Reprod, 1990. 42: 351-65.

50. Maekawa, M., K. Kamimura, and T. Nagano, Peritubular myoid cells in the testis: their structure and function. Arch Histol Cytol, 1996. 59: 1-13.

51. Virtanen, I., et al., Peritubular myoid cells of human and rat testis are smooth muscle cells that contain desmin-type intermediate filaments. Anat Rec, 1986. 215: 10-20.

52. Roosen-Runge, E.C., Motions of seminiferous tubules of rat and dog. Anat Rec, 1951. 109: 413-415.

53. Miyake, K., et al., Evidence for contractility of the human seminiferous tubule confirmed by its response to noradrenaline and acetylcholine. Fertil Steril, 1986. 46: 734-7.

54. Suvanto, O. and M. Kormano, The relation between in vitro contractions of the rat seminiferous tubules and the cyclic stage of the seminiferous epithelium. J Reprod Fertil, 1970. 21: 227-32.

55. Nagano, T., Maekawa, M., Fukuda, Y., Murakami, T., Morphological studies on the wall components of seminiferous and epididymal tubules in rodents in relation to their movements., in perspectives on assisted reproduction., T. Mori, Aono, T., Tominaga, T., and Hiroi, M., Editor. 1994, Ares-Serono Symposia Publications.: Rome, Italy.

56. Fawcett, D.W., L.V. Leak, and P.M. Heidger, Jr., Electron microscopic observations on the structural components of the blood-testis barrier. J Reprod Fertil Suppl, 1970. 10: 105-22.

57. Skinner, M.K., Purification of a paracrine factor, P-Mod-S, produced by testicular peritubular cells that modulates Sertoli cell function. Journal of Biological Chemistry, 1988. 263: 2884-90.

58. Piquet-Pellorce, C., et al., Leukemia inhibitory factor expression and regulation within the testis. Endocrinology, 2000. 141: 1136-41.

59. Dym, M., Basement membrane regulation of Sertoli cells. Endocr Rev, 1994. 15: 102-15.

60. Bustos-Obregon, E., Ultrastructure and function of the lamina propria of mammalian seminiferous tubules. Andrologia, 1976. 8: 179-85.

61. Hadley, M.A. and M. Dym, Immunocytochemistry of extracellular matrix in the lamina propria of the rat testis: electron microscopic localization. Biol Reprod, 1987. 37: 1283-9.

62. Tung, P.S. and I.B. Fritz, Extracellular matrix promotes rat Sertoli cell histotypic expression in vitro. Biol Reprod, 1984. 30: 213-29.

63. Suarez-Quian, C.A., M.A. Hadley, and M. Dym, Effect of substrate on the shape of Sertoli cells in vitro. Ann N Y Acad Sci, 1984. 438: 417-34.

64. Mather, J.P., et al., Effect of purified and cell-produced extracellular matrix components on Sertoli cell function. Ann N Y Acad Sci, 1984. 438: 572-5.

65. Hadley, M.A., et al., Extracellular matrix regulates Sertoli cell differentiation, testicular cord formation, and germ cell development in vitro. J Cell Biol, 1985. 101: 1511-22.

66. Papadopoulos, V. and M. Dym, Sertoli cell differentiation on basement membrane is mediated by the c- fos protooncogene. Proc Natl Acad Sci U S A, 1994. 91: 7027-31.

67. Dym, M., et al., Basement membrane increases G-protein levels and follicle-stimulating hormone responsiveness of Sertoli cell adenylyl cyclase activity. Endocrinology, 1991. 128: 1167-76.

68. Ravindranath, N., et al., Rat Sertoli cell calcium response to basement membrane and follicle- stimulating hormone. Biol Reprod, 1996. 54: 130-7.

69. Dirami G, R.N., Kleinman HK, Dym M., Evidence that basement membrane prevents apoptosis of Sertoli cells in vitro in the absence of known regulators of Sertoli cell function. Endocrinology, 1995. 136: 4439-47.

70. Leydig, F., Zur Anatomie der mannlichen Geschelechtsorgane und Analdrusen ser Saugetiere. Z. Wiss. Zool., 1850. 2: 1-57.

71. Christensen, A.K., Leydig cells, in Handbook of Physiology: Endocrinology, D.W. Hamilton, Greep, R.O., Editor. 1975, American Physiological Society: Washington, D.C.

72. Mori, H. and A.K. Christensen, Morphometric analysis of Leydig cells in the normal rat testis. J Cell Biol, 1980. 84: 340-54.

73. Zirkin, B.R., et al., Testosterone secretion by rat, rabbit, guinea pig, dog, and hamster testes perfused in vitro: correlation with Leydig cell ultrastructure. Endocrinology, 1980. 107: 1867-74.

74. Wing, T.Y., Effects of luteinizing hormone withdrawal on Leydig cell smooth endoplasmic reticulum and steroidogenic reactions which convert pregnenolone to testosterone. Endocrinology, 1984. 115: 2290-6.

75. Hou, J.W., D.C. Collins, and R.L. Schleicher, Sources of cholesterol for testosterone biosynthesis in murine Leydig cells. Endocrinology, 1990. 127: 2047-55.

76. Russell, L.D., et al., Characterization of filaments within Leydig cells of the rat testis. Am J Anat, 1987. 178: 231-40.

77. Papadopoulos, V., et al., Structure, function and regulation of the mitochondrial peripheral-type benzodiazepine receptor. Therapie, 2001. 56: 549-56.

78. Stocco, D.M., StAR protein and the regulation of steroid hormone biosynthesis. Annu Rev Physiol, 2001. 63: 193-213.

Chapter 2

THE SPERM ACROSOME: FORMATION AND CONTENTS

Aïda Abou-Haila And Daulat R.P. Tulsiani
Université René Descartes, Paris, France and Vanderbilt School of Medicine, Nashville, Tennessee, USA

INTRODUCTION

The mammalian spermatozoon is a uniquely shaped cell with a head containing the nucleus and enzyme-filled acrosome, and a flagellum (tail) containing contractile apparatus such as the axoneme, cytoskeletal structures and mitochondria. The shape of the sperm head varies from species to species and usually falls into two categories: a sickle shape in rodents and a paddle shape (spatulate) in several larger species including man. This highly specialized cell, capable of delivering the male genome to the egg, is formed in the testes throughout postpubertal male reproductive life span by a regulated process called spermatogenesis. The final phase of this process, referred to as spermiogenesis, is a continuous process beginning with the formation of the round spermatid and concluding with the release of the spermatozoon into the lumen of the seminiferous tubule. The number of programmed steps needed to transform an ordinary looking round spermatid into a hydrodynamically shaped spermatozoon with a fully developed acrosome varies from species to species and ranges from six to eight steps in man to 19 steps in rats and rabbits.

The formation and organization of the acrosome are perhaps the most fascinating events during spermiogenesis. Despite slow progress, many new details are emerging. The acrosome, a Golgi-derived secretory vesicle, is a sac-like structure filled with a variety of antigens including powerful hydrolytic enzymes. The acrosome contains two sets of components: those that are readily soluble and others that are present in the form of an insoluble matrix. Because of their differential solubility, the acrosomal contents are thought to be released at different rates at the on-set of the acrosome reaction. This prerequisite event allows the acrosome-reacted spermatozoon to penetrate the zona pellucida and fertilize the egg. Thus, an understanding of the acrosome and its contents is necessary for a clear picture of how the organelle prepares itself for a role in early events of fertilization.

The cellular changes during the formation of acrosomic system have been studied and described using biochemical and morphological approaches (1, 2). In this chapter, we review these changes and include our recent studies with anti-acid glycohydrolase antibodies as probes to monitor the formation of the sperm acrosome.

FORMATION OF SPERMATOZOA

Spermatozoa are formed from spermatogonial stem cells throughout male reproductive life by a highly orchestrated process referred to as spermatogenesis. The entire process consists of three sequential phases of proliferation and differentiation. First, spermatogonial stem cells multiply to produce an optimal number of spermatogonia which give rise to primary spermatocytes and also maintain a pool of fresh stem cells. During this phase, cells subdivide into A-type spermatogonia (cells with no heterochromatin in the nucleus) and B-type spermatogonia (cells with heterochromatin in the nucleus). In rats and mice, intermediate (In)-type spermatogonia are also formed (Figure 1). Second, the resulting primary spermatocytes (diploid cells) undergo a lengthy meiotic prophase, followed by two successive divisions: the first division produces two secondary spermatocytes. Each secondary spermatocyte then undergoes a second division to produce four haploid round spermatids. The non-dividing spermatids remain in close association with Sertoli cells during their differentiation into polarized spermatozoa. This last process, referred to as spermiogenesis, is perhaps the most fascinating and involves coordinated changes in intracellular systems including Golgi apparatus, nucleus, mitochondria and centriolar-axonemal complex. These alterations require a series of shape changes from round to elongate as spermatid metamorphoses into a developing spermatozoon. The nucleus elongates, shifts towards the cell surface and takes on a compact and species-specific microanatomy (Figures 2 and 3). The chromatin condenses and testis-specific histones are replaced by spermatozoal proteins (protamines) which are rich in arginine and cysteine. The net result is the formation of a spermatozoon that is devoid of endoplasmic reticulum and Golgi apparatus but contains new structures such as cytoskeletal elements and acrosome.

The process of the transformation of a round spermatid into the spermatozoon is continuous, has several steps, and is completed in 12-14 days in rats and 23 days in man. The progressive steps vary considerably from species to species: there are 19 steps in rats and rabbits, 16 in mice, 14 in monkeys, and 6-8 steps in man.

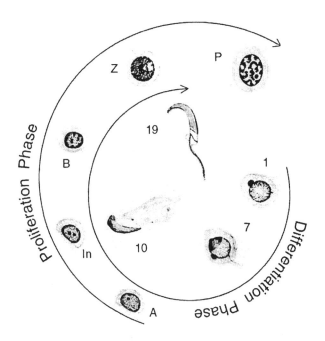

Figure 1. Diagram illustrating spermatogenic cycle of the seminiferous epithelium for rat. The abbreviations for the spermatogenic cells at various stages of the cycle are as follows: A, A-type spermatogonia; In, intermediate-type spermatogonia; B, B-type spermatogonia; Z, zygotene spermatocyte; P, pachytene spermatocyte; and spermatids from stages 1 to 18. The sperm cell (stage 19) is released into the lumen of the seminiferous tubule. Reproduced with permission from Abou-Haila and Tulsiani (3).

The Sperm Acrosome

A well-developed acrosome is a membranous structure covering the anterior portion of the nucleus. There is a large variation in the size and shape of the acrosome from species to species which depends on the morphology of the sperm head. In mammals, the sperm head and its acrosome generally fall into two categories: a sickle-shaped head in rodents, and a skull-cap/paddle-shaped (spatulate) head in several larger species, including man (Figure 4). However, the basic structure and function of the acrosome in all mammals are the same. The acrosome consists of two segments, the acrosomal cap (anterior acrosome) and the equatorial segment (posterior acrosome). The distribution of these two segments differs greatly between species. The acrosomal cap is comprised of the marginal segment (portion that extends beyond the anterior margin of the nucleus) and the principal segment (portion overlying the nucleus). The equatorial segment

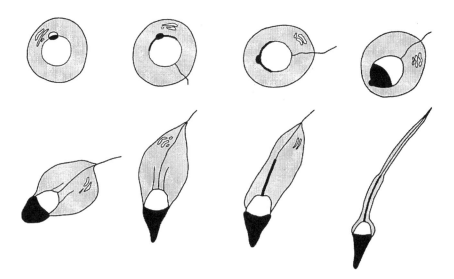

Figure 2. Diagram illustrating eight stages of the transformation of round spermatids into spermatozoa (spermiogenesis) in man. The figure was adopted from Holstein (4).

forms a band that approximately overlies the equator of the head in spatulate spermatozoa and covers much of the lateral surface in sickle-shaped spermatozoa (Figure 4). In most species, the equatorial segment persists until sperm-egg fusion.

A mature acrosome is a sac-like structure with an inner acrosomal membrane close to the nucleus and an outer acrosomal membrane underlying the sperm plasma membrane (Figure 4). It is a Golgi-derived secretory vesicle which resembles the cellular lysosome in many ways. The two organelles originate from the Golgi apparatus. Internal milieu in these vesicles is normally acidic. Finally, both are filled with several common enzymes such as acid glycohydrolases, proteases, esterases, acid phosphatases and aryl sulfatases (5,6). Despite these similarities, the acrosome is different from the lysosome in that it contains several novel components such as acrosin acrogranin, a matrix protein AM67, and several testis-specific serine proteinases (3,7). Because of these differences and its exocytotic properties, the sperm acrosome is considered analogous to a secretory granule. The important features of this secretory vesicles as reported by Burgess and Kelly (8) are : 1) the secretory contents are stored

Round Spermatids

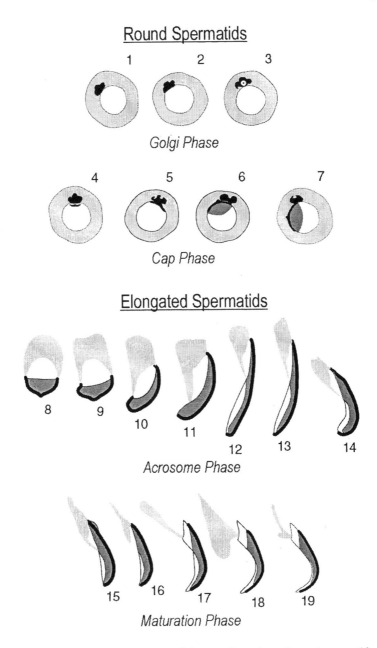

1   2   3

Golgi Phase

4   5   6   7

Cap Phase

Elongated Spermatids

8   9

10   11

12   13   14

Acrosome Phase

15   16   17   18   19

Maturation Phase

Figure 3. Diagram illustrating nineteen stages of the transformation of round spermatids into spermatozoa (spermiogenesis) in the rat. The acrosome is gradually formed during four phases as indicated in the figure. The phases are as follows: Golgi phase, stages 1-3; cap phase, stages 4-7; phase, stages 8-14; and maturation phase, stages 15-19. Reproduced with permission from Abou-Haila and Tulsiani (3).

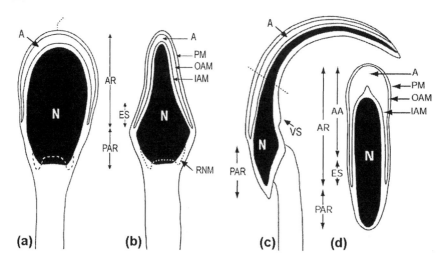

Figure 4. Schematic drawing of human (a,b) and rat (c,d) sperm head. Section through the plane of the nuclear flattening (a), longitudinal sections (a,c), and saggital section (d). The drawings seen in b and d correspond to levels of the plane of sections seen in (a) and (c). A, acrosome; AA, anterior acrosome; AR acrosomal region; ES, equatorial segment; IAM, inner acrosomal membrane; N, nucleus; OAM, outer acrosomal membrane; PAR, postacrosomal region; PM, plasma membrane; RNM, redundant nuclear membrane; VS, ventral spur.

over an extended period of time and are present in a concentrated form; 2) the contents form a dense structure surrounded by inner and outer membranes and the acrosomal granule is stored for several weeks during sperm development in the testis and subsequent maturation in the epididymis; and 3) the organelle undergoes secretion as a result of an external stimulus.

The interior of the acrosome proper is thought to be biochemically and morphologically compartmentalized with specific components present in discrete regions of the organelle. Furthermore, the acrosome contains two sets of components with different solubility, a set of readily soluble components and a set of insoluble (particulate) matrix components. Thus, it has been suggested that the solubility of the acrosomal components determines their function during and after the acrosome reaction. For instance, a soluble component would be released instantly at the on-set of the acrosomal exocytosis, whereas matrix components would remain associated with the acrosome for a longer period of time during the acrosome reaction.

The insoluble matrix components in the acrosome may also act as structural components that allow it to maintain the species-specific shape. In addition, the matrix proteins have been suggested to segregate hydrolases within the acrosome (9). Thus, the acrosome is not just a vesicle filled with enzymes. The high organization within the organelle ensures that its constituents are released in a timely manner as spermatozoa traverse the egg vestments.

constituents are released in a timely manner as spermatozoa traverse the egg vestments.

Formation of the Acrosome

The formation and organization of the acrosomic system takes place during spermiogenesis. Several stages have been identified in man (Figure 2) and rat (Figure 3). In the rat, spermiogenesis has been divided into several sub-phases (steps) based upon changes occurring during the developing acrosome and nucleus. During early stages of spermiogenesis (Golgi and cap phases), the Golgi apparatus is active in synthesizing and delivering proteins/glycoproteins and membrane vesicles required for the maturing acrosome. At this stage, the Golgi complex consists of a prominent system of closely packed tubules and vesicles localized close to one pole of the nucleus. The vesicles that contribute to acrosome formation are derived from the stacks of trans-Golgi network. During the "Golgi phase", numerous proacrosomic granules are formed from the trans-Golgi stacks and accumulate in medullary region. These small granules fuse with each other forming a single acrosomic granule which establishes close contact with the nuclear envelope. During the "cap phase", the spherical acrosomic granule enlarges by the addition of glycoprotein-rich contents into the forming acrosomic system (acrosome and head cap; Figures 5,6). The synthesis of several glycoproteins involved in acrosome formation begins in pachytene spermatocytes and continues throughout spermiogenesis. During these early stages of acrosome formation, there is a close association between the Golgi complex and the forming acrosomal vesicle. Recent evidence suggests the occurrence of a number of small coated vesicles (40-50 nm in diameter) in the Golgi complex that may correspond to coatomer-coated vesicles (COP) which have been observed in somatic cells (10,11). Two of these vesicles, β-COP and clathrin coated, are present in both the Golgi complex and the forming acrosome suggesting that these vesicles may have a role in membrane trafficking during acrosome biogenesis. The developing acrosomic system grows and flattens over the nucleus. The nuclear region which is in close contact with the developing acrosomal vesicle is characterized by a thin layer of condensed chromatin just beneath the nuclear membrane. Finally, during the "acrosomal phase", the acrosomic granule attaches itself to the inner acrosomal membrane and becomes hemispherical. The structure remains distinct throughout the final maturation phase of spermiogenesis and represents the acrosome proper. Additional modifications, such as condensation of the acrosome vesicle and intra-acrosomal modifications of several antigens have been reported as sperm cells undergo maturation in the epididymis. (for details see chapter 6). The

28

mechanism by which the sperm acrosome attains the species-specific shape is far from clear. It has been suggested that flattening and spreading of the

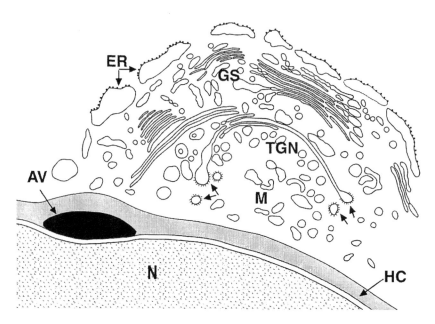

Figure 5. Diagram showing the structure of the hemispherical Golgi apparatus and acrosomic system (acrosome and head cap) at the surface of the nucleus (N) of a step 6 rat spermatid. The acrosomal vesicle (AV) is close to the inner acrosomal membrane lining the nuclear membrane and the head cap (HC) is spread over the surface of the nucleus. The Golgi apparatus shows a cortex composed of several saccules (GS) and a medulla (M) containing membranous vesicular and tubular profiles. A trans-Golgi network (TGN) with coated edges (arrow) is seen in the saccules. Endoplasmic reticulum (ER) is seen at the cis face of the Golgi apparatus.

acrosome over the nucleus may be the result of vesicle fusion at the edges of the forming acrosome coupled with membrane retrieval at the center of the acrosome. According to this model, newly-synthesized Golgi components move from Golgi-to-acrosome during the early stages of spermiogenesis and are responsible for the growth of the acrosome. Over time, the traffic from the Golgi-to-acrosome declines, and there may be a concomitant increase of traffic in the opposite direction (i.e., from acrosome-to-Golgi complex). The spermatids constantly adjust both trafficking routes until the acrosome has flattened and the Golgi apparatus has migrated towards the opposite pole of the germ cell (11). The proposed retrograde and anterograde vesicular transport trafficking pathways which control the growth and shape of the acrosome has yet to be confirmed.

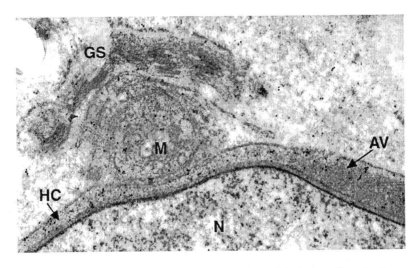

Figure 6. Electron micrograph showing the distribution of β-glucuronidase in the mouse spermatid (step 6). Note the presence of gold particles in the Golgi stacks (GS) and Golgi vesicles in the medulla (M) as well as in the acrosomal vesicle (AV) and the head cap (HC) over the nucleus (N). Original magnification X 24,000.

Acid Glycohydrolases as Markers to Study Acrosome Formation

Glycohydrolases are exo-enzymes that cleave terminal glycosidic residues from the glycan portion of glycoproteins and glycolipids. These enzymes are expressed during spermatogenesis and are localized in the sperm acrosome (see below). We have recently utilized antibodies against two glycohydrolases as probes to monitor the acrosome formation during progressive transformation of spermatids into testicular spermatozoa which contain well-developed acrosome. This was accomplished using immunohistochemical approaches at light and electron microscopic levels. In the former approach, we examined the binding of monospecific immunoglobulin (IgG) to mixed spermatogenic cells prepared by enzymatic disruption of the rat testes. The permeabilized germ cells treated with the primary antibodies were immunostained using fluorescein (FITC)-labeled secondary antibody. Various stages of the acrosome formation were identified by confocal microscopy using Nomarski differential interference contrast optics. Data presented in Figures 7 and 8 demonstrate the distribution of β-D-galactosidase and β-D-glucuronidase, respectively, during spermiogenesis. Both glycohydrolases were expressed in the Golgi apparatus and lysosome-like multivesicular bodies present in diploid spermatogenic cells (pachytene spermatocytes) and early round spermatids. The progressive formation of the acrosome coincided with the decrease in the number of

30

multivesicular bodies. This result suggests that during spermiogenesis the Golgi apparatus is more involved in the formation of acrosomal components.

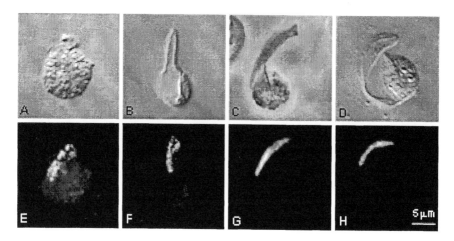

Figure 7. Confocal micrographs showing immunolocalization of β-D-galactosidase in the forming acrosome during spermiogenesis in the rat. The testicular germ cells were immunostained using anti-β-D-galactosidase immunoglobulin (IgG) as primary antibody and FITC-labeled anti-rabbit goat IgG as secondary antibody (3). Various phases of the acrosome formation were photographed with a confocal microscope using Nomarski differential interference contrast optics (A-D). Note the presence of an intense fluorescence in the forming acrosome during progressive differentiation of the spermatids (E-H). The spermatids are from the following stages: Stage 8, A & E; Stage 10, B & F; Stage 14, C & G; and Stage 16, D & H. The absence of flagellum in the elongated spermatids is due to its loss during preparation of the spermatogenic cells by enzymatic disruption of the testis (3).

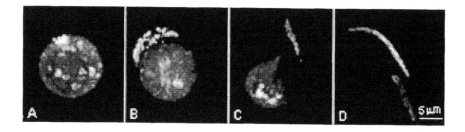

Figure 8. Confocal micrographs showing immunolocalization of β-D-glucuronidase in the forming acrosome during progressive transformation of the rat round spermatid (A) into the spermatozoon (D). Other details are the same as in Figure 7, except that anti-β-D-glucuronidase IgG was used as primary antibody. The spermatids are from the following stages: Stage 4, A; Stage 8, B; and Stage 11, C.

In addition to these two glycohydrolases, male germ cells also contain N-acetyl β-D-glucosaminidase and α-L-fucosidase, the exo-enzymes which cleave terminal β-linked N-acetylglucosamine, and α-linked fucose residues, respectively. Although the biological significance of acrosomal glycohydrolases is still far from clear, a recent study reported abnormal spermiogenesis in dogs suffering from fucosidosis, an inherited storage disorder caused by severe deficiency in lysosomal α-L-fucosidase. It is not known whether male animals and humans with other storage conditions such as galactosidosis or mannosidosis have similar acrosomal abnormalities.

Despite many advances in our understanding of the formation and organization of the acrosome, little is known about the factors that regulate various steps of the complex process. The highly regulated processes of acrosome formation is thought to require sequential activation of many genes expressed in differentiating germ cells (12).

ACROSOMAL CONTENTS AND THEIR FUNCTIONAL SIGNIFICANCE

There is extensive literature on the acrosomal contents and their functional significance in sperm physiology. The following is a brief description of the constituents of the acrosome and their potential role in fertilization.

Glycohydrolases

The acrosome contains a host of hydrolytic enzymes with catalytic and immunological properties similar to the enzymes present within lysosomes. The lysosomal glycohydrolases have both anionic and carbohydrate recognition markers demonstrated to be mannose-6-phosphate (Man6-P). The newly synthesized enzymes are transported to lysosomes by virtue of exposed Man6-P-ligand (13). The ligand is recognized by two types of Man6-P receptors localized in the trans-Golgi cisternae, a 215 kDa cation-independent (CI) and a 46 kDa cation-dependent (CD) type. The two receptors bind to Man6-P ligands and segregate acid glycohydrolases in transport vesicles. These vesicles bud off from the Golgi-associated compartments (transport vesicles) and deliver their enzymes by fusing with prelysosomal (endosome) structures (6,14). The signal required for targeting glycohydrolases to the lysosomes is well documented; however, it is not clear how these enzymes are targeted to the acrosome. Although several glycohydrolases, including α-L-fucosidase, N-acetyl-β-D-glucosaminidase, β-D-galactosidase and β-D-glucuronidase, are expressed in the late spermatocytes and early spermatids, subcellular localization data are available only for the last two enzymes. Both enzymes are present in the Golgi membranes, Golgi-associated vesicles, and lysosome-like multivesicular bodies in the late spermatocytes and early spermatids. This localization suggests that the enzymes are likely phosphorylated and transported to the lysosome-like structures in the late diploid and early haploid cells. Moreover, one of these enzymes (β-D-glucuronidase), localized in the forming/formed acrosome during spermiogenesis has immunological and biochemical properties similar to the lysosomal form of the enzyme present in spermatocytes. Based on the common Golgi apparatus origin of cellular lysosomes and the sperm acrosome, and the fact that mouse germ cells are known to contain CD- and CI- Man6-P receptors, it has been suggested that spermatocytes and spermatids synthesize glycohydrolases which undergo post-translational modifications prior to being transported to either the lysosome or the forming acrosome by virtue of Man6-P receptors (6,15). This suggestion awaits confirmation.

The acrosomal glycohydrolases, like the lysosomal glycohydrolases, have high substrate specificity and will not hydrolyse even a closely related glycosidic linkage (Table 1). These enzymes catalyze hydrolytic cleavage of terminal sugar residues from the glycan portion of glycoprotein and glycolipids. Under *in vitro* or *in vivo* conditions, the hydrolytic enzymes function sequentially in such a way that the product of the first enzymatic cleavage becomes the substrate for the next enzyme (6).

Potential Function of Acrosomal Glycohydrolases

Although the precise role of these enzymes is not yet known, they could be functional in the primed (i.e. capacitated) spermatozoa and/or following the binding of opposite gametes when the bound spermatozoon undergoes the acrosome reaction and releases its contents at the site of sperm binding (14). Spermatozoa penetrate the cumulus cells surrounding the ovulated egg *in vivo*. These cells are held together by an extracellular matrix primarily composed of hyaluronic acid covalently attached to a protein backbone. Hyaluronic acid is a polymer composed of repeating disaccharide units containing glucuronic acid and N-acetylglucosamine in β1,3- and β1,4- linkage (Figure 9). The sperm surface antigen, posterior head-20 (PH-20) with hyaluronidase activity, as well as the acrosomal hyaluronidase which translocates to the plasma membrane after capacitation, have been implicated in hyaluronic acid digestion and sperm penetration through the cumulus oophorus cells. β-D-glucuronidase and β-N-acetylglucosaminidase can further hydrolyse the hyaluronidase product (disaccharide) into monomeric units (14). Two other enzymes implicated in passage through the cumulus matrix are β-D-galactosidase and aryl sulfatase (5,6).

It should be noted that, with the exception of sperm surface PH-20 with hyaluronidase activity, all other glycohydrolases suggested to have a role in digestion and dispersion of the cumulus cells are intra-acrosomal. Thus, any reasonable hypothesis suggesting a role for the acrosomal glycohydrolases must include a possible mechanism(s) which allows the enzymes to be present on the surface of an acrosome-intact spermatozoa and recognize its substrate on cumulus cells. Interestingly, epididymal luminal fluid and seminal fluid are rich in most of the glycohydrolases. Thus, one likely possibility is that these enzymes tightly bind to the sperm surface during epididymal transit and ejaculation and remain bound during interaction of the male gamete with the cumulus mass in the oviduct. A second possibility is that the enzymes present in the acrosome translocate in capacitating spermatozoa and are available on sperm surface as has been reported for hyaluronidase (16) and mannose-binding protein (17). Alternatively, some of the capacitated spermatozoa may undergo spontaneous acrosome reaction releasing the hydrolytic enzymes. These enzymes could disperse the cumulus cells allowing the acrosome-intact spermatozoa a clear passage through the cells.

Table 1. Glycohydrolase Activities Present in Mammalian Sperm Acrosome

Enzyme	Sugar linkage[a]
Hyaluronidase	(GlcNAc-β-Gluc)n
α-L-Fucosidase	Fuc-α-GlcNAc
α-D-Galactosidase	Gal-α-Gal
β-D-Galactosidase	Gal-β-GlcNAc
β-D-Glucuronidase	Glu-β-GlcNAc
β-N-Glucosaminidase	GlcNac-β-(Gal)GalNAc
α-D-Mannosidase	Man-α-Man
β-D-Mannosidase	Man-β-GlcNAc
Neuraminidase	NANA-α-(Gal)GalNAc
Aryl sulfatases A, B and C	Sulfates

[a] GlcNAc, *N*-acetylglucosamine; Gluc, glucuronide; Fuc, fucose; Gal, galactose; GalNac, *N*-acetylgalactosamine; Man, mannose; NANA, *N*-acetylneuraminic acid.

34

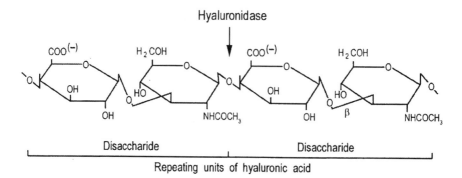

Figure 9. Chemical structure of two repeating (disaccharide) units of hyaluronic acid. The enzyme, hyaluronidase, hydrolyzes β1,4-linkage(s) present between N-acetylglucosaminyl and D-glucuronide residues as indicated.

Proteinases / Proteases

These are proteolytic enzymes which catalyse the breakdown of polypeptide backbone of proteins/glycoproteins. Mammalian sperm acrosome contains several proteinases (Table 2). These include the extensively studied acrosin/proacrosin, the sperm-specific serine protease, to less studied serine proteinases (testis-specific proteinases 1,2,4 and spermin), cysteine proteinases (cathepsins) and a metalloproteinase. The guinea pig sperm acrosome contains dipeptidyl-peptidase II, a 130 kDa protein which is optimally active at an acidic pH of 4.5-5.5 (5,6). In addition, the sperm acrosome is reported to contain a dipeptidyl carboxypeptidase, calpain, and a metalloprotease in the acrosome. Whether all these enzymes are needed for normal functioning of spermatozoa is not known.

Acrosin, a serine-like proteinase, is believed to be present in the acrosome as well as on the surface of capacitated spermatozoa (6). This enzyme is the most extensively studied sperm proteinase implicated in sperm-zona binding and penetration of the zona pellucida. Proteolytic cleavage results in a penetration pathway for the motile hyperactivated spermatozoa through zona pellucida. Since various trypsin inhibitors have been reported to markedly inhibit the sperm binding to, and penetration of, zona pellucida *in vitro*, it has been assumed that a trypsin-like serine protease(s) plays a role in early events of fertilization. However, sperm from mice carrying a targeted mutation of acrosin gene (acrosin-knock-out mice) are still fertile in spite of the absence of acrosin protease activity. This study implies that sperm do not require acrosin to penetrate the zona pellucida and fertilize the egg. Further studies with the mouse spermatozoa lacking the acrosin have revealed that the major role of the protease may be to accelerate the dispersal of the acrosomal constituents during the acrosome reaction since spermatozoa lacking acrosin show delayed fertilization (17). These results suggest that a proteinase(s) other than acrosin may be essential for sperm function. Interestingly, a non-acrosin protease of 42 kD, sensitive to the trypsin/acrosin inhibitor p-aminobenzamidine, has been implicated in the zona penetration step (18). Mouse sperm acrosome also contains four other serine-specific proteases: spermin, TESP1, TESP2 and TESP4 (Table 2). However, the role of these proteases in sperm function, if any, is not known.

Table 2. Proteinase Activities in Mammalian Spermatozoa

Enzyme	Specificity	Localization	References
Acrosin	Serine	Acrosome	6
Acrolysin	Amino proteinase	Acrosome	6
Cathepsin D	Carboxyl	Acrosome	6
Cathepsin L	Cysteine	Acrosome	6
Cathepsin S-like	Cysteine	Spermatozoa	6
Metalloprotease	Not known	Acrosome	6
Spermin	Serine	Acrosome	5
Pz-Protease	Not known	Acrosome	5
TESP1*	Serine	Acrosome	7
TESP2*	Serine	Acrosome	7
TESP4*	Serine	Acrosome	7
Protease 42 kD	Serine**	Acrosome	18

* Testis-specific proteinase
**Sensitive to p-aminobenzamidine, an inhibitor of serine proteinase

Phospholipases

The sperm acrosome from several mammalian species possess two well characterized lipases, phospholipases A2 and C. The former enzyme catalyzes the hydrolysis of phosphatidylcholine into lysophospholipid and cis-double bond (unsaturated) fatty acids. The enzyme is reported to be calcium-dependent and optimally cleaves its substrate between pH 7.4 to 8.0. Phospholipase C is optimally active at pH 6.0. Both enzymes are activated by Ca^{2+} and are thought to be important in the signal transduction pathway which leads to acrosomal exocytosis (19). In addition to these two phospholipases, there is evidence that bovine sperm contains a 115 kDa phospholipase D1. This enzyme is localized in the acrosomal region of the sperm head and occurs in a complex form with protein kinase $C\alpha$ (19).

Phosphatases

Mammalian sperm acrosome contains several phosphatases including acid and alkaline phosphatases. Both have been purified and characterized from the rabbit spermatozoa (5). In addition to these phosphatases, mammalian spermatozoa possess several ATPases believed to have a role in the acrosome reaction (see Chapter 15); however, whether all of the ATPases are associated with the sperm acrosome is not yet known.

Esterases

Non-specific esterases comprise a widely distributed group of enzymes which not only act on ester bonds but also catalyse the hydrolysis of amide bonds and carry out transacylation. Although a total of 11 non-specific esterases isoenzymes have been found in the bull sperm head and tail, none were identified in the acrosome. A corona-penetrating enzyme which disperses corona cells has been extracted from rabbit and human acrosomes. The enzyme, suggested to be esterolytic in nature, has been purified from the testis and appears to be an arylesterase (5). In the guinea pig, a non-specific esterase activity (observed histochemically in the forming/formed acrosome) was recognized at the ultrastructural level on the acrosomal membrane until stage 8 of spermatogenesis and on the plasma membrane after this stage. Based on this localization, the enzyme may modify the sperm plasma membrane. Mouse sperm acrosome also contains low levels of non-specific esterase activity. This activity is not detected in 76% of the acrosome-intact spermatozoa after 8 hours of mating. This is likely due to its loss during passage through the female genital tract or due to the low local level of the enzyme.

Aryl Sulfatases

Aryl sulfatases (A,B,C) are enzymes that remove sulfate from various sulfated glycoconjugates and steroids. Aryl sulfatase activity was demonstrated in acrosomal extracts of the boar, bull, rabbit and ram spermatozoa as a soluble enzyme and localized with the electron microscope in the acrosome of human spermatozoa (5). Aryl sulfatase A, an enzyme which desulfates sulfogalactolipids, has been purified from boar epididymal sperm acrosomes (5) and rabbit testis (20). Boar acrosomal aryl sulfatase A has a pH optimum of 4.2 and has been demonstrated to disperse the cumulus cells of ovulated hamster, rabbit and pig eggs. The enzyme, however, has no effect on the zona pellucida or the oolemma (5).

Other Acrosome-Associated Components

In addition to the above enzymes, mammalian spermatozoa contain calmodulin, a 17 kDa calcium-binding acidic protein, synaptic vesicle protein, synaptotagmin I or a very similar homolog, rab3A/a, a small GTPase, angiotensin II receptor, epidermal growth factor receptor, and soluble N-ethylmaleimide-sensitive factor attachment receptor (SNARE) proteins. All of these components have been reported to be involved in a regulatory role in triggering the acrosome reaction (for details, see chapter 15).

The sperm acrosome also contains cytoskeletal elements between the inner layer of sperm plasma membrane and outer layer of the outer acrosomal membrane. These elements are rich in actin, a protein characteristic of muscle fibers. In capacitated spermatozoa, the actin occurs in filamentous form (F-actin) and has been suggested to provide a physical barrier that prevent the fusion of the plasma membrane and the outer acrosome membrane (6). In response to the increased Ca^{2+} and elevated intrasperm pH during early stages of induction of the acrosome reaction, the F-actin depolymerises to form a soluble monomeric actin (G-form) which disperses, bringing the sperm plasma membrane closer to the outer acrosomal membrane for fusion during the acrosome reaction.

CONCLUSIONS

This book chapter focuses on several aspects of the acrosome, including its formation and organization during sperm development in the testis, its contents, and its functional significance in the prefertilization events. It is important to keep in mind that recent advances in assisted reproductive procedures (*in vitro* fertilization, IVF; and intracytoplasmic sperm injection, ICSI) appear to sideline the importance of the sperm acrosome and its contents. However, the significance of this enzyme-filled organelle should not be undermined during *in vivo* fertilization. It is our hope that the updated information included in this chapter would allow present and future

38

reproductive biologists to devise new strategies to alter the sequence of events during acrosome formation and regulate the sperm function.

ACKNOWLEDGEMENTS

The authors are indebted to Drs. Malika Bendahmane, Thomas N. Oeltman, and Ms. Lynne Black for critically reading this report and to Phillippe Nguyen for assistance with computer graphics. The editorial assistance of Mrs. Loreita Little and Ms. Lynne Black is gratefully acknowledged. The work presented in this report was supported in part by grants HD25869 and HD34041 from the National Institute of Child Health & Human Development.

REFERENCES

1. Clermont Y., Oko R., Hermo L. "Cell Biology of Mammalian Spermiogenesis." In Cell and Molecular Biology of the Testis, C. Desjardins, L.L. Ewing, eds. Oxford: Oxford University Press, 1993, 332-376.
2. Eddy E.M., O'Brien D.A. "The Spermatozoon." In The Physiology of Reproduction, E. Knobil, J.D. Neill, eds. New York: Raven Press, 1994, 29-77.
3. Abou-Haila A., Tulsiani D.R.P. Mammalian sperm acrosome: formation, contents and function. Arch Biochem Biophys 2000; 379:173-182.
4. Holstein A.F., Schirren C. "Classification of abnormalities in human spermatids based on recent advances in ultrastructure research on spermatid differentiation," In The Spermatozoon, D.W. Fawcett, J.M. Bedford, eds. Baltimore: Urban & Schwarzenberg, Inc., 1979, 341-353.
5. Zaneveld L.J.D., de Jonge C.J. "Mammalian Sperm Acrosomal Enzymes and the Acrosome Reaction." In A Comparative Overview of Mammalian Fertilization, B. Dunbar, M. O'Rand, eds. New York: Plenum Press, 1991, 63-79.
6. Tulsiani D.R.P., Abou-Haila A., Loeser C.R., Pereira B.M.J. The biological and functional significance of the sperm acrosome and acrosomal enzymes in mammalian fertilization. Exp Cell Res 1998; 240:151-164.
7. Ohmura K., Kohno N., Kobayashi Y., Yamagata S., Kashiwabara S., Baba T. A homologue of pancreatic trypsin is localized in the acrosome of mammalian sperm and is released during acrosome reaction. J Biol Chem 1999; 274:29426-29432.
8. Burgess T.L., Kelly R.B. Constitutive and regulated secretion of proteins. Annu Rev Cell Biol 1987; 3:243-249.
9. NagDas S.K., Winfrey V.P., Olson G. Identification of hydrolase binding activities of the acrosomal matrix of hamster spermatozoa. Biol Reprod 1996; 5:405-414.
10. Martinez M.J., Geuze H.J., Ballesta J. Identification of two types of beta-cop vesicles in the Golgi complex of rat spermatids. Eur J Cell Biol 1996; 71:137-143.
11. Moreno R.D., Ramalho-Santos J., Sutovsky P., Chan E.K.L., Schatten G. Vesicular traffic and Golgi apparatus dynamics during mammalian spermatogenesis: Implication for acrosome architecture. Biol Reprod 2000; 63:89-98.
12. Hecht N.E. Molecular mechanisms of male germ cell differentiation. BioEssays 1998; 20:555-561.

13. Kornfeld R., Mellman I. The biogenesis of lysosomes. Annu Rev Cell Biol 1989; 5:483-527.

14. Tulsiani D.R.P., Abou-Haila A. Mammalian sperm molecules that are potentially important in interaction with female genital tract and egg vestments. Zygote 2001; 9:51-69.

15. O'Brien D.A., Gabel C.A., Rockett D.L., Eddy E.M. Receptor mediated endocytosis and differential synthesis of mannose-6-phosphate receptors in isolated spermatogenic and Sertoli cells. Endocrinology 1989; 125:2973-2978.

16. Meyers S.A., Rosenberger A.E. A plasma membrane-associated hyaluronidase is localized to the posterior acrosomal region of stallion sperm and is associated with spermatozoal function. Biol Reprod 1999; 61:444-451.

17. Benoff S. Carbohydrates and fertilization: An overview. Mol Hum Reprod 1997; 3:599-637.

18. Yamagata K., Murayama K., Kohno N., Kashiwabra S., Baba T. p-Aminobenzamidine-sensitive acrosomal protease(s) other than acrosin serve the sperm penetration of the egg zona pellucida in mouse. Zygote 1998; 6:311-319.

19. Guraya S.S. Cellular and molecular biology of capacitation and acrosome reaction in spermatozoa. Int Rev Cyt 2000; 199:1-64.

20. Nikolajczyk B.S., O'Rand M.G. Characterization of rabbit testis β-D-galactosidase and aryl sulfatase A: purification and localization in spermatozoa during the acrosome reaction. Biol Reprod 1992; 46: 366-378.

Chapter 3

DUCTUS EPIDIDYMIS

Gail A. Cornwall
Texas Tech University Health Sciences Center, Lubbock, TX USA

INTRODUCTION

"Of the epididymis: This body may be considered as an appendix to the testis, and its name is derived from its being placed upon this organ, as the testes were anciently called didymi. It is of crescenti form; its upper edge rounded, its lower edge thin. Its anterior and upper extremity is called its caput, the middle part its body, and the lower part its cauda."

Sir Astley Cooper
(1830)

Described more than a century ago the epididymis has been the subject of extensive anatomical, histological, biochemical, and molecular analyses. Although spermatozoa are produced in the testis by the process known as spermatogenesis, often overlooked is the fact that testicular spermatozoa in mammalian species are nonfunctional and require transit through the long, convoluted tubule known as the epididymis to acquire progressive motility and the ability to fertilize an egg. Indeed, following early studies which suggested that sperm acquire motility in the epididymis (1), elegant studies by Bedford (2) and Orgebin-Crist (3) showed that the maturation of sperm motility and fertility in the epididymis was not intrinsically due to sperm cells themselves but rather to exposure of spermatozoa to the luminal environment of the epididymis. These observations set the pace for subsequent studies to identify and study the components in the epididymal fluid that are necessary for the sperm maturation process. More recently, studies have also focused on mechanisms of gene regulation in the epididymal cells as well as using gene knockout approaches to study gene function. While significant progress has been made towards identifying specific molecules that are involved in the regulation of epididymal function as well as the maturation of spermatozoa, to date, a full understanding of the molecular events that occur in the epididymis to allow maturation is not known. Further studies aimed at examining epididymal function are critically important, not only to gain a basic knowledge of the epididymis and sperm maturation, but also to facilitate the development of new clinical therapies including the design of male contraceptives as well as treatment for unexplained male infertility.

This chapter will provide brief discussions of epididymal anatomy and histology, functions of the epididymis, region-specific gene and protein expression, regulation by steroid hormones and testicular factors, and gene knockout mouse models. Although birds and reptiles also require an epididymis for sperm maturation, the discussions in this chapter will focus on studies carried out in mammals, primarily in rodents and humans. The goal of this chapter is not to go into great detail on any given topic but to provide the background and scientific basis for subsequent chapters that discuss in greater detail the luminal fluid components of the epididymis and its regulation by testis factors (Chapter 4), sperm motility (Chapter 5), and sperm maturation (Chapter 6).

EPIDIDYMAL ANATOMY AND HISTOLOGY

The epididymis consists of a long, convoluted tubule that is encapsulated by an extension of the tunica albuginea of the testis to form the epididymis proper. The efferent ducts, which arise from the rete testis, converge to form the single epididymal tubule, which if unraveled, can range from several meters in human to up to 60 meters in the boar (4). On the basis of the gross morphology of its tubule, the epididymis is divided anatomically into three regions, the caput (head), corpus, (body), and cauda (tail) (Figure 1). Although the three anatomical regions can be defined in almost all species, the precise boundaries of these regions are not distinct and vary between species. Each epididymal region is characterized histologically by cell type, cellular dimensions and microvilli, as well as by epididymal tubular diameter and surrounding muscle layer. Furthermore, there are functional differences between epididymal regions with early and later maturation events associated with the caput and corpus regions, respectively and sperm storage occurring in the cauda. The most proximal part of the caput region is distinguished from the more distal parts of the caput by a unique tall columnar epithelium with tall microvilli, and therefore this region is termed the initial segment.

The major epididymal epithelial cell is the principal cell, which, according to Robaire and Hermo (5), comprises 80% of the epididymal epithelium in the rat initial segment with a slight decrease along the tubule to approximately 65% of the epithelium in the caudal region (Figure 2). These

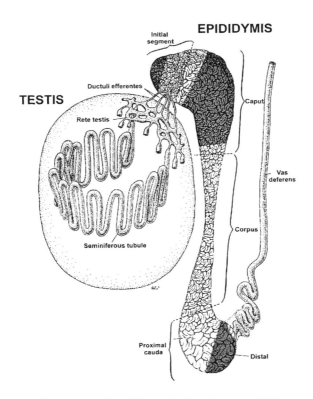

Figure 1. Diagrammatic representation of the testis showing a seminiferous tubule and the rete testis, the ductuli efferentes, the epididymis, and the vas deferens. The shaded regions indicate areas of the different segments of the epididymis, i.e., the initial segment, caput, and distal cauda from which light microscopic photographs were taken for Figure 2. Modified and reprinted with permission from Robaire and Hermo (5), copyright Lippincott, Williams, and Wilkins.

cells exhibit a basal nucleus with apical microvilli. Many small vesicles appear in the apical cytoplasm. The principal cells appear to be the major transporting cells mediating secretion and resorption across the epithelium. Evidence of active absorption by these cells is the presence of microvilli and coated pits at the apical surface and coated vesicles within the apical cytoplasm. Evidence for secretory activity by these cells is less obvious since classical secretory granules are not present. It is thought, however, that the epididymis may utilize an apocrine mechanism of secretion and thus proteins are released from the principal cells by blebbing of the apical plasma membrane (6). In support of this, several studies have described the presence of membrane vesicles within the epididymal lumen (7). Apocrine secretion may be a tissue-specific means to regulate protein function since epididymal proteins would be kept in the neutral pH cytoplasm until secretion into the acidic epididymal lumen.

Throughout the epididymal tubule, the principal cells exhibit regional differences with regard to cell height, length of microvilli, and the tubule with the cells becoming less columnar and more cuboidal in the cauda compared to the caput region (Figure 2). This decrease in cell height allows for the tubular lumen to increase in the cauda epididymis thus allowing for sperm storage (see below).

Basal cells comprise about 20-30% of the epididymal epithelium and are present in all epididymal regions. These cells are located on the basement membrane and interdigitate with the principal cells. These cells are not ciliated. The cytoplasm of these cells contains few organelles but the presence of small pinocytic infoldings of the plasma membrane suggests absorptive activities.

The cells that make up the remaining 10-15% of the epithelium include the narrow, clear, and halo cells. The narrow cells are present only in the initial segment in the adult epididymis, and are tall columnar cells with a dense, elongated nucleus in the upper half of the cell. Although function is not known, the presence of small vesicles in the apical region suggests a role in endocytosis. The narrow cells also are the precursor cells to the clear cells that are present in all regions of the epididymis except the initial segment. These cells are also thought to perform endocytic functions. Halo cells, present in all epididymal regions, have been referred to as lymphocytes but may also be monocytes, precursors to macrophages. The function of these cells is not known but by their description as lymphocytes or macrophages they may perform an immunoprotective role.

Similar to that described in the testis, there is evidence for the presence of a blood-epididymal barrier. Morphological studies of the epididymal epithelium showed the presence of extensive tight junctions between adjacent principal cells (5). Using the tracer lanthanum, Hoffer and Hinton (8) showed that the site of the blood-epididymal barrier was at the luminal surface of the principal cells since lanthanum crossed the blood vessels and the basement membrane of the epididymis, but could not enter the lumen. Other evidence for the presence of a functional barrier is the profound difference in concentrations of specific organic and inorganic components between the epididymal luminal fluid and the blood that is necessary for sperm maturation. Compared to the testis however, fewer studies have been performed examining the blood-barrier in the epididymis and therefore less is known with regard to exactly how tight the barrier is. Functionally, it is logical that the barrier exists in the epididymis as it does in the testis since there are many sperm proteins that the body would recognize as foreign; therefore it is necessary to maintain this immunoprotected site throughout the

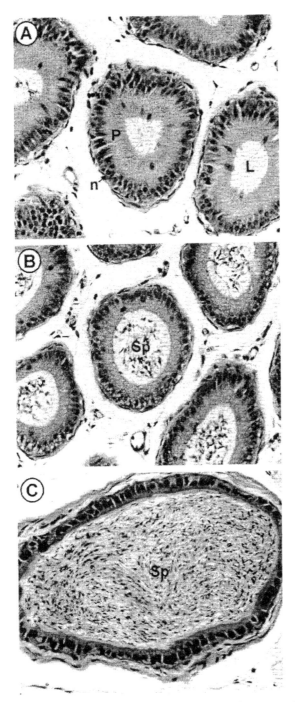

Figure 2. Light microscopic photograph of A) initial segment; B) caput; and C) distal cauda epididymal regions from mouse. All photographs were taken at 40X magnification. P, principal cell; n, nucleus; L, lumen; Sp, spermatozoa.

EPIDIDYMAL FUNCTIONS

Sperm Maturation

The primary function of the epididymis is to create a specialized microenvironment within its lumen to promote the maturation of sperm motility and fertility. Since the presence of the blood-epididymal barrier between epididymal cells prevents molecules from easily passing across the epithelium, the epididymis must utilize its absorptive and secretory activities to create and control its intraluminal environment. A key factor in creating this microenvironment within the lumen is the highly regionalized secretion of ions, organic molecules, and proteins into as well as removal from the lumen. This results in spermatozoa being exposed to a continually changing luminal fluid environment as they migrate from the initial segment to the cauda epididymidis. Exposure of spermatozoa to changing luminal fluid components, described in greater detail below, results in structural, metabolic, and biochemical modifications in the spermatozoa ultimately allowing the maturation of sperm motility and fertility.

Sperm motility and fertility maturation in the epididymis appear to be independent events involving different mechanisms since the actual site where these two events occur in the epididymis are distinct (Figure 3). In all species examined thus far, migration of spermatozoa through the caput region is required for the acquistion of fertilizing ability. However, the site where the first fertilizing spermatozoa are observed differs among species with rodent and human spermatozoa acquiring fertility in the proximal cauda region (5). Furthermore, it appears that the acquisition of fertilizing potential does not coincide with the production of viable offspring. Studies by Orgebin-Crist (3) showed that while spermatozoa from the rabbit corpus will fertilize, viable offspring are not produced. However, when ejaculated rabbit spermatozoa are used, fertilization and development proceeds normally resulting in viable offspring. Additional studies revealed that the loss of embryos appeared to be due to a delay in the fertilization process when corpus spermatozoa were used to fertilize (10).

As sperm progress through the epididymis they also gradually acquire the capacity for progressive motility. The maturation of sperm motility is reflected not only by an increase in the percentage of spermatozoa that are motile but also by qualitative differences in the motility patterns of spermatozoa from different epididymal regions. Spermatozoa removed from the initial segment region and diluted in a physiological buffer exhibit only a weak vibratory motion while spermatozoa from the rat proximal corpus swim in a circular motion (11). Cauda epididymal spermatozoa, however, are like ejaculated spermatozoa and move progressively forward and in straight lines (Figure 3). Detailed studies of sperm flagellar amplitude, frequency, curvature ratio and rate of flagellar beat suggest that the circular motility of immature spermatozoa may be due to an asymmetry of the flagellar beat, whereas the forward progression of cauda spermatozoa is a result of a rotation about the axis of progression (12). Although the mechanisms of sperm motility and motility maturation will be discussed in greater detail in Chapters 5 and 6; respectively, one factor that may contribute to motility maturation is the progressive oxidation of sperm sulfhydryls to disulfide bonds during epididymal transit. It is thought that the formation of disulfide crosslinks in flagellar proteins results in a more rigid sperm flagellum that in turn can affect motility patterns (13).

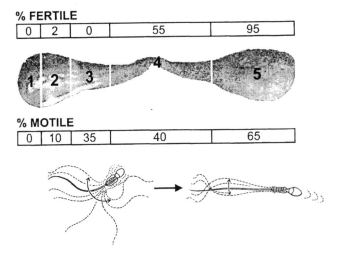

% FERTILE				
0	2	0	55	95

% MOTILE				
0	10	35	40	65

Figure 3. Schematic representation of sperm fertility and motility maturation in the rodent epididymis. 1, initial segment; 2, proximal caput; 3, distal caput; 4, corpus; 5, cauda epididymis. Data taken from Orgebin-Crist (3) and Hinton et al (11).

Sperm Transport

Another function of the epididymis is for the transport of spermatozoa from the testis to the vas deferens. The length of time required for sperm to migrate from region to region as well as through the entire tubule varies among species with a total transport time of approximately 10 days in rodents and a slightly shorter and more variable period of 3-5 days in humans (4). The approximate time required for sperm to migrate from the rat caput to corpus and corpus to cauda is 2-4 days while length of time for sperm to traverse the cauda is longer at 5-6 days (5). Sperm transport time through the cauda epididymidis also can be more variable than in the caput and corpus since it is dependent on ejaculatory frequency. Frequent ejaculations results in accelerated transit times through the cauda epididymis. The mechanisms involved in epididymal sperm transport include the action of cilia, muscular contractions, and hydrostatic pressure (5). Sperm transport in the epididymis appears to be a highly regulated process since following removal of androgens by castration, sperm transport through the rat epididymis increased significantly due to increased contractility while the administration of testosterone reversed this effect (14). Studies also suggest a role for the sympathetic nervous system in sperm transport and storage since following removal of the rat inferior mesenteric ganglion an excessive accumulation of spermatozoa was observed in the cauda epididymis (15).

Sperm Storage

While the caput and corpus epididymal regions are primarily associated with sperm maturation functions, the cauda epididymidis serves as a storage region for functionally mature spermatozoa. This is evident in the rodent epididymis by the presence of densely packed spermatozoa in the lumen of the cauda epididymidis compared to the relatively dilute amounts of spermatozoa in the lumen of the initial segment region (see Figure 2). Depending on the species, spermatozoa can be stored for 30 days and up to several months in the bat epididymis and still remain functional (16). Surprisingly, compared to the rodent, the human cauda epididymidis is not an effective storage site for spermatozoa with the sperm reserves being considerably less than in the rodent. It has been suggested that because sperm transit times through the human epididymis are shorter than in rodents larger sperm reserves in the human cauda epididymis are not required (17).

Absorption and Secretion

Spermatozoa entering the efferent ducts from the testis are highly diluted in testicular fluid and must be concentrated in order for normal maturation to occur in the epididymis. Previously it was thought that the primary site of fluid resorption was the caput epididymidis; however, more recent studies clearly indicate that approximately 90% of the testicular fluid is taken up by the efferent ducts resulting in approximately 8-fold concentration of spermatozoa prior to their entering the initial segment (18). Considering that, like the kidney, the efferent ducts and epididymis are embryonically derived from the mesonephros, it is not surprising that the mechanisms of resorption in the efferent ducts and epididymis are similar to that in the kidney. Studies have identified substantial Na^+-K^+-ATPase activity in the epithelium and thus it appears that a sodium gradient generated by the active sodium pump drives fluid resorption in the efferent ducts. Resorption of testicular fluid also results in a decrease in the pH of fluid from 7.4 in the testis to 6.4 and 6.8 in the caput and cauda epididymidis, respectively (5). The acidification of the luminal fluid is critical for sperm maturation as well as sperm storage and it thought to involve the transport of H^+ and HCO_3^- ions. In support of this, an active Na+/H+ exchanger, carbonic anhydrase and vacuolar proton adenosine triphosphatase (H+-ATPase) have been identified in the epididymal epithelium (19).

Micropuncture techniques have been used to examine changes in the ionic composition of the epididymal luminal fluid from different regions. In the rat, there is a gradual decrease in sodium ions between the caput and cauda while potassium ions increase. Overall, regional differences in the transport of ions result in an epididymal fluid that exhibits decreasing ionic strength from the caput to the cauda epididymidis (5). The osmolarity is maintained in the distal regions of the epididymis by the secretion of organic molecules such as carnitine, glycerylphosphorylcholine, phosphorylcholine, and inositol. Carnitine is present in concentrations in the epididymis that are higher than that found in any other mammalian tissues. While the significance of such high concentrations in the epididymis is not known, carnitine has been proposed to serve as a source of energy for spermatozoa or to be involved in sperm maturation (5). Interestingly, mutant mice that lack carnitine due to a defect in the carnitine transporter gene exhibit epididymal dysfunction characterized by an accumulation of spermatozoa in the proximal epididymis (20). Therefore carnitine may also play important roles in epididymal development or sperm transport. Similar to carnitine, GPC is also present in the epididymal luminal fluid at millimolar concentrations. Both carnitine and GPC are often used as markers of epididymal function in men attending infertility clinics (21).

Other important luminal fluid components include steroids, vitamins, prostaglandins, and peptides such as oxytocin and glutathione. Steroids, which will be discussed further below, include androgens and estrogens, both of which are critical for maintaining normal epididymal function. Vitamins such as retinoids have been shown to be involved in key signaling pathways in the epididymis. The importance of retinoids in the epididymis has been demonstrated by gene knock-out studies which showed that this pathway is essential for maintaining the epididymal epithelium (22). Prostaglandins may play several roles in the epididymis including functions as modulators of epididymal duct contractility thus affecting sperm transport as well as regulators of anion/fluid secretion by the epithelium (23). Similarly, oxtocyin may be involved in sperm transport (24) while glutathione may be important for protecting spermatozoa against oxidative damage (4)

In addition to the transport of water and ions, proteins in the luminal fluid are taken up as well as secreted by the epididymal epithelium. Electron microscopic studies reveal the presence of coated vesicles attached to the cell surface as well as within the cytoplasm of the epithelial cells supporting the mechanism of endocytosis for protein uptake. The epididymal epithelium also exhibits highly regionalized secretory activities as will be discussed below. Combined with the region-dependent removal of proteins from the lumen, these two processes allow for the formation of a luminal fluid environment that is distinct between epididymal regions.

REGIONALIZED GENE AND PROTEIN EXPRESSION

Evidence that proteins secreted by the epididymal epithelium are important for sperm maturation was first demonstrated by Orgebin-Crist and Jahad (25) who showed that androgen-induced maturation of rabbit spermatozoa in cultured epididymal tubules was blocked by inhibition of RNA and protein synthesis. The study of epididymal secretory proteins therefore, has been a major focus for many years as investigators attempt to correlate the appearance and disappearance of particular proteins and their interactions with spermatozoa with changes in sperm function. Investigators have also begun to examine cellular proteins in the epididymis that may serve regulatory or signaling roles. The study of epididymal proteins has been extended to include detailed examinations of genes expressed in a cell and region-dependent manner in the epididymis. As represented schematically in Figure 4, there are distinctive patterns of gene expression in the epididymis with some genes expressed primarily in the initial segment or caput regions, i.e., 5α-reductase, proenkephalin, and thus likely play roles in early maturation events while other genes such as CRISP1 and HE5 are expressed in more distal epididymal regions and may play roles in later maturational events or storage functions. Furthermore, some genes are expressed in a discrete epididymal region, i.e., *cres*, while others exhibit a broader pattern of expression and are expressed by several regions i.e. *Adam7*, E-RABP. To date, the majority of genes that have been examined are expressed by the principal cells. This is not surprising considering that this cell population is the primary cell type in the epididymis and the source of most epididymal secretory proteins. The initial segment/caput is the most metabolically active epididymal region as reflected in Figure 4 by the number of genes expressed in this region.

Genes can also be grouped according to whether they are primarily expressed in the epididymis (reproductive) or are expressed in the epididymis as well as other tissues (somatic) (Figure 4). Genes that are predominantly expressed in the epididymis likely encode proteins with unique functions in the epididymis. Genes expressed in the epididymis as well as nonreproductive tissues may encode proteins that perform the same functions in all tissues or may carry out functions in the epididymis that are distinct from their functions in other tissues. For example, the proenkephalin

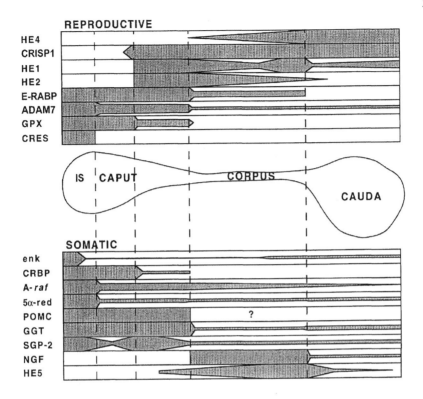

Figure 4. Regional expression of reproductive and somatic genes in the epididymis. IS, initial segment. Modified and reprinted with permission from Cornwall and Hann (26).

gene encodes the opioid peptides met- and leu-enkephalin that function in neural tissue. In the epididymis, larger, atypical enkephalin peptides are present suggesting that these peptides may carry out a unique role in the epididymis.

Outlined in Table 1 is a representative list of genes with known or hypothesized functions in the epididymis. The genes and gene products expressed by the epididymal epithelium are reflective of the many processes that occur in the epididymis that are necessary for sperm maturation. For example, several genes such as DE/CRISP1 and SGP2 encode secretory proteins that bind to spermatozoa during epididymal transit and thus may be directly involved in sperm maturation. Other proteins secreted by the epididymis may directly or indirectly modify existing proteins on the sperm surface. These proteins include the proteases such as matrilysin and CE4 that may be involved in the proteolytic processing of sperm proteins and the cystatin-related protease inhibitor, CRES that may regulate these processing events. Other enzymes modify carbohydrate moieties on existing sperm proteins. Beta galactosidase has been shown in vitro to modify galactosyl residues on sperm surface glycoproteins and thus may perform similar functions in vivo (28). There are also several antioxidant enzymes including gamma glutamyl transpeptidase and superoxide dismutase (E-SOD) whose function may be to protect the spermatozoa from oxidative stress during transit. The Bin1b protein may also serve a protective function in the epididymis. The caput-specific Bin1b protein has been shown to be similar to the antimicrobial peptides, beta-defensins and consequently may play a role in host defense mechanisms (29). The enzyme sulfhydryl oxidase is also expressed in the epididymis in a region-dependent manner and may be involved in the formation of sperm disulfide bonds (30).

Table 1. Gene expression in the epididymis.

GENE	REGION	PROTEIN (kDa)	SECRETED	REGULATION	IDENTITY
ERα	ED, IS	67	no	ND	estrogen receptor
PEA3	IS	62	no	AR/TR	polyomavirus enhancer 3 transcription factor
A-raf	IS	72	no	ND	serine/ threonine kinase
proenk	IS	28	yes	TR	proenkephalin
Cres	IS, pr. caput	14, 19	yes	TR	cystatin-related
Adam7	IS, pr. caput	89	no	AR/TR	A Disintegrin Metalloprotease
B-myc	IS, pr. caput	32	no	AR/TR	transcription factor
BMP8a	IS, pr. caput	45	yes	ND	bone morphogenic/ TGFβ superfamily
5αred	IS, caput	28	no	AR/TR	5α reductase
c-ros	IS, caput	260	no	ND	receptor tyrosine kinase
GGTII-IV	IS, caput	71	yes	AR/TR	γ glutamyl transpeptidase
C/EBPβ	pr. caput	32	no	No	transcription factor
Pem	pr. caput	23	no	AR	homeobox transcription factor
SGP-2	IS, d. caput	73→34, 47	yes	AR	clusterin, ApoJ
B/C E-RABP	d. caput	18.5	yes	AR	retinoic acid binding protein
D/E CRISP1	d. caput	29-32	yes	AR	cysteine-rich secretory protein
GPX	caput	24	yes	AR/TR	glutathione peroxidase
βhex	corpus	63	no	ND	β hexosaminidase
CD52	corpus	26	yes	AR/TR	lymphocyte antigen
CE4	caput, cauda	13	yes	ND	WAP domain
Matrilysin	ED, IS, cauda	28→18	yes	ND	matrix metalloprotease
E-SOD	corpus, cauda	25	yes	ND	superoxide dismutase

ED, efferent duct; IS, initial segment. ND, not determined. AR, androgen-regulated; TR, testis-regulated; pr. caput, proximal caput; d. caput, distal caput. See review by Cornwall et. al. for references (27).

The unique pattern of mRNA expression for each gene in the epididymis is intriguing given that the population of principal cells along the tubule is fairly constant. Therefore, specific control mechanisms must be in place allowing some genes to be expressed by principal cells in one region while neighboring cells do not express the same subset of genes but rather a unique subset of genes. The precise mechanisms that control region and cell-specific gene expression in the epididymis are not known but it is likely that transcriptional initiation may be one such means of regulation. Therefore, it is not surprising that several proteins expressed in a region-dependent manner include signaling and regulatory molecules that may be involved in the regulation of epididymal function. For example, several transcription factors are expressed in a region-dependent manner in the epididymis (Table 1). Specific transcription factors that may be important for caput gene expression include estrogen receptor α, PEA3, C/EBPβ and *Pem*. Gel shift and transfection analyses have demonstrated direct interactions between several of these DNA binding proteins with the promoters of caput expressed genes and so may play key regulatory roles in determining whether these genes are expressed in the epididymis. For example, C/EBPβ protein binds and transactivates the *cres* gene promoter (31) while PEA3 binds to the GGT and GPX gene promoters (32, 33). The B-Myc protein, although belonging to a family of transcriptional regulatory proteins, lacks the DNA binding domain and therefore may regulate gene expression by interactions with other DNA binding proteins (34).

Other regionally expressed genes encode proteins likely involved in signaling pathways and thus these proteins may function by transducing extracellular signals to the nucleus to elicit changes in gene expression. These proteins include the kinases A-*raf* and c-*ros*, BMP8a, Adam7 and E-RABP. Adam7 belongs to a large family of transmembrane proteins with similarity to snake venom proteins. These family members function as metalloproteases or as disintegrins. Because Adam7 protein lacks a consensus metalloprotease active site it is likely that Adam7 function is by its disintegrin domain and therefore may send signals to the nucleus following the binding of luminal factors to its extracellular domain (35). E-RABP is an extracellular retinoic acid binding protein present in the luminal fluid that may be important for the delivery of retinoids to other proteins in the retinoid signaling pathway (36). Glypican, a cell-surface proteoglycan, expressed in the epididymis (Cornwall, unpublished observations) has been shown to bind insulin-like growth factor (IGF2) (37) and thus may be involved in IGF mediated signaling in the epididymis.

REGULATION BY STEROID HORMONES AND TESTICULAR FACTORS

Steroid Hormones

The epididymis is a highly androgen-dependent organ with many of the cellular processes discussed earlier regulated by androgens. Indeed, following the removal of androgens by castration, the epididymis involutes and many of the genes described in Table 1 are no longer expressed resulting in a loss of many epididymal functions. Following the administration of androgens to castrate animals, the epithelium regains its former state with expression of most genes and functions (see below). Metabolic processes including transport of ions and molecules across the epithelium, activity of many glycosidases, proteases, and other enzymes, absorptive and secretory activities, and the functional maturation of sperm motility and fertility are all dependent on the presence of androgens. Although epididymal biosynthesis of androgens has been observed in the epithelium of rodents, most of the androgens present in the epididymis are from testicular fluid, where it is bound to an androgen binding protein or the bloodstream where it is bound to sex hormone-binding globulin (5). Like other steroids, androgens function by binding to an intracellular androgen receptor that then binds to specific DNA response elements in gene promoters and affects gene transcription. In the epididymis testosterone is converted predominantly into 5alpha dihydrotestosterone (DHT) by the enzyme 5α reductase. This enzyme has greater activity in the caput compared to the cauda region.

In addition to androgens, recent studies of the estrogen receptor α showed that estrogens play critical roles in the male reproductive tract. Studies by Hess et al, (see review, 38) showed high levels of estrogen receptors in the efferent ducts and proximal epididymis. Furthermore, in the absence of a functional estrogen receptor α gene, fluid resorption in the efferent ducts was greatly impaired. While it is not clear what the relationship is between androgens and estrogens in their roles in regulating fluid transport, sex steroid hormones are clearly critical molecules in the maintenance and regulation of resorption.

Testicular Factors

The necessity for factors, other than steroid hormones, to maintain epididymal function was first observed by histological analyses. Investigators determined that, following administration of testosterone to castrate animals, all regions of the epididymis except for the initial segment regained cellular morphology similar to that in a noncastrate animal. The finding that the initial segment required factors in addition to androgens for maintenance of morphology and function has been supported in recent years by gene expression studies. Many of the genes (see Table 1) specifically expressed in the initial segment region appear to be no longer expressed following castration as evidenced by the disappearance of mRNA, and remain turned off following the administration of androgen. Other genes expressed in this region regained expression with androgen replacement, but not equivalent to that in the noncastrate animal suggesting that there may be some genes which are entirely dependent on testicular factors for expression, i.e., *cres*, proenkephalin, while other genes require the presence of both androgens and testicular factors, i.e, *Adam7*, GGT. Studies in which the efferent ducts were ligated, thereby preventing the flow of testicular fluid into the epididymis, also showed loss of expression of these genes, suggesting that some component of the testicular fluid is necessary to maintain expression and function in the initial segment.

The observation that the initial segment region, in particular, requires testicular factors for function is intriguing. It is tempting to speculate that the origin of the testicular factors is spermatozoa or factors secreted by Sertoli cells that are in association with spermatozoa. The primary function of these signals from the testis therefore to ensure that the epididymis is expressing important sperm maturation genes and subsequently is ready to receive spermatozoa and begin the maturation process. Along this line of thinking, if spermatogenesis is for some reason disrupted, sperm maturation in the epididymis is not required, and critical genes involved in the initiation of sperm maturation are turned off. In support of this hypothesis, rats administered busulphan to disrupt spermatogenesis, showed a loss of proenkephalin expression in the epididymal initial segment that correlated with the absence of spermatozoa and a regain of expression when spermatogenesis recovered and spermatozoa reentered the epididymis (39).

While the testicular factor(s) involved in maintaining initial segment function are not known, studies suggest that specific growth factors may be involved. Previously, studies had shown that fluid from the rete testis stimulated protein secretion in the rat initial segment region (40). These studies, taken together with the observations that several growth factors including basic fibroblast growth factor (bFGF) are produced by the testis (41) and growth factors have been detected in rete testis fluid (42), prompted Lan and coworkers (43) to test the role of bFGF in the regulation of gamma glutamyl transpeptidase (GGT) activity in the rat initial segment. These studies showed that GGT activity was reduced in the initial segment region following the loss of testicular factors induced by efferent duct ligation and that administration of bFGF but not epidermal growth factor rescued GGT activity to levels similar to that prior to efferent duct ligation. Although further studies examining the regulation of other epididymal proteins by growth factors have yet to be carried out, these studies suggest that growth factors are involved in the testis regulation of epididymal function.

GENE KNOCKOUT MODELS

By far the most powerful approach today to test biological function of a gene is to generate gene knockout mouse models. The specific deletion of a gene and thus its gene product allows the investigator to examine the phenotype of the knockout mouse to determine which biological processes are lost or altered by the absence of the gene. This approach has yielded interesting results, often times unexpected. In particular, thus far most of the gene knockout mouse models with altered epididymal function have been by the deletion of genes expressed in other tissues rather than by the generation of mice lacking a gene specifically expressed in the epididymis. Nonetheless, these mouse models provide valuable tools for the study of epididymal function.

One of the most profound epididymal phenotypes has been generated by the deletion of the c-ros receptor tyrosine kinase gene. Loss of the c-ros kinase resulted in mice lacking the initial segment region and with altered fertility, thus demonstrating that c-ros plays critical roles in epididymal differentiation and emphasizing the importance of initial segment function in sperm maturation (44). In addition to the estrogen receptor α, carnitine transporter, and retinoic acid receptor mouse models described earlier, other knockout models with epididymal phenotypes include deletions of the beta-hexokinase and apolipoprotein B genes. Hexokinase deficient mice had abnormal accumulations of lysosomes in the epididymal epithelial cells thus demonstrating the importance of hexokinase in maintaining the luminal fluid or in endocytosis (45). Mice lacking the apolipoprotein B gene exhibit pronounced fertility defects including reduced sperm motility and lack of fertilizing ability *in vitro* and *in vivo* suggesting that the apolipoprotein B protein plays important roles in epididymal function (46).

CONCLUSIONS

This chapter focuses on broad aspects of epididymal anatomy and histology, function and regulation. The epididymis is an integral part of the male reproductive tract and absolutely critical for the development of functional gametes. However, compared to other organ systems, it is a difficult system to work in due to the lack of appropriate cell lines and in vitro assays to assess sperm function especially in spermatozoa from caput and corpus regions. For these reasons, assigning specific roles in sperm maturation to epididymal secretory proteins has been problematical. Very recently, epididymal cell lines have been developed which hopefully will provide the necessary tools for developing in vitro assays for function as well as for examination of gene promoters (47, 48). Genetic approaches such as transgenic and gene knockout mouse models, briefly discussed in this chapter, have also provided valuable information regarding roles of particular genes in epididymal function. Therefore, it is likely that important insights regarding sperm maturation and epididymal function will be forthcoming.

ACKNOWLEDGEMENTS

The author would like to thank Nelson Hsia for assistance with the computer graphics. The work presented in this chapter was supported in part by grant HD33903 from the National Institute of Child Health and Human Development (G.A.C.).

REFERENCES

1. Young W.C. A study of the function of the epididymis II. Functional changes undergone by spermatozoa during their passage through the epididymis and vas deferens in the guinea pig. J Experimental Biology 1931; 8: 151-162.
2   Bedford J.M. Effect of duct ligation on the fertilizing ability of spermatozoa from different regions of the rabbit epididymis. J Experimental Zoology 1967;166:271-281.
3. Orgebin-Crist M.C. Maturation of spermatozoa in the rabbit epididymis: fertilizing ability and embryonic mortality in does inseminated with epididymal spermatozoa. Annales Biologie Animals Bioch Biophys 1967; 7:373-389.
4. Mann T, and Lutwak-Mann C. "Epididymis and Epididymal Semen", In: Male Reproductive Function and Semen. Springer-Verlag New York, 1981; pp139-159.
5. Robaire B., and Hermo L. "Efferent ducts, epididymis, and vas deferens: Structure, functions, and their regulation". In: Physiology of Reproduction. Volume 1 Eds., E. Knobil and J.D. Neill. Raven Press, New York. 1994; pp.999-1080.
6. Aumuller G.B., Wilhelm B., and Seitz J. Apocrine secretion-fact or artifact? Anat Anz 1999; 181:437-446.
7. Fornes M.W. and De Rosas J.C. Interaction between rat epididymal epithelium and spermatozoa. Anatomical Rec 1991; 231:193-200.
8. Hoffer A.P., and Hinton BT. Morphological evidence for a blood-epididymis barrier and the effects of gossypol on its integrity. Biol Reprod 1984; 30:991-1004.
9. Cyr DG, Robaire B., and Hermo L. Structure and turnover of junctional complexes between principal cells of the rat epididymis. Microsc ResTech 1995; 30:54-66.
10. Orgebin-Crist M.C. Maturation of spermatozoa in the rabbit epididymis: delayed fertilization in does inseminated with epididymal spermatozoa. J Reprod Fertil 1968; 16:29-33.

58

11. Hinton B.T., Dott H.M., and Setchell B.P. Measurement of the motility of rat spermatozoa collected by micropuncture from the testis and from different regions along the epididymis. J Reprod Fertil 1979; 55:167-172.

12. Phillips D.M. Comparative analysis of mammalian sperm motility. J Cell Biol 1972; 53:561-573.

13. Cornwall G.A., Vindivich D., Tillman S., and Chang T.S.K. The effect of sulfhydryl oxidation on the morphology of immature hamster epididymal spermatozoa induced to acquire motility in vitro. Biol Reprod 1988; 39:141-155.

14. Sujarit S., Pholpramool C. Enhancement of sperm transport through the rat epididymis after castration. J Reprod Fertil 1985; 74:497-502.

15. Billups K.L., Tillman S., and Chang T.S.K. Ablation of the inferior mesenteric plexus in the rat: alteration of sperm storage in the epididymis and vas deferens. J Urol 1990; 143:625-629.

16. Setchell B.P., Sanchez-Partida L.G., Chairussyuhur A. Epididymal constituents and related substances in the storage of spermatozoa: a review. Reprod Fertil Dev 1993; 5:601-612.

17. Jones R.C. To store or mature spermatozoa? The primary role of the epididymis. Int J Androl 1999; 22:57-67.

18. Clulow J., Jones R.C., Hansen L.A., and Man S.Y. Fluid and electrolyte reabsorption in the ductuli efferentes testis. J Reprod Fertil Suppl 1998; 53:1-4.

19. Bagnis C., Marsolais M., Biemesderfer D., Laprade R., and Breton S. Na+/H+-exchange activity and immunolocalization of NHE3 in rat epididymis. Am J Physiol Renal Physiol 2001; 280:F426-436.

20. Toshimori, K., Kuwajima M., Yoshinaga K., Wakayama T., and Shima K. Dysfunctions of the epididymis as a result of primary carnitine deficiency in juvenile visceral steatosis mice. FEBS Lett 1999; 446:323-326.

21. Cooper T.G., Yeung C.H., Nashan D., and Nieschlag, E. Epididymal markers in human infertility. J Androl 1988; 9:91-101.

22. Mendelsohn C., Lohnes D., Decimo D., Lufkin T., LeMeur M., Chambon P., and Mark, M. Function of the retinoic acid receptors (RARs) during development (II). Multiple abnormalities at various stages of organogenesis in RAR double mutants. Development 1994; 120:2749-2771.

23. Leung P.S., Chan H.C., Chung Y.W., Wong T.P., and Wong P.Y. The role of local angiotensins and prostaglandins in the control of secretion by the rat epididymis. J Reprod Fertil Suppl 1998; 53:15-22.

24. Harris G.C., Frayne J., Nicholson H.D. Epididymal oxytocin in the rat: its origin and regulation. Int J Androl 1996; 278-286.

25. Orgebin-Crist M.C. and Jahad N. The maturation of rabbit epididymal spermatozoa in organ culture: inhibition by antiandrogens and inhibitors of ribonucleic acid and protein synthesis. Endocrinology 1978; 103:46-53.

26. Cornwall G.A., and Hann S.R. Specialized gene expression in the epididymis. J Androl 1995; 16:379-383

27. Cornwall G.A., Lareyre J.-J., Matusik R.J., Hinton B.T. and Orgebin-Crist M.C. "Gene Expression and Epididymal Function" In: The Epididymis, B.Robaire and B. Hinton, eds., Plenum Press. (In press).

28. Tulsiani D.R.P., Skudlarek M.D., Araki Y., and Orgebin-Crist, M.C. Purification and characterization of two forms of beta-D-galactosidase from rat epididymal luminal fluid: evidence for their role in the modification of sperm plasma membrane glycoproteins. Biochem J 1995; 305:41-50.

29. Li P., Chan H.C., He B., So S.C., Chung Y.W., Shang Q., Zhang Y.D., Zhang Y.L. An antimicrobial peptide gene found in the male reproductive tract system of rats. Science 2001; 291:1783-1785.

30. Chang T.S.K., and Zirkin B.R. Distribution of sulfhydryl oxidase activity in the rat and hamster male reproductive tract. Biol Reprod 1978; 17:745-748.

31. Hsia N., and Cornwall G.A. CCAAT/enhancer binding protein β regulates expression of the cystatin-related epididymal spermatogenic (cres) gene. Biol Reprod 2001;65: 1452-1461 .

32. Lan Z.-J., Palladino M.A., Rudolph D.B., Labus J.C., and Hinton B.T. Identification, expression and regulation of the transcriptional factor polyomavirus enchancer activator 3 (PEA3) and its putative role in regulating the expression of gamma glutamyl transpeptidase mRNA-IV in the rat epididymis. Biol Reprod 1997; 57:186-193.

33. Drevet J.R., Lareyre, J.-J., Schwaab V., Vernet P., and Dufaure J.-P. The PEA3 protein of the Ets oncogene family is a putative transcriptional modulator of the mouse epididymis-specific glutathione peroxidase gene gpx5. Mol Reprod Develop 1998; 49:131-140.

34. Cornwall G.A., Collis R., Xiao Q., Hsia N., and Hann S.R. B-myc, a proximal caput epididymal protein, is dependent on androgens and testicular factors for expression. Biol Repro. 2001; 64:1600-1607.

35. Cornwall G.A. and Hsia N. ADAM7, a member of the ADAM (a disintegrin and metalloprotease) gene family is specifically expressed in the mouse anterior pituitary and epididymis. Endocrinology 1997; 138:4262-4272.

36. Lareyre J.-J., Zheng W.-L., Zhao G.-Q., Kasper S., Newcomer M.E., Matusik R.J., Ong D.E., and Orgebin-Crist M-.C. Molecular cloning and hormonal regulation of a murine epididymal retinoic acid-binding protein. Endocrinology 1998; 139:2971-2981.

37. Pilia G., Hughes-Benzie R.M., MacKenzie A., Baybayan P., Chen E.Y., Hub Cao A., Forabosco A., and Schlessinger D. Mutations in GPC3, a glypican gene, cause the Simpson-Golabi overgrowth syndrome. Nat Genet 1996; 12:241-247.

38. Hess R.A., Bunick D., and Bahr J. Oestrogen, its receptors and function in the male reproductive tract: a review. Mol. Cell. Endocrinology 2001; 178:29-38.

39. Garrett, J.E, Garrett SJ., Douglass J. A spermatozoa-associated factor regulates proenkephalin gene expression in the rat epididymis. Mol Endocrinol 1990; 4:108-118.

40. Sujarot S., Jones R.C., Setchell B.P. Chaturapanich G., Lin M., Clulow J. Stimulation of protein secretion in the initial segment of the rat epididymis by fluid from the ram rete testis. J Reprod Fertil 1990; 88:315-321.

41. Skinner M.K. "Secretion of growth factors and other regulatory factors". In: The Sertoli Cell, 1st ed. Russell L.D., Griswold M.D. (eds). Clearwater: Cache Press; 1993; pp.237-248.

42. Brown K.D., Blakeley D.M., Henville A., Setchell B.P. Rete testis fluid contains a growth factor for cultured fibroblasts. Biochem Biophys Res Commun 1982; 105:391-397.

43. Lan, Z.J., Labus, J.C., and Hinton B.T. Regulation of gamma-glutamyl transpeptidase catalytic activity and protein level in the initial segment of the rat epididymis by testicular factors: role of basic fibroblast growth factor. Biol Reprod 1998; 58: 197-206.

44. Sonnenberg-Riethmacher E., Walter B., Riethmacher D., Godecke S., and Birchmeier C., The c-ros kinase receptor contrls regionalization and differentiation of epithelial cells in the epididymis. Gene Develop 1996; 10:1184-1193.

45. Adamali H., Somani I.H., Huang J.Q., Mahuran D., Gravel R.A., Trasler J.M., and Hermo L. Abnormalities in cells of the testis, efferent ducts, and epididymis in juvenile and adult mice with beta-hexosaminidase A and B deficiency. J Androl 1999; 20:779-802.

46. Huang L.S., Voyiaziakis E., Chen H.L., Rubin E.M., and Gordon J.W. A novel role for apolipoprotein B in male infertility in heterozygous apolipoprotein B knock-out mice. Proc Natl Acad Sci USA 1996; 93:10903-10907.

47. Araki Y., Suzuki K., Matusik R.J., Obinata M., and Orgebin-Crist M.C. Immortalized epididymal cell lines from transgenic mice harboring temperature-sensitive SV40 large T antigen gene. J Androl Suppl 2001; p.195.

60

48. Telgmann R., Brosens J.J., Kappler-Hanno K., Ivell R., and Kirchhoff C. Epididymal epithelium immortalized by simian virus 40 large T antigen: a model to study epididymal gene expression. Mol Human Reprod 2001; 7:935-945.

Chapter 4

THE TESTICULAR AND EPIDIDYMAL LUMINAL FLUID MICROENVIRONMENT

Carmen M. Rodríguez and Barry T. Hinton
University of Virginia Health System, Charlottesville, VA, USA

INTRODUCTION

From their humble beginnings as spermatogonia and throughout their development, spermatozoa are continually exposed to a specialized luminal fluid microenvironment. The composition of this fluid is unique to the ducts of the male reproductive tract and is distinctly different from blood plasma. Moreover, the composition of the testicular and epididymal luminal fluid is not constant but changes dramatically from one region to the next. Remarkably, the changes in the luminal microenvironment occur even along very small distances in the reproductive tract. These changes have been assumed to reflect the needs of spermatozoa as they undergo spermatogenesis in the testes and maturation in the epididymis. However, with the exception of some proteins, few components of the microenvironment have been shown to directly play a role in either testicular or epididymal function. This chapter will review some of the key components of the testicular and epididymal luminal fluid milieu and briefly discuss how this microenvironment is formed. Special attention will be given to the role of transport proteins in regulating the luminal fluid microenvironment. The information presented in this chapter will be primarily from data gathered from studies of the rat male reproductive tract; although wherever possible, data from other species including the human will be included. For more information, the reader is encouraged to consult some of the previously published reviews (1,2).

THE TESTICULAR LUMINAL FLUID MICROENVIRONMENT

Fluid Secretion and Composition

The Sertoli cells form tight junctions that divide the seminiferous epithelium into a basal and an adluminal compartment and form the basis for the blood-testis barrier. A function of the blood-testis barrier is to create a proper environment for meiosis and spermiogenesis to occur (2). Therefore, the tight junctions enable the Sertoli cells to form a specialized luminal fluid microenvironment that is necessary, not only for the development of the germ cells but also for the transport of spermatozoa to the rete testis, efferent ducts and epididymis. Investigators have examined the manner by which fluid is secreted from the testis as well as the fluid composition and mechanisms involved in the formation of the testicular luminal fluid microenvironment.

The first measurement of fluid secretion from the testis was made by Baillie (1) who compared the weights of testes that had been ligated to those that had not been ligated. The increased weight observed in the ligated testis suggested that fluid was secreted by the seminiferous tubules. These studies were later confirmed by Barack, Smith, Setchell and Gustaffson (1). Fluid secretion measurements from a single seminiferous tubule determined that the rat testis secretes approximately 0.5nl fluid/cm/min, which is equivalent to 80 µl fluid/h/testis. The flow rate along a seminiferous tubule is 1 µl/h, which when considering the approximately 30 seminiferous tubules in the rat testis, equals a fluid flow from the testis of about 30 µl/h (1). Fluid secretion increases substantially from approximately 30 to 60 days of age and may be regulated by hormones (1). In addition, fluid secretion from the testis can be affected by the presence and/or absence of germ cells as well as by a number of drugs. For example, acetazolamide, a carbonic anhydrase inhibitor known to decrease fluid secretion in the kidney, also decreases testicular fluid secretion in the rat and ram (1).

The composition of seminiferous tubule fluid (STF) is distinctly different from that of blood plasma. For example, the concentration of potassium in STF is 10 times higher compared to blood plasma whereas the sodium concentration is lower (Table 1). With a pH of approximately 7.3 (1), the osmolality of STF is isosmotic in the rat but hyperosmotic in the hamster (3). In addition to ions, STF also contains inositol (1.8 mM), protein (6 mM), testosterone (40-155 ng/ml) and dihydrotestosterone (1.0-1.5 ng/ml).

Seminiferous tubular fluid (originally called "free flow fluid") was thought to be a mixture of primary fluid, presumably secreted by the Sertoli cells, and rete testis fluid. However, an alternative hypothesis proposed by

Setchell and co-workers suggested that the seminiferous epithelium secretes a potassium bicarbonate-rich fluid, which is later modified to a sodium chloride-rich environment as STF enters the rete testis (1). Moreover, the concentration of other components such as protein and steroids are also modified as STF enters the rete. If this hypothesis is correct, then rete

Table 1: Approximate concentration (mM) of ions within seminiferous tubule fluid (STF) of the rat and rete testis fluid (RTF) of the rat, rabbit, ram, bull, boar, monkey and wallaby. Data from Setchell et al., (2).

Fluids	Na	K	Ca	Mg	Cl	PO$_4$	HCO$_3$
STF							
Rat	108	50	0.44	1.2	120	<0.1	20
RTF							
Rat	143	14	0.81	0.39	140	<0.1	21
Rabbit	147	7.8	0.8	0.5			
Ram	121	11.2	1.0	0.4	128	0.025	8
Bull	133	9.1	0.4	0.4	122	0.017	7
Boar	116	8.8	1.2	-	134	0.22	-
Monkey	136	7.4	-	-	-	-	-
Wallaby	118	14	-	-	137	-	-

testes fluid (RTF) is not representative of fluid that is found within the seminiferous tubule and, therefore, associated with germ cell development. This hypothesis further suggests that the epithelium of the rete testis is far more active in regulating the luminal fluid microenvironment than was previously thought. However, the role of the rete testis in regulating the luminal fluid microenvironment has not been investigated.

The ionic composition of rete testis fluid (RTF) has been studied in a number of species (Table 1). RTF contains some of the same components present in STF including inositol (2.5 mM), protein (1 mM), testosterone (22-46.5 ng/ml) and dihydrotestosterone (1.9-32.7 ng/ml). RTF also contains pregnenolone, progesterone, dehydroepiandrosterone, 5-androstene-3, 17-dione, 5α- androstan-3α, 17β-diol, 5α- androstan-3β, 17β-diol, 5-androstene-3β, 17β-diol, and estrogens. Presumably, the tight junctions located between the cells of the rete testis play a role in maintaining the luminal fluid milieu in this region.

STF and RTF also contain proteins that are found in blood plasma. Sertoli cells secrete a number of proteins including transferrin, ceruloplasmin, saposin (SGP-1, testibumin), clusterin (SGP-2), androgen binding protein, α2-macroglobulin, growth factors including FGF, TGF, interleukins, as well as a number of proteases and protease inhibitors (4-6).

The exact role of many of these proteins during spermatogenesis and spermiogenesis is not known. Interestingly, growth factors secreted by Sertoli cells into the seminiferous tubule appear to play a role in the regulation of gene expression in the epididymis (7). Hence, some of the proteins secreted by Sertoli cells may well have functions in tissues that lie distal to the testis.

EFFERENT DUCT LUMINAL FLUID MICROENVIRONMENT

Fluid Reabsorption and Composition

"During the time the ingredients of the semen are propelled through the very long ducts of the testicles, the semen is elaborated in their cavities in such a way that what was watery and ash-like in the testicles becomes milk and thick in the epididymis"

deGraaf (1668)

DeGraaf (8) was remarkably observant when he noticed that fluid leaving the testis was watery and ash-like but became milky and thick as it passed out of the epididymis. One of the principle functions of the efferent ducts and the epididymis is to reabsorb water resulting in fluid containing a high concentration of spermatozoa. The concentration of sperm within the rat RTF is approximately 10^7 sperm/ml but increases to greater than 10^9 sperm/ml in cauda fluid (9).

Recently, there has been considerable interest in the role of the efferent ducts in the regulation of fluid flow across its epithelium. Net fluid fluxes across the epithelium of the efferent ducts are quite impressive and in one species, the quail, the rate of fluid flux (approx. 100 μl cm^{-2} h^{-1}) is almost equivalent to that of the rat renal proximal tubule (90-190 μl cm^{-2} h^{-1}). Between 50-96% of the fluid leaving the rete testis of several species is reabsorbed by the efferent ducts (10).

The dramatic movement of water out of the efferent ducts results in a luminal fluid composition within the rat coni vasculosi of 144 mM Na$^+$, 5.7 mM K$^+$, 113 mM Cl$^-$, 2.7 mM Mg^{2+}, 2.2 mM Ca^{2+}, and an osmolality of 303 mOsm/kg water (10). The driving force for active solute transport across the efferent ducts is the Na$^+$-K$^+$ ATPase located on the basolateral membranes. However, the passive movement of sodium through the paracellular pathway may exceed that of the sodium pump. Also, it appears that the Na+-H+ exchanger plays a pivotal role in the movement of sodium across the epithelium, and together with carbonic anhydrase provides a mechanism for the reabsorption of bicarbonate. Fluid phase endocytosis also contributes to the movement of water and solutes across the efferent duct epithelium.

Recent evidence suggests that estrogen may play a role in the regulation of fluid transport across the epithelium of the efferent ducts. The efferent ducts of estrogen receptor alpha knock-out (αERKO) mice display defects in water transport and the males are infertile (11). Although further studies are needed to define the role of estrogen in the male reproductive system, estrogen receptors are likely to modulate the expression of genes involved in the regulation of fluid resorption in the efferent ductules. Other factors such as vitamin D, oxytocin, progesterone, and inhibin may also regulate efferent duct epithelial function. Clearly, the regulation of water transport across the efferent duct epithelium is quite complex.

THE EPIDIDYMAL LUMINAL FLUID MICROENVIRONMENT

The epididymal microenvironment is carefully regulated from one epididymal region to the next such that spermatozoa come into contact with a distinct luminal fluid milieu at the appropriate time of maturation. Of great interest to investigators is how the epididymal epithelium generates and regulates an ever-changing microenvironment along a duct that is approximately 3 m in length in the rat and 6 m in the human. The epididymal luminal fluid microenvironment is formed by 1) fluid entering the epididymis from the testis; 2) de novo synthesis, secretion and absorption of proteins by the epididymal epithelium; and 3) the movement of molecules and ions in and out of the epididymis.

Fluid Composition

Protein

The protein composition of epididymal fluid has been studied extensively by a number of investigators. Therefore, protein composition will only be briefly reviewed and the reader is referred to Chapter 3 (Regionalized Gene and Protein Expression) for further information.

The protein concentration within the luminal fluid collected from the proximal regions of the rat epididymis is approximately 25 μg/μl, which is about 25 times the protein concentration found within rete testis fluid. The measured protein concentration within the luminal fluid does not change dramatically from the proximal to the distal regions of the epididymis.

However, if water reabsorption is taken into account, assuming that protein is neither degraded nor absorbed, the calculated concentration of protein within cauda luminal fluid is greater than 100 µg/µl. The difference between the calculated concentration and the measured concentration is likely due to protein endocytosis by the epididymal cells, protein degradation and/or association of secreted epididymal proteins with the maturing spermatozoa. Spermatozoa are exposed to a highly dynamic protein microenvironment that is constantly changing from one region to the next. New proteins are secreted into distinct epididymal regions whereas others are absorbed. The net protein composition found within the distal epididymal regions is a mixture of testicular proteins, de novo synthesis and secretion of epididymal proteins, modified testicular and epididymal proteins and sperm-associated proteins that have become dissociated from sperm.

The epididymal luminal fluid contains a repertoire of proteins including: protease inhibitors, acrosome stabilizing proteins, multiple enzymes (for example, proteases, gamma-glutamyl transpeptidase, glycosyltransferases, glycosidases, glutathione peroxidase), protective proteins such as the defensins, growth factors (FGF for example), proteins involved in sperm-egg binding, and sperm-immobilizing agents (immobilin). Some of these proteins may have specific functions within the epididymis, whereas others are used by spermatozoa when they reach the female reproductive tract.

Ions, water and organic solutes

The epididymal luminal fluid microenvironment contains a number of ions, organic solutes and macromolecules. Studies by Levine and Marsh demonstrated that the ionic composition of epididymal fluid changes as this fluid progresses from the caput to the cauda (Fig. 1) (12). These investigators also estimated the osmotic deficit present across the epididymis. The osmotic deficit is the difference, in mM, between the measured osmolality and the concentration of osmotically active ions. When the total concentrations of organic solutes and ions were compared to the osmolality, Levine and Marsh noticed that there was an estimated osmotic deficit of 150 mM, 225 mM and 250 mM for the caput, corpus and cauda, respectively. Levine and Marsh postulated that the epididymis had to secrete organic solutes to make up the deficit. Indeed, studies demonstrated that epididymal luminal fluid contains high levels of organic solutes including inositol, L-carnitine, glutamate, taurine, hypotaurine, sialic acids, glycerophosphorylcholine, phosphorylcholine, and a number of other amino acids. The presence of these organic solutes appears to make up over 50% of the calculated osmotic deficit. Nonetheless, an osmotic deficit of greater than 50 mM is still present in the corpus and cauda suggesting that

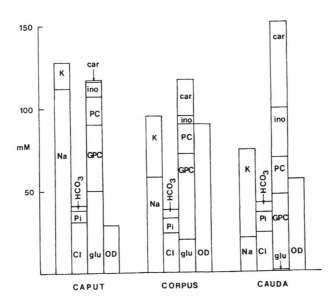

Figure 1: Mean concentration of ions and organic solutes within the luminal fluid of the rat caput, corpus and cauda epididymidis. Na - sodium; K- potassium; Cl - chloride; Pi - inorganic phosphate; HCO3 - bicarbonate; car - L-carnitine; ino - inositol; PC - phosphorylcholine; glu - glutamate; GPC - glycerylphosphorylcholine; OD - osmotic deficit. Reproduced from Hinton (12) with permission from the American Society of Andrology.

additional solutes are secreted into the lumen in these regions. However, other factors could also account for the estimated osmotic deficit observed in the corpus and cauda epididymides. The measured concentration of each solute, distal to its site of secretion, is underestimated if water reabsorption is taken into account. For example, the measured concentration of inositol in rete testis fluid and the vas deferens is 2.49 and 49.17 mM, respectively (13). However, if water reabsorption from the rete testis to the vas deferens is taken into account, calculated concentration of inositol within the vas deferens luminal fluid would be 107.3 mM. Hence, the difference (107.3 - 49.17 = 58.13 mM) suggests that the epididymal epithelial cells are reabsorbing much of the secreted solutes. Therefore, further secretion of inositol into the epididymal duct would not have to be postulated to account for the osmotic deficit. The bidirectional movement of inositol and other solutes is likely to be necessary to maintain the appropriate physiological concentration of that solute within the epididymal lumen. Presumably, specific transporting processes are responsible for the movement of solutes out of the epididymal lumen. Alternatively, metabolism by either spermatozoa and/or the epididymis would also account for the difference.

The role of organic solutes in epididymal function is not known although there is plenty of speculation. Many of the organic solutes present in the epididymal lumen may play a role in the osmoregulation of sperm and/or the epididymal epithelium. Organic solutes are ideal molecules to regulate water movement across epithelia because they are nonperturbing to a cell's interior (14). In contrast, ions are considered to be perturbing because cells cannot tolerate high intracellular concentration of ions. Interestingly, the osmolality of cauda fluid from several hibernating bat species is over 1000 mOsm/kg water (15), but the solute(s) responsible for maintaining this extraordinary high osmolality is/are not known. The kidney utilizes solutes such as glycerophosphorylcholine (GPC) and inositol to move water in and out of cells (16,17). Taurine, which is also found in relatively high concentrations in epididymal luminal fluid, may also be a prime candidate for the regulation of water across the epididymal epithelium (12). Osmoregulation appears to play a critical role in male fertility as demonstrated by the phenotype of the tyrosine kinase, c-Ros, knock-out mouse. The male mice have underdeveloped initial segments, sperm flagellar defects and are infertile (18). The sperm flagellar defects are thought to be the consequence of the inability of spermatozoa to regulate their volume (19). These defects in turn impair sperm motility resulting in infertility. Water reabsorption along the epididymal duct also results in dramatic changes in the concentration of several organic solutes. In addition to playing a role in osmoregulation, other solutes such as L-carnitine have been postulated to play a role in the acquisition of sperm motility. L-carnitine has also been implicated in the protection of oxidative stress in cultured neurons and may well have a similar role in the epididymis.

Ions and organic solutes may also play different roles in different species. For example, the luminal fluid of the human vas deferens contains many of the organic solutes found in the same tissue of other species; however, the concentration is almost an order of magnitude lower (20). Though much lower than in other species, the concentrations of organic solutes in the human vas deferens may still be considered to be high. Nevertheless, such differences in concentrations may denote differences in function and warrant further study.

The elaborate tight junctional network present between the epididymal epithelial cells is critical for the maintenance of the specialized luminal fluid and forms the basis of the blood-epididymis barrier. Permeability studies illustrate that the epididymis does not readily allow molecules to cross its epithelium (21-22). Therefore, the epididymis utilizes a series of specific cellular pathways and transporters to generate a specialized luminal fluid microenvironment. Figure 2 (23) summarizes the movement of ions, water, organic solutes and protein across different regions of the rat epididymis.

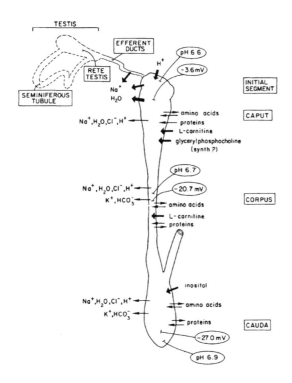

Figure 2: Summary of the movements of ions, water, organic solutes and macromolecules across different regions of the rat epididymis. Transepithelial potential difference and intraluminal pH are also shown. Thickness of arrows reflects degree of movement. Reproduced from Hinton and Turner (23) with permission from the American Physiology Society.

Transport Systems

The flow of ions, water, organic solutes and macromolecules across epithelia is regulated by specific transport systems. Despite their importance in regulating the luminal fluid microenvironment, the transporters involved in the movement of molecules across the testicular and epididymal epithelia are just beginning to be characterized.

Studies have established that the efferent ducts reabsorb the majority of the rete testis fluid. Osmotic gradients and hydrostatic pressure differences can drive water across cells via the transcellular and paracellular pathways. The paracellular pathway entails the movement of water through the apical junctions. In contrast, the transcellular pathway consists of water movement through the lipid bilayer or via the use of transport proteins (Fig. 3) (24). Indeed, substantial quantities of water pass directly through the lipid bilayer. However, in tissues such as the kidney, water channels known as aquaporins are critical transporters of water across membranes. In view of the importance of aquaporins in fluid transport, investigators quickly turned to examine the expression of these proteins in the efferent ducts and epididymis. Aquaporin-1 (AQP-1) was localized to the apical brush border membrane of efferent ducts (25,26) and estrogens were implicated as regulators of AQP-1 expression during development (26). Surprisingly, AQP-1 knock-out mice are fertile (10) suggesting that either the paracellular pathway or other water channels or transporters may also be involved in water movement in the efferent ducts. Aquaporin-9 was localized in the apical region of the efferent duct and principal cells of the epididymis as

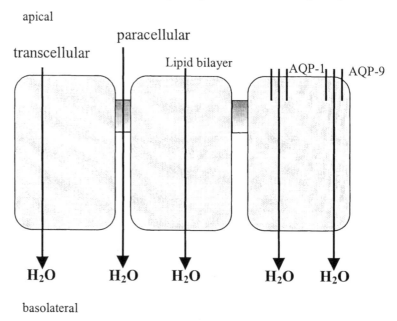

Figure 3: Diagram of the transcellular and paracellular pathways involved in water transport in the efferent ducts. The transcellular pathways include the movement of water through the lipid bilayer as well as movement via water channels or transporters. AQP-1 and AQP-9 have been localized to the apical region of the efferent ducts. Diagram adapted from Verkman (24).

well as the Leydig cells in the testis (27,28). In addition to the movement of water, AQP-9 has high permeability for solutes such as carbamides, polyols, purines and pyrimidines and may also be involved in the movement of these solutes within the male reproductive system.

The pH of the intraluminal fluid within the seminiferous tubule becomes progressively more acidic as it travels along the epididymis. Acidification of epididymal fluid and maintenance of a low bicarbonate concentration have been suggested to maintain spermatozoa in an immotile, quiescent state. Sperm motility is triggered, during ejaculation, when the epididymal fluid is alkalinized by the secretions of the accessory sex glands. Manipulation of luminal pH has been suggested as a possible way to regulate male fertility (29). Therefore, a number of studies have attempted to identify the transport systems responsible for luminal acidification and bicarbonate absorption.

Investigators proposed that an apical Na^+/H^+ exchanger was involved in proton (H^+) secretion in the epididymis (30). In fact, immunohistochemistry studies have identified a number of Na^+/H^+ exchangers including NHE1, NHE2 and NHE3 (31,32). Not surprisingly, these transporters are differentially expressed along the different regions of the epididymis. NHE1 was localized to the basolateral side of epididymal cells along the entire epididymis and may be involved in HCO_3^- secretion. NHE2 is present on the apical side of the epithelium of the caput, corpus and cauda and may be involved in Na^+ reabsorption (31). NHE3 is highly expressed on the apical membrane in all cells of the caput (32).

An additional mechanism involved in proton (H^+) secretion was proposed when a vacuolar H^+-ATPase was identified in selected cells within the epididymis and vas deferens (33,34). In the vas deferens, a bafilomycin-sensitive H^+ATPase is responsible for a large fraction of the measured proton secretion (34). The apical vacuolar H^+-ATPase in conjunction with a basolateral $Na(HCO_3)$ cotransporter were proposed to mediate the transepithelial bicarbonate absorption. A Na^+/HCO_3^- cotransporter (NBC) protein was localized to the basolateral region of the principal cells with highest expression observed in the proximal regions of the epididymis (35). An NBC isoform, NBC3, was later found to be highly expressed on the apical membrane of narrow cells in the caput epididymis and light clear cells in the corpus and cauda epididymidis (32). The co-localization of NBC3 with the vacuolar H^+-ATPase in specialized acid-secreting cells in the epididymis prompted investigators to propose a model by wherein these two transporters are in close association (32). Studies have suggested that some transporters may associate with other transport proteins to regulate their functions. For example, the Na^+/H^+ exchanger regulatory factor, NHE-RF, may bind and regulate the function of the cystic fibrosis transmembrane conductance regulator (36).

The cystic fibrosis transmembrane conductance regulator (CFTR) is a cAMP-activated chloride channel that also participates in the regulation of the luminal fluid microenvironment of the epididymis. Present in the apical membrane of the principal cells of the efferent duct (37), CFTR may prevent excessive dehydration of the lumen by providing a counterbalance to the absorptive activities that predominate in the efferent ducts (37). Mutations in the CFTR gene are responsible for cystic fibrosis, an autosomal recessive disease observed primarily in caucasians. Men with cystic fibrosis display abnormal epididymides and vasa deferentia (38). Whether these abnormalities are the results of morphological defects arising during development or the lack of proper chloride transport is unclear. The low levels of CFTR expression in the epithelium of the post-natal human epididymis were suggested to indicate that this organ was more sensitive to CFTR dysfunction than other developing organs (39). However, CFTR knock-out mice appear to have normal testicular and epididymal development (40).

In addition to transporting ions, a number of transport systems are responsible for the transport of organic solutes. An active transport system comprised of both a basolateral and an apical transporter was suggested to be involved in the transport of L-carnitine across the epididymal epithelium (41). L-carnitine is transported against a concentration gradient greater than 2000-fold, with intraluminal concentrations reaching as high as 50 mM in the cauda epididymides of the rat (42). Studies in our laboratory have identified a transporter that may be responsible for L-carnitine transport in the epididymis. OCTN2, a member of the organic cation transporter family, is expressed in the rat epididymis in a region-dependent manner with highest expression observed in the regions previously shown to be involved in L-carnitine uptake (43). Moreover, OCTN2 protein is localized to the basolateral region of epididymal cells suggesting that this protein is likely responsible for the transport of L-carnitine across the epididymal epithelium (43). Mutations in OCTN2 are responsible for primary carnitine deficiency in the juvenile visceral steatosis (jvs) mouse (44) and these mice were found to display a number of epididymal abnormalities (45).

Clearly, transport systems are critical controllers of the luminal fluid microenvironment of the epididymis. However, transport systems are also being characterized in the testis and other regions of the male reproductive tract. Table 2 lists some of the transport proteins that have been identified in the testis, efferent ducts, epididymis and vas deferens. Many of these transporters may serve as potential targets of male fertility. By targeting testicular and epididymal transporters, investigators may be able to further define the role of the luminal components in sperm development and maturation. Such studies may provide fundamental information for the treatment of certain forms of male infertility and for the development of a male contraceptive.

CONCLUSIONS

Fluid homeostasis is critical for normal development and function of the male reproductive system particularly the testis and epididymis. Alterations in fluid balance can have adverse effects and may be the source of certain forms of male infertility. The tight junctional networks in the testis and epididymis ensure that spermatozoa are exposed to a specialized

Table 2: Transport proteins identified in the testis, efferent ducts, epididymis and vas deferens

Transporter	Function	Tissues expressed or localized	Reference
Aquaporin 1 (AQP-1)	Water channel	Efferent ducts, epididymis	(25,26)
Aquaporin 7 (AQP-7)	Water channel	Developing testis, adult testis	(46,47,48)
Aquaporin 8 (AQP-8)	Water channel	Developing testis	(48,49)
Aquaporin 9 (AQP-9)	Water channel	Apical region of efferent ducts, principal cells of epididymis, Leydig cells	(28)
Anion exchanger 2 (AE2)	Cl/HCO$_3$ exchanger	Epididymis	(50)
Cystic fibrosis transmembrane conductance regulator (CFTR)	cAMP-regulated chloride channel	Efferent ducts, vas deferens, developing epididymis	(37,39)
H$^+$ATPase	Proton pumping H$^+$ATPase	Efferent ducts, epididymis, vas deferens	(29,33,34)
Concentrative Nucleoside transporter 2 (CNT2)	Nucleoside transporter	Epididymis	(51)
Equilibrative Nucleoside transporter (ENT1)	Nucleoside transporter	Epididymis	(51)
Equilibrative Nucleoside transporter 2 (ENT2)	Nucleoside transporter	Epididymis	(51)
Glutx1/Glut8	Glucose transporter	Testis and epididymis	(52,53,54)
Na$^+$/dicarboxylate cotransporter (SDCT1)	Sodium dicarboxylate cotransporter	Initial segment and proximal regions of epididymis	(55,56)
Na$^+$-driven Cl-HCO$_3$ exchanger (NDCBE1)	Cl/HCO$_3$ exchanger	Testis	(57)
Na$^+$/H$^+$ exchanger 1 (NHE1)	Na$^+$/H$^+$ exchanger	Basolateral surface of epididymis	(31)
Na$^+$/H$^+$ exchanger 2 (NHE2)	Na$^+$/H$^+$ exchanger	Apical surface of epididymis	(31)
Na$^+$/H$^+$ exchanger 3 (NHE3)	Na$^+$/H$^+$ exchanger	Testis	(32)
NBC	Na$^+$/HCO$_3^-$ cotransporter	Basolateral surface of proximal epididymis	(35)
NBC3	Na$^+$/HCO$_3^-$ cotransporter	Apical surface of caput epididymidis	(32)
Novel organic cation transporter (OCTN1)	Organic cation transporter	Epididymis	(43)
Novel organic cation/carnitine transporter (OCTN2)	Organic cation/carnitine transporter	Distal caput, corpus, proximal cauda epididymides	(43)
Novel organic cation transporter (OCTN3)	Organic cation transporter	Mouse testis	(43,58)
Testis anion transporter 1 (Tat1)	Anion transporter	Human testis	(59)

luminal fluid microenvironment at their appropriate time of development and maturation. These junctional complexes restrict the movement of molecules across the seminiferous and epididymal epithelium allowing transport systems to become critical controllers of the luminal fluid

74

microenvironment. Proteins involved in the transport of water, ions, solutes and other macromolecules are quickly beginning to be characterized. The advent of new molecular biology tools such as those providing possible tight reversible regulation of genes (60), may provide the means to further define the role of transport proteins in providing the proper luminal fluid microenvironment required for sperm development and maturation.

ACKNOWLEDGEMENTS

The authors acknowledge the support from The Rockefeller Foundation / Ernst Schering Research Foundation, NIH-NICHD HD 32979, Training Grant NIH T-32-HD07382 and the NICHD/NIH through cooperative agreement (U54 HD28934) as part of the Specialized Cooperative Centers Program in Reproduction Research.

REFERENCES

1. Hinton BT and Setchell BP. Fluid secretion and movement. In: The Sertoli Cell Eds Russell LD and Griswold MD, Cache River Press, Clearwater, FL 1993; pp 249-267.
2. Setchell BP, Maddocks S and Brooks DE. Anatomy, vasculature, innervation, and the fluids of the male reproductive tract. In: The Physiology of Reproduction. eds: E Knobil and JD Neill, Raven Press, 1994; pp1063-1175.
3. Johnson AL and Howards S. Hyperosmolality in intraluminal fluids from hamster testis and epididymis: a micropuncture study. Science 1977; 195:492-493.
4. Griswold MD. Protein secretion by Sertoli cells: General considerations. In: The Sertoli Cell, eds LD Russell and MD Griswold, Cache River Press FL. 1993; pp 195-200.
5. Skinner MK. Secretion of growth factors and other regulatory factors. In: The Sertoli Cell, eds LD Russell and MD Griswold, Cache River Press FL, 1993; pp 237-247.
6. Sylvester, SR. Secretion of transport and binding proteins. In : The Sertoli Cell, eds LD Russell and MD Griswold, Cache River Press, Clearwater, FL, 1993; pp 201-216.
7. Lan ZJ, Labus JC and Hinton BT. Regulation of gamma-glutamyl transpeptidase catalytic activity and protein level in the initial segment of the rat epididymis by testicular factors. Role of basic fibroblast growth factor. Biol Reprod 1998; 58:197-206.
8. de Graaf R 1668 Tractatus de Virorum Generationi Inservientibus. Hackiana, Lugduni Batavorum. Translated Jocyln HD and Setchell BP. Regnier de Graaf on the human reproductive organs. J Reprod Fertil 1972; Suppl 17.
9. Turner TT. Resorption versus secretion in the rat epididymis. J Reprod Fertil 1984; 72:509-514.
10. Hess RA. The efferent ductules; structure and functions. In: The Epididymis: From molecules to clinical practice, a comprehensive survey of the efferent ducts, the epididymis and the vas deferens. Edited by B Robaire and BT Hinton. Kluwer Academic/Plenum Publishers, 2002; pp 49-80.
11. Hess RA, Zhou Q and Nie R. The role of estrogens in the endocrine and paracrine regulation of the efferent ductules, epididymis and vas deferens. In: The Epididymis: From molecules to clinical practice, a comprehensive survey of the efferent ducts, the epididymis and the vas deferens. Edited by B Robaire and BT Hinton. Kluwer Academic/Plenum Publishers, 2002; pp 317-337.

12. Hinton BT. The testicular and epididymal luminal amino acid microenvironment in the rat. J Androl 1990; 11:498-505.

13. Hinton BT, White RW and Setchell BP. Concentrations of myo-inositol in the luminal fluid of the mammalian testis and epididymis. J Reprod Fertil 1980; 58:395-399.

14. Yancey PH. Clark ME. Hand SC. Bowlus RD. Somero GN. Living with water stress: evolution of osmolyte systems. Science 1982; 217:1214-1222.

15. Crichton EG, Hinton BT, Pallone TL and Hammerstedt RH. Hyperosmolality and sperm storage in hibernating bats: prolongation of sperm life by dehydration. Am J Physiol 1994; 267:1363-1370.

16. Nakanishi T, Turner RJ and Burg MB. Osmoregulatory changes in myo-inositol transport by renal cells. Proc Natl Acad Sci USA 1989; 86:6002-6006.

17. Nakanishi T and Burg MB. Osmoregulation of glycerylphosphocholine content of mammalian renal cells. Am J Physiol 1989; 257:C795-C801.

18. Sonnenberg-Riethmacher E, Walter B, Riethmacher D, Godecke S and Birchmeier C. The c-ros tyrosine kinase receptor controls regionalization and differentiation of epithelial cells in the epididymis. Genes Develop 1996; 10:1184-1193.

19. Yeung CH, Wagenfeld A, Nieschlag E and Cooper TG. The cause of infertility of male c-ros tyrosine kinase receptor knockout mice. Biol Reprod 2000; 63:612-618.

20. Hinton BT, Pryor JP, Hirsch AV and Setchell BP. The concentration of some inorganic and organic compounds in the luminal fluid of the human ductus deferens. Int J Androl 1981; 4:457-461.

21. Cooper TG and Waites GMH. Investigation by luminal perfusion of the transfer of compounds into the epididymis of the anesthetised rat. J Reprod Fertil 1979; 56:159-164.

22. Hinton BT and Howards SS. Permeability characteristics of the epithelium in the rat caput epididymidis. J Reprod Fert 1981; 63:95-99.

23. Hinton BT and Turner TT. Is the epididymis a kidney analogue? Am Physiol Soc 1988; 3:28-31.

24. Verkman AS. Lessons on renal physiology from transgenic mice lacking aquaporin water channels. J Am Soc Nephrol 1999; 10:1126-35.

25. Brown D, Verbavatz JM, Valenti G, Lui B and Sabolic I. Localization of the Chip28 water channel in reabsorptive segments of the rat male reproductive tract. Eur J Cell Biol 1993; 61:264-273.

26. Fisher JS, Turner KJ, Fraser HM, Saunders PTK, Brown D, and Sharpe RM. Immunoexpression of Aquaporin-1 in the efferent ducts of that rat and marmoset monkey during development, its modulation by estrogens, and its possible role in fluid resorption. Endocrinology 1998; 139:3935-3945.

27. Elkjær M-L, Vajda Z, Nejsum LN, Kwon T-H, Jensen UB, Amiry-Moghaddam M, Frøkiær J and Nielsen S. Immunolocalization of AQP9 in liver, epididymis, testis, spleen and brain. Biochem Biophys Res Commun 2000; 276:1118-1128.

28. Pastor-Soler N, Bagnis C, Sabolic I, Tyszkowski R, McKee M, Van Hoek A, Breton S and Brown D. Aquaporin 9 expression along the male reproductive tract. Biol Reprod 2001; 65:384-393.

29. Brown D, Smith PJS and Breton S. Role of V-ATPase-rich cells in acidification of the male reproductive system. J Exp Biol 1997; 200:257-262.

30. Au CL and Wong PYD. Luminal acidification by the perfused rat cauda epididymidis. J Physiol (Lond) 1980; 309:419-427.

31. Cheng Chew SB, Leung GPH, Leung PY, Tse CM and Wong PYD. Polarized distribution of NHE1 and NHE2 in the rat epididymis. Biol Reprod 2000; 62:755-758.

32. Pushkin A, Clark I, Kwon TH, Nielsen S and Kurtz I. Immunolocalization of NBC3 and NHE3 in the rat epididymis: colocalization of NBC3 and the vacuolar H+-ATPase. J Androl 2000; 21:708-720.

33. Brown D. Lui B. Gluck S. Sabolic I. A plasma membrane proton ATPase in specialized cells of rat epididymis. Am J Physiol 1992; 263:C913-916.

34. Breton S, Smith PJ, Lui B, Brown D. Acidification of the male reproductive tract by a proton pumping (H+)-ATPase. Nature Medicine 1996; 2:470-472.

35. Jensen LJ, Schmitt BM, Berger UV, Nsumu NN, Boron WF, Hediger, Brown D, Breton S. Localization of sodium bicarbonate cotransporter (NBC) protein and messenger ribonucleic acid in rat epididymis. Biol Reprod 1999; 60:573-579.

36. Wang S, Raab RW, Schatz PJ, Guggino WB and Li M. Peptide binding consensus of the NHE-RF-PDZ1 domain matches the C-terminal sequence of cystic fibrosis transmembrane conductance regulator (CFTR). FEBS Lett 1998; 427:103-108.

37. Leung GPH, Gong XD, Cheung KH, Cheng-Chew SB and Wong PYD. Expression of cystic fibrosis transmembrane conductance regulator in rat efferent duct epithelium. Biol Reprod 2001; 64:1509-1515.

38. Wong PYD. CFTR gene and male fertility. Mol Hum Reprod 1998; 4:107-110.

39. Tizzano EF, Chitayat D and Bushwald M. Cell-specific localization of CFTR mRNA shows developmentally regulated expression in human fetal tissues. Hum Mol Genet 1993; 2:219-224.

40. Reynaert I, van der Schueren B, Degeest G, Manin M, Cuppens H, Scholte B and Cassiman J-J. Morphological changes in the vas deferens and expression of the cystic fibrosis transmembrane conductance regulator (CFTR) in control, ΔF508 and knock-out CFTR mice during postnatal life. Mol Reprod Dev 2000; 55:125-135.

41. Yeung CH, Cooper TG and Waites GM. Carnitine transport into the perfused epididymis of the rat: regional differences, stereospecificity, stimulation by choline, and the effect of other luminal factors. Biol Reprod 1980; 23:294-304.

42. Hinton BT, Snoswell AM, Setchell BP. The concentration of carnitine in the luminal fluid of the testis and epididymis of the rat and some other mammals. J Reprod Fertil 1979; 56:105-111.

43. Rodríguez CM, Labus JC and Hinton BT. The organic cation/transporter, OCTN2, is differentially expressed in the adult rat epididymis. *Unpublished*

44. Lu K, Nishimori H, Nakamura Y, Shima K, Kuwajima M. A missense mutation of mouse OCTN2, a sodium dependent carnitine cotransporter, in the juvenile visceral steatosis mouse. Biochem Biophys Res Commun 1998; 252:590-594.

45. Toshimori K, Kuwajima M, Yoshinaga K, Wakayama Y, Kenji Shima. Dysfunctions of the epididymis as a result of primary carnitine deficiency in juvenile visceral steatosis mice. FEBS Lett 1999; 446:323-326.

46. Suzuki-Toyota F, Ishibashi K and Yuasa. Immunohistochemical localization of a water channel aquaporin 7 (AQP7), in the rat testis. Cell Tissue Res 1999; 295:279-285.

47. Ishibashi K, Kuwahara M, Gu Y, Kageyama Y, Tohsaka A, Suzuki F, Marumo F, and Sasaki S. Cloning and Functional Expression of a New Water Channel Abundantly Expressed in the Testis Permeable to Water, Glycerol, and Urea. J Biol Chem 1997; 272: 20782-20786.

48. Calamita G, Mazzone A, Bizzoca A and Svelto M. Possible involvement of aquaporin-7 and −8 in rat testis development and spermatogenesis. Biochem Biophys Res Commun 2001; 288:619-625.

49. Calamita G, Mazzone A, Cho YS, Valenti G and Svelto M. Expression and localization of the aquaporin-8 water channel in rat testis. Biol Reprod 2001; 64:1660-1666.

50. Jensen LJ, Stuart-Tilley AK, Peters LL, Lux SE, Alper SL and Breton S. Immunolocalization of AE2 anion exchanger in rat and mouse epididymis. Biol Reprod 1999; 61:973-980.

51. Leung GP, Ward JL, Wong PY, Tse CM. Characterization of nucleoside transport systems in cultured rat epididymal epithelium. American Journal of Physiology - Cell Physiology. 2001; 280(5):C1076-82.

52. Doege H. Schurmann A. Bahrenberg G. Brauers A. Joost HG. GLUT8, a novel member of the sugar transport facilitator family with glucose transport activity. J Biol Chem 2000; 275:16275-16280.

53. Ibberson M. Uldry M. Thorens B. GLUTX1, a novel mammalian glucose transporter expressed in the central nervous system and insulin-sensitive tissues. J Biol Chem 2000; 275:4607-4612.

54. Rodríguez CM, Katz E, Du XQ, Hinton BT, Charron M. Expression of Glutx1, a member of the facilitative glucose transporter family, in the rat testis and epididymis. *Unpublished*

55. Chen X-Z, Shayakul C, Berger UV, Tian W, Hediger MA. Characterization of a rat Na+-dicarboxylate cotransporter. J Biol Chem 1998; 273:20972-20981.

56. Jervis KM and Robaire B. Dynamic changes in gene expression along the rat epididymis. Biol Reprod 2001; 65:696-703.

57. Grichtchenko II, Choi I, Zhong X, Bray-Ward P, Russell JM and Boron WF. Cloning, characterization, and chromosomal mapping of a human electroneutral Na(+)-driven Cl-HCO3 exchanger. J Biol Chem 2001; 276:8358-8363.

58. Tamai I, Ohashi R, Nezu JI, Sai Y, Kobayashi D, Oku A, Shimane M and Tsuji A. Molecular and functional characterization of organic cation/carnitine transporter family in mice. J Biol Chem. 2000; 275:40064-40072.

59. Touré A, Morin L, Pineau C, Becq F, Dorseuil O and Gacon G. Tat1, a novel sulfate transporter specifically expressed in human male germ cells and potentially linked to Rho GTPase signaling. J Biol Chem 2001; 276:20309-20315.

60. Cronin CA, Gluba W and Scrable H. The lac operator-repressor system is functional in the mouse. Genes and Development 2001;15:1506-1517.

Chapter 5

SPERM MOTILITY: PATTERNS AND REGULATION

Srinivasan Vijayaraghavan
Kent State University, Kent, OH, USA

INTRODUCTION

Spermatozoa are the only cells destined to be exported from the body. Thus, motility is an essential property of fertile spermatozoa. It enables ejaculated spermatozoa to traverse the female reproductive tract and reach the site of fertilization; it is also essential for penetration of the outer investments of the oocyte including the zona pellucida. This chapter will describe the following aspects of mammalian sperm motility: structure of the flagellum, development of motility in the epididymis, changes in sperm movement patterns following ejaculation and during fertilization, biochemical mechanisms regulating sperm kinetic activity, methods to quantitate sperm motility, and a brief outline of the methods available to alter sperm motility *in vitro*.

STRUCTURE OF THE FLAGELLUM

The site of motility is the flagellum. Details of the structure of the flagellum have accumulated from extensive studies of not only spermatozoa but also of simple ciliated organisms such as protozoa (1-3). The structure responsible for kinetic activity of the flagellum is the axoneme. The axoneme consists of a central pair of microtubules surrounded by nine outer doublets (Figure 1). This 9+2 pattern of the axial filament is a characteristic of almost all forms of eukaryotic cilia and flagella. Each outer doublet is made up of two tubules, designated as A and B. The A microtubule is composed of 13 protofilaments. The B tubule which is attached to the A microtubule is made up of 10 protofilaments. The central pair, designated C1 and C2, is made up of 13 protofilaments each. The microtubules are composed of α and β tubulin. Each A tubule has attached to it an inner and

80

an outer dynein arm. The dynein arms extend towards the B tubule of the adjacent doublet. Flagellar dynein is a multi-subunit protein complexed with ATP hydrolyzing activity responsible for the conversion of chemical energy into kinetic activity. The 9+2 axonemal filament is held together by cross-linking protein structures. The central pair, C1 and C2, is connected to each other by regular bridges, resembling the steps of a ladder. The central pair is also surrounded by a central fibrous sheath. Each pair of the outer doublets is joined to the adjacent doublet by nexin links. Radial spokes are the cross-linking structures projecting from each A microtubule towards the central sheath.

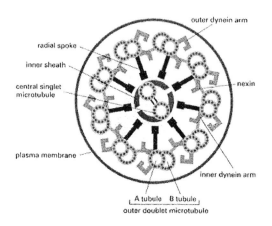

Figure 1. Structure of microtubules in a flagellum. Diagram of the parts of the axoneme showing the dynein arms and the cross linking protein structures. In mammalian spermatozoa additional structures are present (see Figure 2).

Axonemal kinetic activity is due to ATP-dependant sliding of the nine doublet microtubules relative to one another (2,4,5). The sliding forces are manifested as bending since the doublets are held together by cross-linking protein structures. Limited proteolysis of demembranated spermatozoa results in the loss of these cross-linking protein structures of the axoneme. Addition of ATP to such proteolytically modified spermatozoa results in the telescoping of the microtubules. In other words, there is no bending motion when the cross-linking protein structures are absent; the microtubules slide past one another. The dynein arms are important in sliding. Demembranated flagella, prepared under conditions resulting in the loss of dynein arms, lack kinetic activity. Sliding activity can be restored following reconstitution of the dynein arms. Further confirmation of the role of dynein arms in sliding came from studies with mutant *Chlamydomonas* defective in motility.

In mammalian spermatozoa, the 9 + 2 axonemal filament is surrounded by nine outer dense fibers (6-8). The fibers appear in association with a complementary outer doublet but are usually separate from the outer doublets in mature spermatozoa (Figure 2). The shape and thickness of the outer dense fibers varies in different species. These fibers do not contain tubulin or ATPase activity and thus their role in the creation of bending movements appears unlikely. The exact role of the outer dense fibers in the kinetic activity of flagella is unknown.

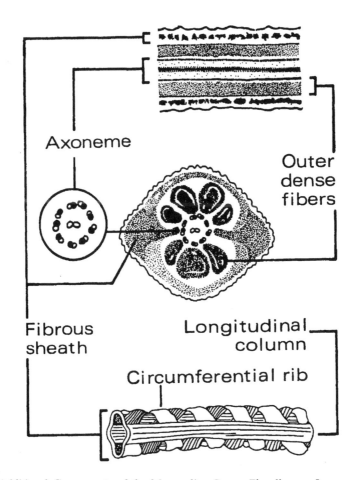

Figure 2. Additional Components of the Mammalian Sperm Flagellum. In mammalian spermatozoa, the 9+2 microtubule structure is surrounded by 9 outer dense fibers. In the principal piece, which is the region below the mid-piece, a fibrous sheath surrounds the outer dense fibers. Reproduced from (8).

Mammalian spermatozoa have several tightly coiled mitochondria around the flagellum in a region posterior to the nucleus called the mid-piece or the middle piece (Figure 3). The number of mitochondria, arranged end to end at the mid-piece, varies among species ranging from 15 in human to about 350 in rat spermatozoa. Below the mid-piece is the principal piece of the flagellum that is surrounded by a fibrous sheath. This sheath is made up of semicircular ribs that join to form two lateral columns along the length of the principal piece. The mitochondria and the fibrous sheath are covered by the plasma membrane. As with outer dense fibers the exact structure of the fibrous sheath varies greatly between species. The fibrous sheath terminates before the end of the flagellum. The end piece of the tail has only a central pair of microtubules under the plasma membrane. The role of the fibrous sheath in flagellar activity is not known.

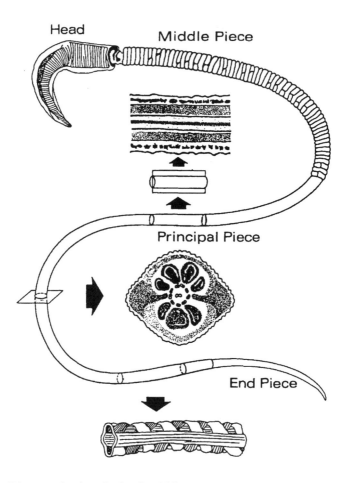

Figure 3. Diagram showing the head, middle piece, principal piece and end piece of a mammalian spermatozoon. Reproduced from reference (8).

MOTILITY DEVELOPMENT IN THE EPIDIDYMIS

Spermatozoa exiting from the testis are immotile and unable to fertilize the egg. These functions develop during their passage through the epididymis (9-11), a period varying from 6 to 14 days, depending on the species. In almost all mammalian spermatozoa, motility and fertilizing ability develop in the proximal and distal caudal regions of the epididymis (Figure 4). Maturation in the epididymis is a complex combination of processes taking place within spermatozoa and through interaction of the spermatozoa with environment of the androgen sensitive epithelium and its secretions.

During epididymal maturation, spermatozoa undergo alterations in structure and metabolism. Structural changes include modifications in the plasma membrane and increases in disulfide linkages affecting the stability of the outer dense fibers, fibrous sheath and mitochondria. Changes in the activities of metabolic enzymes and in the composition of membrane phospholipids and proteins also take place. Marked increases in glycolysis and respiration rates accompany the development of motility. It is not known whether this increase in metabolic rate is an independent event or a consequence of the energy demand from motility.

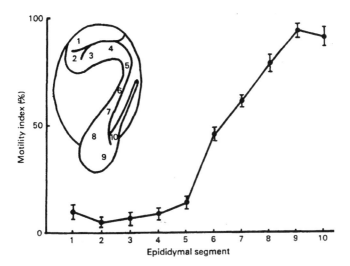

Figure 4. Motility development in the epididymis. Motility index of spermatozoa removed from successive regions of the ram epididymis. Motility was measured after a 10 minute incubation at 37°C in a Krebs-Ringer bicarbonate buffer. Maximum motility is displayed by spermatozoa from regions 8 and 9 of the cauda epididymidis. Reproduced from (12).

Motility of testicular and immotile caput epididymal spermatozoa demembranated by Triton X-100 extraction can be activated in the presence of ATP and cAMP (13-15). This suggests that the structural machinery required for motility exists in immature spermatozoa. This observation implies that cAMP levels in immotile spermatozoa are limiting. Other studies with developing epididymal spermatozoa have shown that, in addition to cAMP, calcium and intrasperm pH also play important roles in the regulation of sperm kinetic activity. As spermatozoa pass through the epididymis cyclic AMP levels and pH increase, while intracellular total and free calcium levels decrease (16-18). It is not known how the changes in the levels of these mediators are effected. However, these observations have provided us with approaches to alter motility *in vitro*. Thus, elevation of intrasperm cAMP levels, with phosphodiesterase inhibitors such as caffeine and isobutyl methyl xanthine, leads to motility initiation (19). Changes in calcium and pH, depending on the conditions, can stimulate or inhibit sperm motility (17). Under optimum conditions, motility induced *in vitro* in immotile caput epididymal spermatozoa can resemble that of motile mature caudal spermatozoa.

EJACULATED SPERMATOZOA AND HYPER-ACTIVATED MOTILITY

Mature spermatozoa stored in the cauda epididymidis are apparently immotile. The reasons for this appear to be species dependent. In the rat and hamster a protein, "immobilin," is responsible for suppressing motility (20). Immobilin increases the viscosity of epididymal fluid suppressing sperm movement. Dilution of the viscous fluid with a physiological buffer restores motility. The metabolism of these mechanically immobilized sperm, as measured by their respiration rates, is not different from that of motile rat caudal epididymal spermatozoa. A completely different mechanism operates in the bull. Bovine caudal epididymal spermatozoa are held immotile during their storage due to acidity of the epididymal fluid (21-22); the pH of caudal epididymal fluid is around 5.5. Lactate present in the epididymal fluid enters the sperm plasma membrane as lactic acid acidifying the sperm cytoplasm. It appears that a low intracellular pH inhibits dynein ATPase activity. Dilution of the immotile caudal spermatozoa in a physiological buffer causes alkalinization of the cytoplasm resulting in vigorous motility. It is interesting that intracellular pH is also used as a means of regulating sperm motility in some invertebrate species. Sea urchin spermatozoa are activated upon contact with sea water by alkalinization of their intracellular pH caused by a plasma membrane sodium/proton exchanger. The nature of the inhibitory mechanism by which human spermatozoa may be held immotile in the epididymis is unknown.

Spermatozoa are vigorously motile following ejaculation. This suggests that dilution by seminal plasma or that some component of seminal plasma serves to inactivate sperm immobilizing factors. Seminal plasma contains a complex mixture of secretions of the male reproductive tissues (23). A glycolysable substrate, such as glucose or fructose, and bicarbonate ion that activates sperm adenylyl cyclase is also present in semen. There is no doubt that motility, along with high sperm density, are generally positive indicators of fertile semen. However, there is considerable uncertainty about the lower limits of motility and sperm number.

Following ejaculation, spermatozoa undergo a process called capacitation before they can bind to and fertilize the ovum. In a wide variety of species, one of the features of capacitated spermatozoa is hyperactivated motility (24-27). Hyperactivated motility is variously described as 'whip lash', 'dancing', and 'serpentine' movements. Detailed analysis of flagellar beat reveals that hyperactivated motility is characterized by a high amplitude, asymmetric waveform of the tail (Figure 5). The asymmetric tail beat results in a non-linear progression. Hyperactivated motility is thought to be required for penetration of the outer investments of egg. Because of this potential significance in fertilization the detection of the ability of spermatozoa to undergo hyperactivation could be of clinical value in predicting fertility of semen samples. It is believed that elevation of cAMP and calcium levels mediate the onset of hyperactivated sperm motion.

BIOCHEMICAL MECHANISMS REGULATING MOTILITY

ATP hydrolysis by the inner and outer dynein results in microtubule sliding, flagellar bending, and motility. Microtubule sliding velocity is thought to be regulated by phosphorylation of yet unidentified axonemal proteins. Net protein phosphorylation is a result of the action of protein kinases and the opposing action of protein phosphatases. As noted above, cAMP appears to be a key regulator of motility (9). Most of the known actions of cAMP in cells are mediated by protein kinase A. Inactive protein kinase A is a tetramer consisting of two regulatory and two catalytic subunits. Binding of cAMP to the regulatory subunits results in the dissociation and activation of the catalytic subunit. The activated catalytic subunit, when intrasperm cAMP levels are sufficiently high, is thought to phosphorylate proteins leading to motility and metabolic activation of spermatozoa.

As described above definitive studies on the role of cAMP in motility come from experiments with demembranated spermatozoa (13-15). Mature motile spermatozoa have higher cAMP levels than immotile and immature spermatozoa. Thus, in demembranated caudal and ejaculated spermatozoa, ATP alone can sustain motility if sperm were motile prior to removal of the membrane. However, demembranated immotile testicular and caput epididymal spermatozoa require cAMP, in addition to ATP, for motility. Consistent with these studies is the observation that membrane permeable cAMP analogs and phosphodiesterase inhibitors induce motility in immature spermatozoa and stimulate motility of mature spermatozoa. Enzymes of cyclic nucleotide metabolism, adenylyl cyclase and phosphodiesterase, are present in spermatozoa. The soluble sperm adenylyl cyclase is unique compared to the membrane bound enzyme found in somatic cells (28). The sperm enzyme appears not to be coupled for activation or inhibition by trimeric G-proteins as is the case in hormone responsive somatic cells. The sperm enzyme is sensitive to activation by bicarbonate ions (29,30). Activation of the bicarbonate-sensitive sperm adenylyl cyclase and the

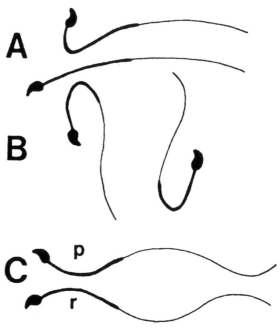

Figure 5. Patterns of flagellar beat in hyperactivated hamster spermatozoa. A, two sperm showing the waveform of the principal bend in the proximal mid-piece; B, reverse bends at the distal mid-piece are propagated along the flagellum followed by principal mends of very low curvature. This asymmetric waveform results in non-linear motion resembling a star-shaped or figure eight trajectory; C, sinusoidal symmetric principal bends (p) and the reverse bends (r) results in linear trajectories. Reproduced from (27).

resultant increase in cAMP levels are thought to explain the requirement for this anion in sperm suspension buffers used in *in vitro* fertilization.

If there is an involvement of a protein kinase in a cellular response, the response is most likely modulated by a protein phosphatase. Inclusion of protein phosphatases in the inactivation buffer of demembranated spermatozoa prevents motility induction. A protein phosphatase is found to be tightly associated with the *Chlamydomonas* axoneme. Treatment of immotile spermatozoa with protein phosphatase inhibitors results in a dramatic induction of motility. These observations highlight the importance of protein phosphatases in flagellar kinetic activity.

Somatic cell protein phosphatases can be divided into four subtypes, PP1, PP2A, PP2B, and PP2C, depending on their substrate specificity, metal ion requirement and inhibitory characteristics (31). The predominant protein phosphatase in spermatozoa is the PP1 subtype. Sperm contain a testis specific isoform of PP1: PP1γ2 (32-34). This isoform contains a carboxyl terminus extension not found in the somatic cell isoform PP1γ1. Antibodies against the carboxyl terminus peptide cross-react with PP1γ2 in spermatozoa from a wide range of species - mouse, rat, bull, monkey and human. Since the carboxyl terminus is not required for catalytic activity, the seeming conservation of this extension is intriguing. It is possible that the unique amino acid sequences in the protein could be involved in enzyme localization and regulation in spermatozoa. In addition to PP1γ2, relatively lower activity levels of the serine/threonine phosphatase subtypes PP2A and PP2B are also present in spermatozoa. Mechanisms by which protein kinases and phosphatases are regulated are likely to be key aspects of the biochemical mechanisms regulating sperm motility.

METHODS FOR QUANTITATING SPERM MOTILITY

Methods for quantitating motility range from simple microscopic observation to complex computerized automated analysis (27). Microscopic observation by an experienced investigator, even though subjective, has significant value and is still used for semen evaluation. Motility quantitation techniques can be classified as indirect or direct. Indirect techniques, such as turbidometry and laser techniques, use the optical properties of the sperm suspension to estimate percent motility and velocity distribution. Direct techniques involve observation of individual spermatozoa. Such a direct quantitative analysis is made following cinematographic or video recordings of the sperm sample under a microscope. Analysis of these images can be manual or computerized. One of the manual techniques, for example, involves the use of a transparent grid placed on a monitor to follow sperm head motion through a specified number of video frames. Such a relatively simple, but labor intensive, manual procedure can provide reliable head motion parameters such as percent motility and velocity. Computerized motility analysis techniques have now become more prevalent (35-37). These motility analysis procedures, CASMA or CASA (Computer Assisted Sperm Motility Analysis or Computerized Assisted Sperm Analysis), consist of a video-recording device, image accession and digitization in the

88

computer, and image analysis software. The positions of the head of spermatozoa, tracked through a specified number of video frames, can be used to compute a number of motility parameters. These include direct parameters such as percent motility, straight line, curvilinear, and average velocities. Other derived parameters such as beat cross frequency, turn angle and linearity can also be measured. While motility quantitation is of great value as a research tool, the utility of these procedures in clinical andrology as a predictor of sperm function is still being investigated.

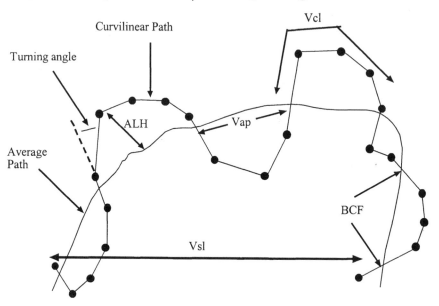

Figure 6. An idealized tracing of the head positions of a sperm tracked by a computerized analysis system. Vsl: Straight line (end-to-end) velocity. Vcl and Vap: are curvilinear and average path velocities. ALH: average amplitude of lateral head movement. BCF: Beat cross frequency. Adopted from (27).

PHARMACOLOGICAL METHODS TO ALTER SPERM MOTILITY *IN VITRO*

There is considerable clinical and economic interest in the development of safe and reliable methods to maintain optimum motility in sperm preparations used in *in vitro* fertilization and artificial insemination. Spermatozoa are notorious for their lack of sensitivity to hormones and other signaling molecules that activate somatic cells. However, a number of pharmacological additives have been shown to stimulate sperm kinetic activity. As noted earlier, phosphodiesterase (PDE) inhibitors, by virtue of their ability to elevate cAMP levels, are stimulators of motility. The PDE inhibitors that have been used are caffeine, theophylline, and pentoxyfylline. All these compounds, used at millimolar levels, stimulate motility. The effect of these compounds on *in vitro* fertilization rates, however, is variable.

Yet another class of compounds more potent in stimulating motility are analogues of adenosine (38). The effect of these compounds on sperm motility was first discovered in this author's laboratory. The most potent of these analogues are 2'-deoxyadenosine and 2-chloroadenosine. A relatively new compound, 2'-deoxy-2-chloroadeonsine, is effective in stimulating bovine sperm motility at low micromolar levels. This compound, commercially called Cladribine, is used in the treatment of certain types of leukemia. The mechanism by which these nucleoside analogues, including 2'-deoxyadenosine, stimulate sperm motility is unknown. While these compounds appear beneficial to sperm function paradoxically they are cytotoxic to somatic cells in culture and *in vivo*.

Another approach to stimulate motility is to increase protein phosphorylation by compounds that inhibit protein phosphatase activity. Indeed the protein phosphatase inhibitors, okadaic acid and calyculin A, stimulate sperm motility at nanomolar levels (32). These are the most potent stimulators of motility known. However, a clinical use for these phosphatase inhibitors is unlikely because they are potent tumor promoters. The development of compounds capable of specifically inhibiting sperm protein phosphatase could be useful in this regard. It is quite likely that an increased understanding of the biochemical mechanisms regulating male gamete function should lead to rational approaches to optimize sperm function *in vitro* and in procedures used in managed reproduction.

CONCLUSIONS

The structure of the highly differentiated sperm cell is suited for two essential properties: motility and the ability to fertilize the ovum. Its relatively simple metabolic functions compared to a somatic cell is a reflection of the fact that it is the only cell designed to function outside the body. Testicular spermatozoa following spermiation are immotile and cannot bind to the egg. The capacity for motility and fertilization ability develops during their passage through the epididymis. Motility initiation is regulated by the intracellular factors: pH, cyclic AMP and calcium. These factors induce changes in sperm protein phosphorylation by altering the activities of protein kinases and phosphatases. Studies on regulation of sperm motility and protein phosphorylation have resulted in the identification of a number of procedures to alter sperm motility *in vitro*. These include the use of cAMP, phosphodiesterase and protein phosphatase inhibitors, bicarbonate anion, and analogues of the nucleoside adenosine. New studies

90

on the biochemical mechanisms underlying regulation of sperm motility are not only of basic importance but also have practical implications in the development of rational approaches to manipulate sperm function *in vitro* and *in vivo*.

REFERENCES

1. Gibbons I.R. Cillia and flagella of eukaryotes. J Cell Biol 1981; 91:107s-124s
2. Satir, P. "Basis of Flagellar Motility in Spermatozoa: Current Status." In The Spermatozoon, D.W. Fawcet, M. Bedford, eds. Urban and Schwarzenberg, 1979. pp 81-90
3. Porter M.E., Sale W.S. The 9+2 axoneme anchors multiple inter-arm dyneins and a network of kinases and phosphatases that control motility. J Cell Biol 2000; 151:F37-F42
4. Gibbons I.R. Introduction: dynein ATPase. Cell Motility 1982; Supplement 1:87-93
5. Lindemann C.B., Kanous K.S. A model for flagellar motility. Int Rev of Cytology 1987; 173:1-73
6. Fawcett D.W. The mammalian spermatozoon. Dev Biol 1975; 44:394-436
7. Fawcett D.W. A comparative view of sperm ultrastructure. Biol Reprod 1970; Supplement 2:90-127
8. Eddy E.M., O'Brian D.A. "The Spermatozoon," In Physiology of Reproduction, E. Knobil, J.D. Neill, eds. NY: Raven Press, 1993, pp 29-78
9. Bedford J.M. "Maturation, Transport, and Fate of Spermatozoa in the Epididymis." In Handbook of Physiology and Endocrinology, Volume V, 1975, pp 303-313
10. Orgebin-Crist M.-C. Sperm maturation in rabbit epididymis. Nature 1967; 216:816-818
11. Bedford J.M., Hoskins D.D. "The Mammalian Spermatozoon: Morphology, Biochemistry and Physiology." In Marshall's Physiology of Reproduction, G.E. Lamming, ed. Edinburgh, London, Melbourne, New York: Churchill Livingstone, 1990, pp 379.
12. Pariset C.C., Feinberg J.M.F., Dacheux J.L., Weinman, S.J. Changes in calmodulin level and cAMP-dependent protein kinase activity during epididymal maturation of ram spermatozoa. J Reprod Fert 1985; 74:102-112.
13. Lindemann C.B., Kanous K.S. Regulation of mammalian sperm motility. Arch Androl 1989; 23:1-22
14. Yeung C.H. Effects of cyclic AMP on the motility of mature and immature hamster epididymal spermatozoa studied by reactivation of the demembranated cells. Gamete Res 1984; 9:99-114
15. Mohri H., Yanagimachi R. Characteristics of motor apparatus in testicular, epididymal and ejaculated spermatozoa. Exper Cell Res 1980; 127:191-196
16. Hoskins D., Stephens D., Hall M. Cyclic adenosine 3':5'-monophosphate and protein kinase levels in developing bovine spermatozoa. J Reprod Fert 1974; 37:131-133
17. Vijayaraghavan S., Critchlow L.M., Hoskins D.D. Evidence for a role for cellular alkalinization in the cyclic adenosine 3',5'-monophosphate-mediated initiation of motility in bovine caput spermatozoa. Biol Reprod 1985; 32:489-500
18. Vijayaraghavan S., Hoskins D.D. Changes in the mitochondrial calcium influx and efflux properties are responsible for the decline in sperm calcium during epididymal maturation. Mol Reprod Dev 1990; 25:186-194
19. Hoskins D., Hall M., Munsterman D. Induction of motility in immature bovine spermatozoa by cyclic AMP phosphodiesterase inhibitors and seminal plasma. Biol Reprod 1975; 13:168-176

20. Usselman M.C., Cone R.A. Rat sperm are mechanically immobilized in the caudal epididymis by "immobilin," a high molecular weight glycoprotein. Biol Reprod 1983; 29:1241-1253

21. Acott T.S., Carr D.W. Inhibition of bovine spermatozoa by caudal epididymal fluid: ii. interaction of pH and a quiescence factor. Biol Reprod 1984; 30:926-935

22. Carr D.W., Acott T.S. Inhibition of bovine spermatozoa by caudal epididymal fluid: i. studies of a sperm motility quiescence factor. Biol Reprod 1984; 30:913-925

23. Mann, T., Lutwak-Mann, C. "Male Reproductive Function and the Composition of Semen. In MaleRreproductive Function and Semen. NY: Springer Verlag, 1981, pp 1-34.

24. Yanagimachi R. "Mammalian Fertilization." In The Physiology of Reproduction, E. Knobil, J.D. Neill, eds. NY: Raven Press Ltd., 1994, pp 189-378.

25. Katz D.F., Yanagimachi R. Movement characteristics of hamster and guinea pig spermatoz oa upon attachment to the zona pellucida. Biol Reprod 1981; 25:785-791

26. Katz D.F., Yanagimachi R., Dresdner R.D. Movement characteristics and power output of guinea-pig and hamster spermatozoa in relation to activation. J Reprod Fert 1978; 52:167-172

27. Katz D.F. Characteristics of sperm motility. Ann NY Acad Sci 1991; 637:409-423

28. Buck J., Sinclair M.L., Schapal L., Cann M..J., Levin L.R. Cytosolic adenylyl cyclase defines a unique signaling molecule in mammals. Proc Natl Acad Sci USA 1999; 96:79-84

29. Chen Y., Cann M.J., Litvin T.N., Iourgenko V., Sinclair M.L., Levin L.R., Buck J. Soluble adenylyl cyclase as an evolutionarily conserved bicarbonate sensor. Science 2000; 289:559-560

30. Okamura N., Tajima Y., Onoe S., Sugita Y. Purification of bicarbonate-sensitive sperm adenylylcyclase by 4-acetamido-4'-isothiocyanostilbene-2,2'-disulfonic acid affinity chromatography. J Biol Chem 1991; 266:17754-17759

31. Cohen, P. "Classification of Protein-Serine/Threonine Phosphatases: Identification and Quantitation in Cell Extracts." In Methods in Enzymology, Vol. 201, Chapter 33, T. Hunter, B.M. Sefton, eds. NY: Academic Press, Inc., 1991.

32. Vijayaraghavan S., Stephens D.T., Trautman K., Smith G.D., Khatra B., da Cruz E., Silva E.F., Greengard P. Sperm Motility development in the epididymis is associated with decreased glycogen synthase kinase-3 and protein phosphatase 1 activity. Biol Reprod 1996; 54:709-718

33. Smith G.D., Wolf D.P., Trautman K.C., da Cruz E., Silva E.F., Greengard P., Vijayaraghavan S. Primate sperm contain protein phosphatase 1, a biochemical mediator of motility. Biol Reprod 1996; 54:719-27.

34. Smith G.D., Wolf D.P., Trautman K.C., Vijayaraghavan S. Motility potential of macaque epididymal sperm: the role of protein phosphatase and glycogen synthase kinase-3 activities. J Androl 2000; 20:47-53.

35. Stephens D.T., Hickman R., Hoskins D.D. Description, validation and performance characteristics of a new computer automated sperm motility analysis system. Biol Reprod 1988; 38:577-586

36. Stephens, D.T., Hoskins, D.D., Controls of Sperm Motility: Biological and Clinical Aspects. Boca Raton: CRC Press, Inc., 1990, pp 251-260.

37. Mortimer S.T. A critical review of the physiological importance and analysis of sperm movement in mammals. Hum Reprod Update 1997; 5:403-439

38. Vijayaraghavan S., Hoskins D.D. Regulation of bovine sperm motility and cyclic adenosine 3', 5'-monophosphate by adenosine and its analogues. Biol Reprod 1986; 34:468-477

Chapter 6

TESTICULAR AND EPIDIDYMAL MATURATION OF MAMMALIAN SPERMATOZOA

Kiyotaka Toshimori
Department of Anatomy and Reproductive Cell Biology, Miyazaki Medical College, Miyazaki, Japan

INTRODUCTION

Mammalian spermatozoa are highly differentiated haploid cells which are unable to synthesize proteins *de novo* post-testicularly. These unique cells undergo morphological and biochemical changes in components as diverse as the plasma membrane, cytoplasm (the cytoplasmic droplet) and certain internal components such as the nucleus, the acrosome, and the cytoskeletal elements (perinuclear theca, outer dense fibers and fibrous sheath). Most of these components are primarily organized into developing germ cells and nascent spermatozoa during spermatogenesis, and the initial maturational modifications occur in the testis, which is typically referred to as testicular maturation. Spermatozoa that have left the testis undergo further modifications in the epididymis referred to as epididymal maturation (1, 2). These changes are commensurate with the physiological and functional events continuously induced in developing spermatids and maturing spermatozoa. Thus, spermatozoa present in the testis and proximal epididymis are immotile and immature (infertile), while spermatozoa that reach the distal epididymis are motile and mature (fertile). In this context, the maturation of spermatozoa is defined as the process of achieving the fertilizing ability. Such maturational events are prerequisite for spermatozoa to undergo the final maturational events (capacitation and the acrosome reaction) that are inevitable steps prior to penetration into the oocyte.

In the last four decades, various maturational events in the testis and epididymis have been analyzed, and have been extensively reviewed (recent reviews; 3-16). In this chapter, typical events associated with testicular and epididymal maturation of mammalian spermatozoa are described; the maturation of the nucleus, the acrosome, the plasma membrane, as well as some flagellar components.

STRUCTURE AND COMPONENTS OF SPERMATOZOA

A spermatozoon consists of a head and a flagellum, and is completely covered by the plasma membrane, on which distinct surface domains are present (Figure 1). The head is divided into two major regions (domains); the acrosomal region and the postacrosomal region. In general, the nucleus occupies nearly the entire head. The acrosome, a cap-like structure located in the anterior-most (rostral) region, contains a variety of hydrolytic enzymes such as serine-like protease acrosin and hyaluronidase as well as a number of other functional proteins. The acrosome is segregated into two major domains, the anterior acrosome (acrosomal cap) and posterior acrosome (equatorial segment). Functionally, the anterior acrosome is involved in the acrosome reaction, while the equatorial segment is involved in the fusion with the oocyte. The perinuclear theca (PT) surrounds nearly all of the nucleus, extending from the rostral region (perforatorium) throughout to the basal region of the head. The head is demarcated from the flagellum by a diffusion barrier, a constricted zonule of the plasma membrane called the posterior ring.

The flagellum is divided into three major regions (domains); the middle piece (MP), the principal piece (PP) and the end piece (EP). The MP is separated from the PP by the MP-PP juncture, another constricted zonule of the plasma membrane called the annulus or Jansen's ring. The neck region is sometimes regarded as an independent region called the connecting piece. The neck structure is complex, consisting of the following structures; the distal-most part of the nucleus with a redundant nuclear envelope, the basal plate, the capitulum, the proximal centriole, the transitional connecting centriole and the segmented columns. The basal plate is situated in the implantation fossa. The capitulum covers the proximal centriole, which is docked in the centrosome. Amorphous pericentriolar materials such as centrin and pericentrin surround the centrosome. The distal centriole is transformed into the transitional connecting centriole, which is further transformed into the centrally placed axoneme. The axoneme extends distally throughout the flagellum. The axoneme represents the machinery associated with motility and consists of microtubules (alpha and beta tubulins) and dyneins containing ATPase. The segmented columns are transformed into outer dense fibers (ODFs) and are surrounded by the proximal part of mitochondrial sheath. The MP region consists of the axoneme, the ODFs (nine in number) and the mitochondrial sheath. Mitochondria generate adenosine triphosphate (ATP) as an energy source for movement of the flagellum. The PP region consists of the axoneme, the ODFs (reduced to seven in number) and the fibrous sheath

95

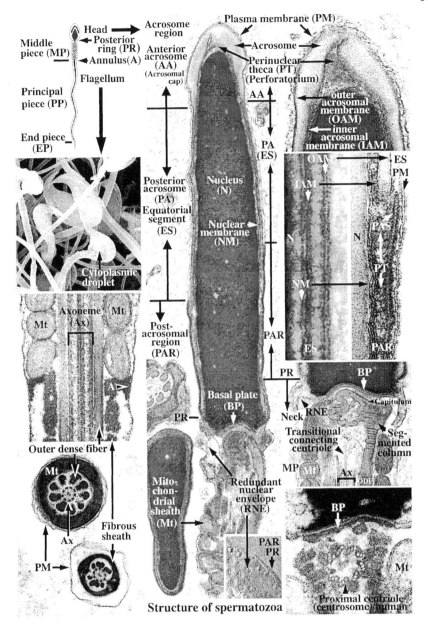

Figure 1. The structure of mammalian spermatozoa. A schematic drawing (top, left) showing the various sperm domains. The components located in each domain are shown in electron microscopic images. A scanning electron micrograph (EM) shows the cytoplasmic droplet in the flagellum (below the scheme). The other components described in this chapter are shown as transmission EM images. A freeze fracture image shows the neck region with the redundant nuclear envelope (inset, bottom center). The undulating plasma membrane over the anterior acrosome region (top, right) is labeled with a polyene antibiotic, filipin, to show the cholesterol-rich nature.

(FS). The FS is localized beneath the plasma membrane, encircling the ODF. The FS consists of two longitudinal columns connected by semicircular ribs. The EP region contains only residues of the axoneme. These fundamental structures or elements (matrices) consist of specific proteins and associating molecules. The PT, ODF and FS are cytoskeletal elements which are specifically differentiated in spermatozoa.

GENERAL ASPECTS OF TESTICULAR MATURATION

A drastic change in cell shape and structure, i.e., metamorphosis, occurs in spermatids (17). Spermatids are derived from diploid spermatocytes (primary and secondary), and are produced as haploid cells after meiosis. The nuclear chromatins of spermatids are condensed in the late elongating spermatid stage, and transcription activity is terminated. Notwithstanding the nuclear condensation, sperm constituent proteins are initially generated in spermatocytes and/or spermatids and are integrated into structures or associated elements in the nascent heads and flagella. Therefore, the constituent proteins are programmed so as to be synthesized in a timely manner and are then transported and organized into structures and elements. Thus, testicular maturation requires strict coordination between molecular events and cell organelle behavior. For example, the Golgi apparatus is transformed into the acrosome, in which many functional proteins are destined to be stored in the early spermatids (see details for Section 5). Microtubules act to form the sperm cell shape and to temporarily store proteins required for the formation of structures and elements. The distal centriole is transformed to the axoneme during overall spermiogenesis. The plasma membrane establishes the surface domains, where specific proteins (molecules) are distributed. Consequently, at the end of differentiation, upon spermiation, spermatozoa emerging from the germinal epithelium discard most of the cytoplasm and cell organelles, and contain only small amounts of residual cytoplasm called the cytoplasmic droplet (CD). The CD contains practically no cellular machinery necessary for protein synthesis, and Sertoli cells phagocytose the discarded cytoplasm. Thus, spermatozoa become highly differentiated and transcriptionally silent cells during testicular maturation (Figure 2).

GENERAL ASPECTS OF EPIDIDYMAL MATURATION

Spermatozoa released from the germinal epithelium are transported to the rete testis, efferent ductules, and are then collected into an epididymal duct in the epididymis. The epithelia of the rete testis and the efferent ductules consist of cuboidal cells and columnar cells (non-ciliated and ciliated cells), respectively. The epithelium of the efferent ductules reabsorbs approximately 90% of the testicular fluid, an estrogen-dependent process. If the function is impaired, then the efferent ductules are dilated, and spermatozoa cannot enter the epididymal duct, which eventually causes infertility in the mouse (18).

The length of the epididymal duct varies, depending on the species (5-6 m in men and 80 m in horses). Such a long duct is convoluted in the epididymis. Spermatozoa pass through the epididymal duct in about 1-2 weeks depending on the species; approximately 3-7 days in most mammals including human. The epididymis is generally divided into three functionally distinct regions based on the secretory activity of the

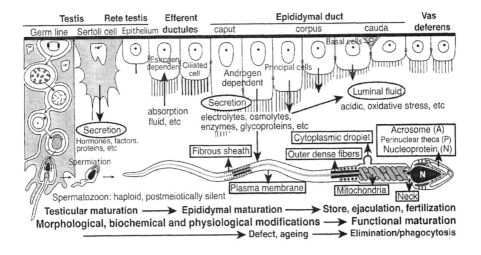

Figure 2. A schematic drawing of the relationship between maturing spermatozoa and lining epithelia in the testis and epididymis.

duct epithelium; caput, corpus and cauda epididymidis (Figure 3). The epididymal epithelium consists of columnar cells bearing microvilli or stereocilia (principal cells or secretory cells). The epithelial height varies in regions: high in the proximal region, but low in the distal region. Epididymal epithelial cells are involved, not only in secretions but also in phagocytosis of the cell debris and dead spermatozoa. The basal cells are sporadically distributed between the columnar cells. In rodents, clear cells are distributed in the epithelium.

In general, the proximal part of the epididymis is actively involved in sperm maturation. For example, both the number of spermatozoa (spermatocrit) and protein concentration in the fluid are quite low in the proximal caput (about 15 % and 35 mg/ml, respectively), but rise in the distal caput (about 85 % and 60 mg/ml, respectively), and then gradually decrease in the cauda (about 30 % and 20-30 mg/ml, respectively) in the stallion. Such a remarkable distinction in secretion and activity are under the control of androgen and other testicular factors. The inhibition of the secretory activity leads to a loss of fertilizing ability of spermatozoa. An optimal environment is necessary for spermatozoa to mature and survive in the epididymis.

The epididymal fluid is acidic; approximately pH 6.2 - 6.5 in the caput and pH 6.8 - 6.9 in the cauda. The fluid contains a wide variety of molecules: 1) transudates from the blood, 2) secretions from the testis (Sertoli products) and the duct epithelium, 3) lytic products of molecules in the fluid and spermatozoa, and 4) metabolic products of spermatozoa. Therefore, the composites are electrolytes (Na^+, K^+, Cl^-), hormones (both androgen which is bound to androgen binding protein and estrogen), enzymes (transferases, cathepsin, glutathione peroxidase, prostaglandin synthase, H^+ V-ATPase), and many other proteins (lactoferrin, clusterin, cholesterol transfer protein, albumin), carnitine, vitamine D, inositol, steroid, glycerol, glycerylphosphorylcholine, sialic acid, etc. Many of these molecules are related to epididymal maturation or are involved in the fertilization process at a later stage (not discussed here). Glutathione peroxidase and lactoferrin have potential roles of antioxidants. Such luminal microenvironments are effectively isolated from the blood by the epithelial tight junctions (blood-testis barrier and blood-epididymis barrier). Thus, spermatozoa become fertilization- competent cells and are stored until ejaculation, bathed in the epididymal fluid.

MATURATION OF THE NUCLEUS

The nuclear shape and nucleoproteins of spermatozoa are stabilized during testicular and epididymal maturation, undergoing remodeling in the nuclear chromatin.

Testicular Maturation of the Nucleus

The spermatid nucleus is condensed, and is completely remodeled in shape and size. According to the nuclear shape and chromatin texture, growing spermatids can be classified into three groups; early spermatid with a round nucleus (round spermatid), intermediate spermatid with an elongating nucleus (elongating or condensing spermatid) and mature spermatid with a condensed nucleus (mature spermatid). During nuclear condensation, the nucleosomal histones (somatic and testis- specific isoforms) of the meiotic germ cell DNA are replaced, first by intermediate proteins called transition proteins, and then by small arginine-rich protamines. Minas that encode transition proteins and protamines are synthesized in round spermatids, stored as cytoplasmic ribnucleoprotein particles for up to a week, and finally are translated into elongated spermatids during spermiogenesis. Histone acetylation and ubiquitination facilitate chromatin transformation from the nucleosome to the nucleoprotamine. The binding of protamines to DNA is under the control of a phosphorylation-dephosphorylation mechanism. Thus, the spermatid chromatin becomes highly condensed in late spermiogenesis.

During chromatin condensation, the nuclear pore complexes move toward the nascent neck region and are removed from the main region of the nucleus covering the head. Consequently, the complexes are retained as the redundant nuclear membrane in the neck of spermatozoa (Figure 2). The transcription activity of germ cells is terminated, generally several days before the completion of spermatogenesis. Therefore, protein synthesis is stringently regulated at several steps such as transcription, translation and/or post-translation in spermatogenic cells (19).

The change in nuclear shape is coordinated with the formation of the acrosome and perinuclear theca. These formations are harmonized with the behavior of cytoskeletal filaments such as manchette and actin. The manchette is a half-mitotic spindle-like microtubule assembly surrounding the caudal region of the nucleus, and transiently appears only in elongating spermatids.

Epididymal Maturation of the Nucleus

During epididymal transit, the nuclei of spermatozoa become slightly reduced in size. Sperm chromatins become more stable, and the sulfhydryl cross- link (-SH) between thiol groups in adjacent protamine chains are converted into disulfide cross-link (-S-S-). Therefore, the number of –S-S- cross-links reflects the stability of the nucleoprotein in spermatozoa. The paternal genome is protected from external environment hazards such as acidity and oxidative stress that may cause nuclear damage. An impaired genome is implicated in male infertility. However, the details of the mechanisms involved in these processes are unclear: the issues of how the somatic chromatin is transformed into the testis specific histones, how the genome is impaired or protected, and how the impaired genome is detected are currently unclear.

MATURATION OF THE ACROSOME

The acrosome contains infrastructures and spatially and biochemically distinct subdomains (subcompartments) within itself. The acrosomal contents are segregated into these subdomains and closely interact with the infrastructures. The acrosome contains not only hydrolytic enzymes to dissolve the extracellular matrices of the egg, but specific proteins required for the various steps of the sperm-egg interaction as well. Acrosomal organization is remodeled during testicular and epididymal maturation, and the constituent proteins undergo maturational modifications.

Structure and Components of the Acrosome

The mature acrosome is a cap-like structure enclosed with limited membranes, the inner acrosomal membrane (IAM) overlying the nucleus and the outer acrosomal membrane (OAM) underlying the plasma membrane (Figure 1). The anterior acrosome contains several domains (subcompartments), wherein the acrosomal proteins are segregated. For example, many acrosomal proteins such as acrosin and SP-10 (HS-63 antigen) are distributed over the entire anterior acrosome, and some other proteins are confined to the subdomains: for instance, a zona pellucida penetration-facilitating protein MC101 (acrin 3) is confined in the outer region (cortex) beneath the OAM (Figure 3e). The guinea pig acrosome contains three subcompartments; the electron-light, -intermediate and – dense region, called M1, M2 and M3, respectively. A member of the complement 4- binding protein family AM 67 (mouse sp56) localized in the M1 region, AM 50 (p50, apexin, Narp, or NPTX2) in the M3 region, proacrosin in the M2 and M3 regions, and autoantigen 1 (mouse Tpx-1, human TPX1, or CRISP-2) in all these regions. Such compartmentalizations are found in many mammals (mouse, hamster, guinea pig, rabbit, bovine and hedgehog).

The hamster acrosome has a structural framework consisting of detergent- stable materials (AM29 and AM22), referred to as the acrosomal lamina-matrix complex (20). Proacrosin/acrosin and N-acetyl-glucosaminidase specifically bind to the acrosomal matrix. Guinea pig acrosomal matrix protein AM50 also has an affinity for proteases. Thus, the acrosome consists of soluble proteins and insoluble (particulate) proteins called acrosomal matrix proteins. The former includes hyaluronidase, dipeptidyl peptidase, CRISP-2, and acrosome reaction- relating protein MN7 (acrin 1) etc, while the latter includes AM22, AM29, AM50, AM67, proacrosin-binding protein sp32, zona-binding protein MC41 (acrin 2) etc.

Testicular Maturation of the Acrosome

Drastic remodeling of the acrosome occurs in the late stages of acrosome biogenesis. It is known that some acrosomal proteins are redistributed into the anterior acrosome in this stage, and some other proteins are specifically localized to the restricted region. For example, AM50 is localized to the ventral matrix of the apical segment, and MN9 (equatorin) is confined to the equatorial segment. Some proteins such as AM22, SM29, SP-10 and MN7 (acrin 1) are further modified during epididymal maturation. The major matrix proteins, AM22 and AM29, are derived from the common precursor protein (40 kDa), processed and localized to the acrosome cap region, adhering to the OAM. Human sperm SP-10, which is initially distributed over the entire acrosome in the round spermatid phase, becomes

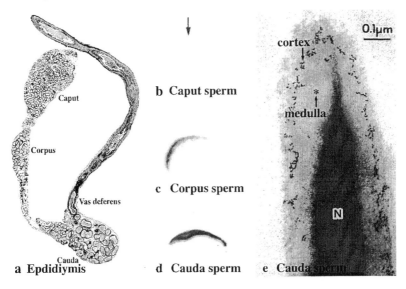

Figure 3. a, Mouse epididymis. b-d, Increment of the antigenicity of MC101 (acrin 3) during epididymal maturation. e, Localization of MC101 at the cortex region within the acrosome.

localized to the acrosomal membranes, and then undergoes proteolytic processing during its passage into the initial segment or caput epididymis. The mouse acrosomal membrane protein, cyritestin (ADAM3, tMDC I), is a Cyrn gene product and a member of the metalloproteinase-disintegrin family, ADAM (a disintegrin and metalloprotease) or MDC (metalloprotease-like, disintegrin-like cyritestine-rich domain). Cyritestin is initially organized as an integral IAM transmembrane during spermiogenesis, and is then post-translationally processed, decreasing in molecular size (from 110 kDa to 55 kDa) by removal of the N- terminal half during epididymal transport. In contrast, the 1D4 protein loses its antigenicity.

Acrosome remodeling appears to be affected by the behavior of periacrosomal substances such as the perinuclear theca in developing spermatids or the cytoskeletal filaments (microtubules and actin) in spermatids and somatic Sertoli cells.

Epididymal Maturation of the Acrosome

The change in acrosome size and structure is clear in some mammals (guinea pig, rabbit, and some primates). Acrosomal proteins also are reduced in molecular weight and their antigenicity modified in many mammalian spermatozoa (mouse, rat, rabbit, guinea pigs, boar, and humans). For example, in addition to the above proteins, acrosomal glycoconjugates become reduced in molecular weight via modification of side chains containing D-galactose and N- acetyl-D-galactosamine residues. Guinea pig MN7 (acrin 1) becomes reduced in molecular size, and is redistributed within the acrosome. In contrast, the antigenicity of mouse MC101 (acrin 3) increases (Figure 3b-d).

Thus, the maturational modifications of acrosomal proteins can be attributed to deglycosylation and/or proteolytic processing. The physicochemically distinct affinities to the substructures or the matrix and maturational changes are thought to reflect the differential release of the acrosome contents during the acrosome reaction. For example, mouse M42, which is processed from the precursor form to the mature form and develops the capacity for the zona-induced acrosome reaction, is relatively diffusible and involved in the acrosome reaction. AM22 and AM29 proteins remain associated with the OAM after the acrosome reaction. MN7 (acrin 1), MC41 (acrin 2) and MC101 (acrin 3), which are independently transported, distributed and modified during the maturation processes, are differentially released from the acrosome and involved in different steps of the sperm-zona pellucida interaction, the acrosome reaction, firm binding to the zona pellucida and penetration through the zona, respectively. Thus,the mature form of cyritestin remains on the acrosome-reacted spermatozoa and is involved in binding to the zona pellucida during fertilization.

MATURATION OF THE PLASMA MEMBRANE

The surface domains on the plasma membrane are primarily formed during spermiogenesis, and surface distinction becomes clear during epididymal maturation. Epididymal fluid contains a wide variety of secretions and transudates from the blood. Some of these molecules specifically bind to the sperm surface. For example, sialic acid residues bind the traversing spermatozoa, producing negative charge on the surface. Membrane cholesterols, which stabilize membrane fluidity, are reduced in quantity as they are trapped by albumins. Therefore, the cholesterol/phospholipid molar ratio in the plasma membrane is sequentially changed. Accordingly, membrane stability and composition of the proteins are modified. Thus, sperm plasma membrane molecules are not static within the membrane domains. Various events during sperm plasma membrane maturations have been analyzed and reviewed (recent reviews; 5, 6, 13, 14, 21-23).

Based on recent findings, possible mechanisms involved in sperm plasma membrane remodeling are as follows; 1) uptake or binding of new proteins (molecules) to the membrane, and 2) modification of pre-existing proteins on the membrane. In both cases, it is likely that further mechanisms also operate; 1) elimination or removal, 2) masking, 3) limited proteolysis, and 4) redistribution or migration within or to on adjacent domain. As a result of these modifications, some proteins may lose their antigenicity, while the antigenicities of other proteins, including newly bound proteins, may be exposed to new domains. Accordingly, the immunostaining pattern on the sperm surface could be changed. Based on these possible mechanisms, typical maturation cases are described below.

Uptake or Binding of New Proteins (Molecules)

Glycoproteins, e.g., sialoglycoproteins, are secreted from the epididymal epithelium and bound to the sperm plasma membrane as stated above. Prostasome- like particles in the bull epididymal fluid transfer an epididymis-secreted protein p25 to the plasma membrane. The rat sperm fusion-related protein DE (AEG) is inserted into the dorsal region of the periacrosomal plasma membrane, and later, DE migrates to the equatorial segment. Other epididymal proteins that could be inserted into the sperm plasma membrane include the group of GPI-anchored sperm plasma membrane proteins; CD52/HE5 (gp20, CAMPATH-1), and CD55 (DAF) and CD59 (protectin).

Modifications of Pre-existing Proteins

Membrane glycoproteins are primarily synthesized as large precursor proteins in the testis, and then undergo biochemical modifications. This is mainly the result of deglycosylation or proteolysis.

Glycans on the Plasma Membrane: Some glycan-modifying enzymes such as glycosidases (glycohydrolases) and glycosyltransferases (fucosyltransferase, galacosyltransferase and sialyltransferase) are present in the epididymal fluid or associated with the sperm plasma membrane. The concentration level of these glycosyltransferases is high in the caput spermatozoa, but low in the cauda spermatozoa. These glycan-modifying enzymes modify the surface glycan chains. A deficiency of these enzymes is found in male infertility patients.

Beta 1,4-galactosyltransferase (GalTase) is expressed on the entire surface of the spermatocyte, and then coalesces onto the surface of the dorsal region of the acrosome during testicular and epididymal maturation. GalTase functions, not only as a receptor for extracellular oligosaccharide ligands of ZP3 (a glycoprotein that consists of the most external layer of the zona pellucida), but also as an inducer of signal transduction. Rat sperm plasma membrane alpha-D-mannosidase is converted from an enzymatically inactive or less active precursor form (135 kDa and 125 kDa) in the testis into an active mature form (115 kDa), localized to the periacrosomal region. This occurs during epididymal maturation.

In the head: Proteins of the ADAM or MDC family undergo the maturational modifications. A heterodimeric glycoprotein (alpha and beta), fertilin (PH-30, ADAM 1 & 2), undergoes limited proteolysis on the extracellular domains by the aid of protease present on the sperm surface or in the environment (24). Initially, fertilin is synthesized as a large precursor pro-protein in primary spermatocytes, and is localized to the entire head in testicular spermatozoa. The fertilin alpha subunit (ADAM1) is processed in the testis, while the pre-beta subunit (ADAM2), pre fertilin, undergoes proteolytic cleavage in the epididymis to become mature form (25) (Figure 4). In accordance with proteolysis, the immunostaining pattern is shifted to the postacrosomal region. In this case, the external peptide domains are continuously cleaved off in a stepwise manner, and the newly exposed functional domain interacts with the surrounding molecule(s). Such a limited proteolysis continues up to the point of sperm-egg interaction, and the mature form of fertilin is thought to play dual functions in sperm-egg binding and fusion via integrin/disintegrin-like interactions. Fertilin is also present in monkey spermatozoa.

In another case, guinea pig PH-20 (SPAM1) is anchored to the sperm plasma membrane as well as the IAM by a glycerophosphatidyl inositol. This molecule has multifunctional activities such as hyaluronidase, zona binding and cell signalling. PH-20 is expressed in round spermatids, and is localized in the acrosomal membrane (PH-20AM) and the entire surface of testicular spermatozoa (PH- 20PM). PH-20 becomes reduced in molecular size during epididymal maturation, and is simultaneously distributed to the posterior head (PH-20PM). This modification is due to the deglycosylation of N-linked oligosaccharides. PH-20PM migrates to the IAM of the anterior acrosome (PH-20IAM) after the acrosome reaction. This event is found in spermatozoa of many mammals (mouse, guinea pig, bull, horse, and human).

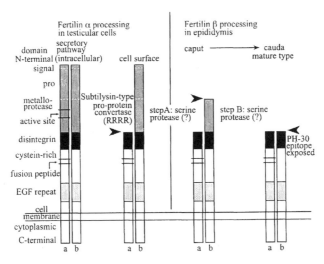

Figure 4. A proposed model for sequential proteolytic events involved in the processing of fertilin alpha/beta (ADAM1/ADAM2, PH-30). Modified from Lum and Blobel (25).

Zonadhesin, which is homologous to the von Willebrand factor and mucins, is expressed on the peri-acrosomal plasma membrane, and the precursor protein is proteolytically processed to become new, 105 kDa and 45 kDa proteins in the boar. The complement (C) regulatory protein CD59 binds to the entire surface during sperm maturation and is thought to be required for protection from immune attack during the transit of spermatozoa through the male and female reproductive tracts in the human.

In the flagellum: Flagellar proteins belonging to the immunoglobulin superfamily, CE9, MC31 and basign, also undergo limited proteolysis (26-28). These proteins migrate from the principal piece to the middle piece across the annulus subsequent to modification. CE9 is endoproteolytically processed near the N terminus at Arg 74 and redistributed from the posterior tail to the anterior tail during its passage through the proximal caput. The redistribution of CE9 (23-33 K) has been examined by the photobleaching technique; the redistribution was found to be time-, temporal- and energy dependent.

In the case of MC31, a homologue of CE9, the precursor protein is initially synthesized in spermatocytes and strongly expressed on the Golgi apparatus and plasma membrane at around stage XIII-II of rat spermatogenesis (Figure 5). The synthesis is maintained at a quite low level during the round spermatid phase and is intensively reexpressed in the early elongating spermatid to be integrated into the plasma membrane of the nascent principal piece. MC31 is then slightly reduced in molecular size from approximately 45-35 kDa to approximately 35-25 kDa during epididymal maturation, and concomitantly shifts to the middle piece as in the case of CE9. Thus, plasma membrane proteins are not organized into the correct or final site when spermatozoa are formed in the testis. Subsequently, biochemical modifications mainly due to deglycosylation and/or limited proteolysis modify the large precursor proteins (immature type) to the reduced-size proteins (mature type). During the maturation process, the surface proteins move to the correct site to play their destined roles. Some of the plasma membrane proteins are further redistributed from the flagellar regions to the head regions or vice versa (Table 1). If this is lateral diffusion, unknown mechanisms may be operative since the proteins need to cross over a strong diffusion barrier, the posterior ring. In any case, specific roles ascribed to the proteins (molecules) thus modified include protection, motility, passage through the cumulus oophorus, recognition of and binding to the zona pellucida, the acrosome reaction, attachment to and fusion with the oolemma, and modulation of relevant molecules.

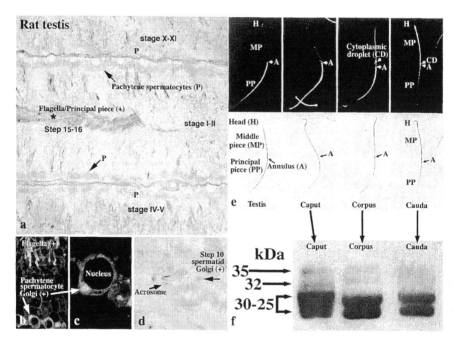

Figure 5. Maturational modification of an immunoglobulin superfamily protein MC31 (CE9) in the rat(a-d). Initial expression in the testis; a and d, immunohistochemistry images, b and c, confocal laser micrscope images. e, Modification of MC31 in the epididymis; the immunostaining pattern changes from the middle piece (MP) to the principal piece (PP). f, Reduction in molecular weight; MC31 is reduced from (35 kDa) in the caput to 30-25 kDa in the cauda epididymidis. The reduction in molecular size also occurs for another immunoglobulin superfamily protein, basigin (28).

Other Maturation-Associated Flagellar Components

Other flagellar components also undergo maturational changes. Among these, events associated with three components are described; the cytoplasmic droplet (CD), outer dense fibers (ODF) and the fibrous sheath (FS).

Cytoplasmic Droplet (CD)

Only a small amount of the cytoplasm, the cytoplasmic droplet (CD), is retained in spermatozoa. The CD is generally located near the neck region in the testis and caput spermatozoa, while it moves toward the annulus as spermatozoa move toward the cauda epididymidis. Estrogen is produced through P450 aromatase in the CD during epididymal passage. Spermatozoa are likely to develop the ability of regulatory response in order to adjust to the osmotic changes through the CD during epididymal maturation. However, if excessive residual cytoplasm remains, then sperm DNA is likely to be damaged by oxidative stress, the origin of which is the excess of residual cytoplasm (29).

Outer Dense Fibers (ODF) and Fibrous Sheath (FS)

The ODF proteins assemble in a proximal-distal direction along the length of the axoneme, while FS proteins are assembled in the opposite direction. These assemblies then overlap within the periaxonemal cytoplasm. The ODF proteins appear to be temporarily stored in granulated bodies of the cytoplasmic lobe during the assembly, while the FS proteins are randomly distributed throughout the cytoplasm. Later, intermolecular disulfide bonds are formed, stabilizing the ODF and FS during epididymal maturation. The ODF proteins reach maturity in the epididymis, undergoing protein degradation via the ubiquitin system.

The FS has long been thought to serve as a mechanical component responsible for flagellar flexibility (5). Proteolysis occurs in the case of some groups of the FS proteins during epididymal maturation. Two families of proteins are of interest. One is a family of cAMP-dependent protein kinase anchoring proteins (AKAP) and the testis A-kinase anchoring protein (TKAP). The other is a family of glycolytic enzymes that include glyceraldehyde 3-phosphate dehydrogenase-s (GAPDS) and hexokinase type 1-s (HK1-S). Since the former group functions as the anchoring site for cAMP-dependent protein kinase, the AKAP and TKAP family proteins are implicated in the signaling pathway in spermatozoa. The latter group, GAPDS and HK, is involved in energy production or motility. Therefore, the FS develops the ability to serve as a scaffold for functional proteins that regulate varieties of sperm functions leading to fertilization during the maturation process.

Table 1. Examples of the Modification of Plasma Membrane Proteins in Mammalian Spermatozoa

Protein	When modified	Redistribution	Ref.
PH-20	Epididymal passage	Whole cell to posterior head	1
SPAM 1 (mouse PH-20)	Epididymal passage	Whole cell to anterior and posterior head	2
PH-30 (fertilin)	Epididymal passage	Whole head to posterior head	3
AH-50	Epididymal passage	Whole head to anterior head	3
Surface galactosyl transferase	Epididymal passage	Anterior head over acrosome to more restricted acrosome cap	4
CE9	Epididymal passage	Posterior tail to anterior tail	5
MC31	Epididymal passage	Posterior tail to anterior tail	6
Basigin	Epididymal passage	Posterior tail to anterior tail	7
PT-1	Capacitation	Posterior tail to whole tail	8
2B1 (rat PH-20)	Capacitation	Tail to anterior head	9
PH-20	Acrosome reaction	Posterior head to inner acrosomal membrane on anterior head	1
Basigin	Capacitation/Acrosome reaction	Tail to head	7

Modified from Myles and Primakoff (1997) Biol Reprod, 56; 320-327. References: 1. Phelps and Myles (1987) Dev Biol, 123; 63-72. 2. Deng et al. (1999) Mol Reprod Dev, 52; 196-206. 3. Phelps et al. (1990) J Cell Biol, 111; 1839-1847. 4. Scully et al. (1987) Dev Biol, 124; 111-124. 5. Petruszak et al. (1991) J Cell Biol, 114; 917-927. 6. Toshimori et al. (1992) Mol Reprod Dev, 32; 399-408. 7. Saxena et al. (2002) Reproduction, 123; 435-444. 8. Myles and Primakoff (1984) J Cell Biol, 99; 1634-1641. 9. Jones et al. (1990) Dev Biol, 139; 349-362.

Other Maturation-Associated Factors

Ubiquitin system and carnitine are also involved in sperm maturation.

Ubiquitin-System

Ubiquitin is instrumentally involved in controlling a variety of biological events such as cell cycle progression and protein degradation/recycling, etc. Ubiquitinated polysubstrates are ligated with polyubiquitinated chains and targeted to cellular organelles, such as lysosomes and proteosomes for degradation. In these processes, the polyubiquitinated chains are cleaved by specific ubiquitin hydrolases. The process of ubiquitin attachment to target proteins is activated through enzymatic pathways involving enzymes such as ubiquitin-ligases. A variety of elements involved in the ubiquitin system have been detected in the testis and epididymis, and the general activity of the ubiquitin system is relatively high during spermiogenesis. The ubiquitin system has been implicated in various steps of protein degradation/recycling, repairing and elimination during spermatogenesis (e.g., nuclear condensation), epididymal maturation (e.g., centrosome degradation) and fertilization (e.g., sperm mitochondrial degradation) (30).

Carnitine

Carnitine is an essential cofactor in the transport of long chain fatty acids inside the mitochondria and their subsequent beta-oxidation (producing acetyl- CoA). The concentration of carnitine in the epididymal duct fluid dramatically increases from 1 mM in the rete testis to 60 mM in the cauda; the concentration is 2000 times higher than that in the blood. If carnitine is deficient, spermatozoa cannot move toward the cauda epididymis, and the duct of proximal epididymis is dilated or ruptured, leading to the production of anti-sperm autoantibody.

CONCLUSIONS

Typical events of the testicular and epididymal maturation of mammalian spermatozoa are described in this chapter. Such maturational events include a variety of biologically important phenomena, all of which are required for successful fertilization *in vivo*. Future studies will undoubtedly lead to an understanding of the mechanisms involved in sperm maturation at the molecular level. It is hoped that the updated information included in this chapter will help to decipher the complex biology of sperm maturation and provide insights into the control of spermatogenesis, fertilization and their diagnostic criteria.

ACKNOWLEDGEMENTS

The author is indebted to Drs. Kazuya Yoshinaga, Ichiro Tanii, Dinesh K. Saxena, Tadasuke Oh-oka, and Chizuru Ito for their cooperation. I also am appreciative to Mr Yasunori Fujii and Miss Yuko Kiyotake for their

110

technical assistance. The work presented in this review was supported by grants from the Ministry of Education, Science, Sports and Culture of Japan.

REFERENCES

1. Bedford J.M. "Maturation, transport and fate of spermatozoa in the epididymis." In: Handbook of Physiology, Sec. 7, Vol. 5, R.O. Greep, E.B. Astwood, eds. Washington DC: American Physiological Society, 1975; 303-317.
2. Orgebin-Crist M.C., Danzo B.J., Davies J. "Endocrine control of the development and maintenance of sperm fertilizing ability in the epididymis." In: Handbook of Physiology, Sec. 7, Vol. 5, R.O. Greep, E.B. Astwood, eds. Washington DC: American Physiological Society, 1975; 319-338.
3. Robaire B., Hermo L. "Efferent ducts, epididymis, and vas deferens: structure, functions, and their regulation." In: The Physiology of Reproduction, E. Knobil, J.D. Neill et al., eds. New York: Raven Press, 1988; 999-1080.
4. Setchell B.P., Sanchez-Partida L.G., Chairussyuhur A. Epididymal constituents and related substances in the storage of spermatozoa: a review. Reprod Fertil Dev 1993; 5:601-612.
5. Eddy E.M., O'Brien D.A. "The spermatozoon." In: The Physiology of Reproduction, 2nd edition, E. Knobil, J.D. Neill, eds. New York: Raven Press, 1994; 29-77.
6. Yanagimachi R. "Mammalian fertilization." In: The Physiology of Reproduction, 2nd edition, E. Knobil, J.D. Neill, eds. New York: Raven Press; 1994; 189-317.
7. Hinton B.T., Palladino M.A. Epididymal epithelium: its contribution to the formation of a luminal fluid microenvironment. Microsc Res Tech 1995; 30:67-81.
8. Oko R.J. Developmental expression and possible role of perinuclear theca proteins in mammalian spermatozoa. Reprod Fertil Dev 1995; 7:777-797.
9. Krichhoff C., Pera I., Derr P., Yeung C.H., Cooper T. The molecular biology of the sperm surface. Post-testicular membrane remodelling. Adv Exp Med Biol 1997; 424:221-232.
10. Aitken R.J., Vernet P. Maturation of redox regulatory mechanisms in the epididymis. J Reprod Fertil Suppl 1998; 53:109-118.
11. Cooper T.G. Interactions between epididymal secretions and spermatozoa. J Reprod Fertil Suppl 1998; 53:119-136.
12. Dacheux J.L., Druart X., Fouchecourt S., Syntin P., Gatti J.L., Okamura N., Dacheux F. Role of epididymal secretory proteins in sperm maturation with particular reference to the boar. J Reprod Fertil Suppl 1998; 53:99-107.
13. Jones R. Plasma membrane structure and remodelling during sperm maturation in the epididymis. J Reprod Fertil Suppl 1998; 53:73-84.
14. Toshimori K. Maturation of mammalian spermatozoa: modifications of the acrosome and plasma membrane leading to fertilization. Cell Tissue Res 1998; 293:177-187.
15. Tulsiani D.R.P., Orgebin-Crist M.-C., Skudlarek M.D. Role of luminal fluid glycosyltransferases and glycosidases in the modification of rat sperm plasma membrane glycoproteins during epididymal maturation. J Reprod Fertil Suppl 1998; 53:85-89.
16. Jones R.C. To store or mature spermatozoa? The primary role of the epididymis. Int J Androl 1999; 22:57-67.
17. Fawcett D.W. The mammalian spermatozoon. Dev Biol 1975; 44:394-436.
18. Hess R.A., Bunick D., Lee K.H., Bahr J., Taylor J.A., Korach K.S., Lubahn D.B. A role for oestrogens in the male reproductive system. Nature 1997; 390:509-512.
19. Hecht N.B. Molecular mechanisms of male germ cell differentiation. Bioessays 1998; 20:555-561.

20. Olson G.E., Winfrey V.P., NagDas S.K. Acrosome biogenesis in the hamster: ultrastructurally distinct matrix regions are assembled from a common precursor polypeptide. Biol Reprod. 1998; 58:361-370.
21. Blobel C.P. Functional processing of fertilin: evidence for a critical role of proteolysis in sperm maturation and activation. Rev Reprod 2000; 5:75-83.
22. Primakoff P., Myles D.G. The ADAM gene family: surface proteins with adhesion and protease activity. Trends Genet 2000; 16:83-87.
23. Cuasnicu P.S., Ellerman D.A., Cohen D.J., Busso D., Morgenfeld, M.M., Da Ros V.G. Molecular mechanisms involved in Mammalian gamete fusion. Arch Med Res 2001; 32:614- 618.
24. Evans J.P. Fertilin beta and other ADAMs as integrin ligands: insights into cell adhesion and fertilization. Bioessays 2001; 23:628-639.
25. Lum L., Blobel C.P. Evidence for distinct serine protease activities with a potential role in processing the sperm protein fertilin. Dev Biol 1997; 191:131-145.
26. Phelps B.M., Myles D.G. The guinea pig sperm plasma membrane protein, PH-20, reaches the surface via two transport pathways and becomes localized to a domain after an initial uniform distribution. Dev Biol 1987; 123:63-72.
26. Cesario M.M., Bartles J.R. Compartmentalization, processing and redistribution of the plasma membrane protein CE9 on rodent spermatozoa. Relationship of the annulus to domain boundaries in the plasma membrane of the tail. J Cell Sci 1994; 107:561-570.
27. Toshimori K., Yoshinaga K., Tanii I., Wakayama T., Saxena D.K., Ohoka T. Protein expression and cell organelle behavior in spermatogenic cells. Kaibogaku Zasshi 2001; 76:267-279.
28. Saxena D.K., Oh-oka T., Kadomatsu K., Muramatsu T., Toshimori K. Behaviour of a sperm surface transmembrane glycoprotein basigin during epididymal maturation and its role in fertilization. Reproduction 2001; 123:435-444.
29. Aitken R.J., Krausz C. Oxidative stress, DNA damage and the Y chromosome. Reproduction 2001; 122:497-506.
30. Sutovsky P., Moreno R., Ramalho-Santos J., Dominko T., Thompson W.E., Schatten G. A. Putative, ubiquitin-dependent mechanism for the recognition and elimination of defective spermatozoa in the mammalian epididymis. J Cell Sci 2001; 114:1665-1675.

Chapter 7

GLYCOSYL PHOSPHATIDYL INOSITOL (GPI) ANCHORED MOLECULES ON MAMMALIAN SPERMATOZOA

Ben M.J. Pereira, Parul Pruthi and Ramasare Prasad
Indian Institute of Technology, Roorkee, Uttaranchal, India

INTRODUCTION

Sperm cells, like somatic cells, are surrounded by a plasma membrane that participates in early events of fertilization including capacitation, sperm-egg interaction and the acrosome reaction. Thus, an understanding of the composition and organization of sperm membrane is important to ascertain its role in the fertilization process. The sperm plasma membrane is predominantly composed of phospholipids and cholesterol. A change in the relative concentrations of these two components determines the permeability, fluidity, rigidity and functional state of the gamete. A mosaic of other molecules including proteins is scattered and embedded in this structure (1). Spermatozoa require a large number of proteins to become competent and fertilize the egg. Unfortunately, the cells have no known synthetic activity and the number of molecules that can be physically accommodated at any given time is severely restricted by their miniscule size. One strategy by which spermatozoon overcomes this limitation is to choose a type of anchorage/mechanism that economizes on the number of surface molecules.

IMPORTANCE OF THE GPI ANCHOR

We will begin this chapter by projecting the basic patterns of membrane anchorage and importance of the glucosyl phophatidyl inositol (GPI) anchored molecules. Of all types of attachments of bio-molecules to the membrane surface, the GPI anchor is of great interest since it is present only on the outer leaflet of the lipid bi-layer (Figure 1). This ensures that the

114

molecules associated with a GPI anchor are capable of unhindered movement on the membrane surface. The question whether the GPI anchorage offers stability to the proteins is difficult to address since it depends on many factors including the nature of the surrounding lipids, the nature and amount of the neighboring membrane proteins and other factors such as accessibility to proteolytic enzymes (2). Several proteins attached to membranes through GPI anchors are released from the cell surface by the action of specific enzymes such as phospholipases C or D (3); however, a chemical modification like fatty acid acylation of inositol can make glycoinositol phospholipids resistant to the phospholipase cleavage. The GPI anchor may also accept proteins from the environment provided they contain the desired signal sequence. Lastly, the amphipathic nature of GPI-anchored proteins makes it possible to transfer from one cell to another by the assistance of vesicles and phospholipid transfer proteins (4,5). Thus, the GPI type of anchorage can offer special structural advantages to sperm membrane proteins.

Cell Exterior

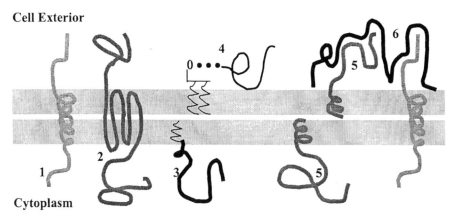

Cytoplasm

Figure 1: Diagram illustrating various ways by which proteins are anchored in the lipid bilayer of the sperm plasma membrane. 1, Single-pass transmembranal protein; 2, Multi-pass transmembranal protein; 3, Protein linked through fatty acid (myristyl and farnesyl anchors); 4, GPI-anchored protein; 5, Protein inserted through hydrophobic interactions; 6, Peripheral protein

STRUCTURE OF GPI-ANCHORED PROTEINS

A typical GPI-anchored protein consists of a glycan bridge (Man α-1,2 Man α-1,6 Man α-1,4 GlcN) between phosphatidyl-ethanolamine and phospho-inositol; the phospho-ethanolamine is attached to the C-terminus of the protein via amide linkage. The fatty acid linked to phosphatidylinositol is inserted in one leaflet of the membrane lipid bi-layer while the glycan core and the attached protein is exposed (Figure 2). The glycan core of GPIs

(GlcN-Man-Man-Man) is remarkably conserved throughout evolution; although it is modified in many cell types/organisms by the addition of side chains (R_1-R_3) at specific locations (6).

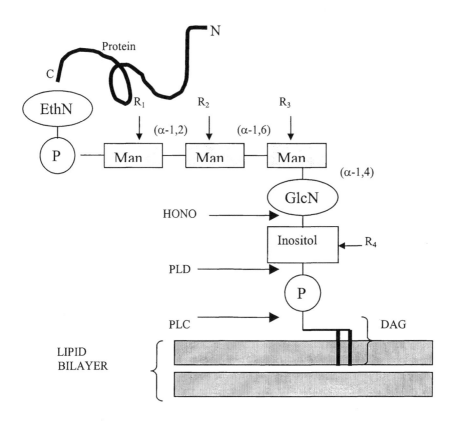

Figure 2: The core structure of a typical GPI-anchored protein showing the various sites of cleavage by chemical and enzymatic methods. R_1 is usually Man, R_2 is ethanolamine phosphate, R_3 can be mono, oligo, polysaccharide or ethanolamine phosphate and R_4 is palmitate. P, phosphate group; GlcN, glucosamine; Man, mannose; EthN, ethanolamine; HONO, nitrous acid; PLC, phospholipase C; PLD, phospholipase D; DAG, diacylglycerol.

BIOSYNTHESIS OF GPI-ANCHORED PROTEINS

No significant work has been carried out on the biosynthesis of GPI anchors of sperm proteins. Thus, much of the knowledge on the biosynthetic pathway of GPI proteins is derived from model studies involving parasites, yeast, and mammalian cell lines (7). It should be noted that spermatozoa leaving the testis lack endoplasmic reticulum (ER), the intracellular compartment crucial for the biosynthetic machinery. In view of the numerous molecules required, the large number of sperm produced and their small size, it is difficult to envision that their entire protein requirement is met during development in the testis. An alternative option is that several molecules including GPI-anchored proteins are synthesized at several sites along the reproductive tract and then acquired by the sperm through specialized mechanisms (8).

The GPI Anchor Precursor

The biosynthesis of the core anchor is initiated in the cytoplasmic face of the ER and completed in its lumen. Basically, sugar residues (glucosamine, mannose) and phosphoethanolamine are sequentially added on to phospha- tidylinositol forming a GPI anchor precursor (7). The formation of the GPI anchor precursor is a multi-step process requiring enzymes such as glycosyl synthase, glycosyl transferase, inositol acyl transferase, deacytylase, ethanolamine phosphodiesterases and mannosyl transferases (Figure 3).

Transfer of Protein to GPI Anchor

Proteins destined for GPI anchorage are synthesized following the same generalized plan consisting of transcription and translation processes similar to other proteins. However, the protein is synthesized as a nascent precursor with two specific signal sequences. One is the N-terminal signal peptide sequence that is responsible for its translocation into the ER lumen during synthesis, and the second is a small hydrophobic C-terminal stretch of 15-20 amino acids that holds the protein in the ER membrane. As soon as the protein synthesis is completed, it undergoes processing in the ER lumen (Figure 4). This involves cleavage of the nascent protein in such a way that the hydrophobic portion is retained on the membrane while the rest is transferred to the GPI anchor. The C-terminal amino acid of the free protein to which the GPI anchor gets attached is termed the W-site and always has a small side chain. The second and third amino acids are referred to as W+1 and W+2. This is followed by a sequence of amino acids (5-7 hydrophilic and 12-20 hydrophobic residues) characteristic of an attachment signal. The attachment of the protein to its anchor requires two gene products Gaa1p and Gpi8p, which together form a complex with transamidase activity. It has been suggested that the enzyme transamidase binds to the attachment signal of the nascent peptide, cleaves the carboxyl end of the W-site amino acid and then guides the released peptide to the GPI anchor to complete the transamidation reaction (9).

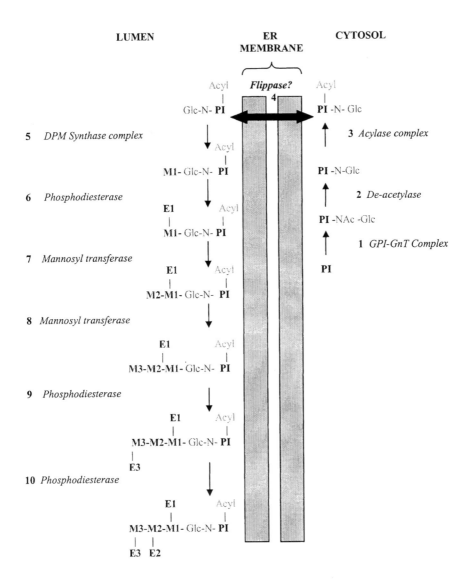

Figure 3: Schematic representation of GPI anchor biosynthesis in the ER. The biosynthesis is initiated in the cytoplasmic side of the ER membrane (steps 1-3). Both GlcN and the acyl group are attached at different positions of the inositol residue on PI. This intermediate then flips to the luminal side of the ER (step 4). The transfer of three mannosyl residues and three ethanolamine residues assembles the rest of the glycan backbone on the PI intermediate (steps 5-10). The sequence of these transfers may vary in different cells. The enzymes involved are shown in italics. PI, phosphatidylinositol; GlcN, glucosamine; GlcNAc N-acetylglucosamine; M1, M2, M3 mannose residues; E1 E2 E3 phosphoethanolamine residues;

Modifications in Golgi and Transfer to Plasma Membrane

The addition of sugars to the GPI protein (N-/O-glycosylation) that takes place in the Golgi apparatus gives rise to the mature GPI-anchored protein, which finds its way to the plasma membrane surface. The details of sorting and exporting the GPI protein from the ER to the plasma membrane have been adequately reviewed (2). Briefly, protein sorting is governed by a

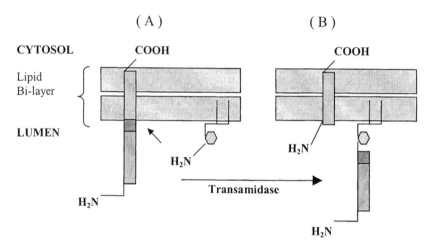

Figure 4: Mechanism of protein transfer to the GPI anchor by transamidation reaction. ⬡, glycan bridge; ▪, attachment site. (A) Nascent protein precursor and GPI anchor attached separately to the luminal side of ER membrane. (B) Freed protein after cleavage is transferred to GPI anchor by transamidation.

combination of different factors, which include GPI addition, transcytosis, N- and O-glycosylation and oligomerization. Some GPI- anchored proteins are believed to partition into specialized membrane microdomains known as DIGs (detergent insoluble glycolipids) or lipid rafts (10). In the sperm of sea urchins, GPI-anchored proteins and transducer proteins have been co-localized in the DIGs, implicating an organized receptor activation and signal transduction. Whether such a mechanism occurs in mammalian sperm is not yet known.

GPI-ANCHORED PROTEINS IN GERM CELLS AND SPERMATOZOA

The total number of GPI-anchored proteins that have been reported so far in various cells is quite large. Several GPI-anchored proteins have been detected in the male reproductive tract, some of which have an indirect effect on the developing spermatozoa. For example, a GPI-anchored Ceruloplasmin found in Sertoli cells transports iron across the blood testis barrier and makes it available for sperm during its development. Other GPI proteins are synthesized in the epithelial cells of the male sex accessory organs and transferred to the sperm surface (11). This chapter summarizes only those GPI proteins detected in the testicular germ cells and spermatozoa (Table 1).

MECHANISM(S) OF GPI PROTEIN TRANSFER TO SPERM MEMBRANES

GPI-anchored proteins on the sperm plasma membrane have attracted the same attention as other integral components. The distribution/ translocation of this class of protein on the sperm during development, maturation, storage, capacitation, sperm-egg interaction and the acrosome reaction is a clear indication of their importance in reproduction (11). However, the mechanism(s) through which GPI proteins are acquired by spermatozoa is far from clear. The observation that GPI-anchored proteins are acquired by sperm post-testicularly implies that there must be other mechanisms involved.

Several regions of the male reproductive tract are known to actively support the function of the spermatozoa by secreting biomolecules that are incorporated into the cell surface. The epididymis is one such organ that has been shown to secrete maturation-associated proteins, which are acquired by the cauda spermatozoa. Other antigens common to lymphocytes, like the complement restriction factors found in the epithelial cells of the epididymis, vas deferens, seminal vesicle and prostate are also transferred to the sperm surface. Even GPI-anchored ectoenzyme, 5' nucleotidase, is transferred to spermatozoa post-testicularly. Some of these antigens have been detected in secretory vesicles found in the luminal/seminal fluid. These vesicles (prostasomes) have been shown to fuse with spermatozoa under specified conditions. In addition, phospholipid-transfer proteins that can act as specific carriers for single phospholipids between membranes are known to exist in almost all eucaryotic systems (5). These facts make it possible to conceptualize a cell-to-cell transfer of GPI-anchored proteins. It is likely that GPI proteins are synthesized by the epithelial cells lining the male reproductive tract (epididymis/seminal vesicles/prostate), secreted into luminal/seminal fluids, transported by secretory vesicles and ultimately incorporated into spermatozoa by a process of fusion (4).

Another possibility to be examined is that the GPI anchor on the sperm surface remains the same, but the proteins are replaced. This hypothesis stems from experiments demonstrating that desired proteins can be transferred to the GPI anchor on membranes when linked to specific signal sequences. When 29 residues of the C-terminus of a GPI-anchored protein, decay-accelerating factor (DAF), were fused to the C-terminus of a secretory protein, the fusion protein was directed to the cell surface by means of a GPI anchor (9). Thus, if proteins with these specific signal sequences are synthesized and secreted by the epithelial cells lining the reproductive tract, they could be picked up by sperm membranes with GPI anchor; provided that the appropriate enzymatic machinery for this transfer is also available in the seminal fluid.

Table 1: Consolidated list of GPI-anchored proteins found in germ cells and spermatozoa

GPI protein/ Homologue	Species	Location	Function
HemT-3 transcript (22kD protein)	Mice	Early spermatocytes	Germ cell differentiation
Sperm agglutination antigen 1(SAGA-1)	Human	Sperm	Immunological reactions
CD52/CAMPATH/ HE5 / B7 /MB7/ Gp20-sialoglycoprotein	Human, Rat, Mice, Monkey	Sperm, Epididymis Seminal vesicle, lymphocytes	Fertilization, sperm maturation, Immunoprotection
CD55/ DAF	Human, Rat	Sperm, Epididymis	Protection from complement attack, Interaction with egg.
CD59/ Protectin/ MACIP	Human, Rat Mice	Sperm, Prostate Epididymis	Prevents formation of membrane attack complex, Complement receptor
CD73/ Ecto 5'nucleotidase	Human, Ox	Sperm, Seminal vesicle	Adhesion, Receptor
PH20/ 2B1 glycoprotein/ Hyaluronidase	Human, rat, Guinea pig, Monkey	Sperm	Sperm penetration through cumulus Secondary sperm-Zona binding
Alpha mannosidase	Boar	Sperm	Mannose receptor, Catalysis, Fertilization
SmemG/RB7/CE5	Rat, Dog	Sperm	?

Kooyman et al. (12), using several experimental models, have demonstrated that the transfers of GPI proteins occur under physiological conditions. The investigators have gone a step further by detailing a procedure that could effect the transfer of GPI-linked proteins from one cell type to another in either an *in vivo* or *in vitro* system. Based on these inputs, we have formulated a hypothetical model for the replacement of the protein component in sperm GPI-anchored molecules (Figure 5). Several lines of evidence support this model. First, GPI anchors are recycled in eucaryotic cells. Second, GPI proteins participate in potocytosis. Third, spermatozoa undertake continuous remodeling of both proteins and lipids on membranes (1). Finally, there are reports that spermatozoa acquire new GPI-anchored proteins post-testicularly.

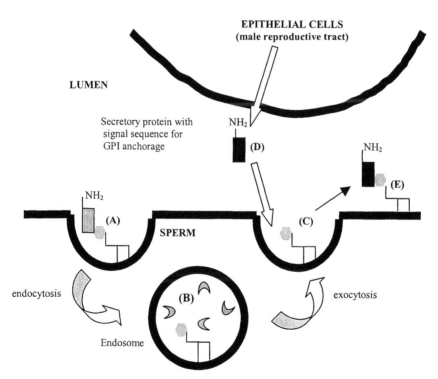

Figure 5: Hypothetical model for the replacement of sperm membrane proteins attached to GPI anchors. (A), GPI-anchored protein is synthesized by testicular germ cells and inserted en-block into sperm membrane; (B), After the protein's function has been served, it is degraded within the sperm; (C), The GPI anchor may be recycled and returned as part of sperm membrane remodeling; (D), Epithelial cells of the male reproductive tract synthesize and secrete proteins with signal sequence required for GPI anchorage; (E), The GPI anchor picks up and accommodates this new protein.

FUNCTIONAL RELEVANCE OF GPI-ANCHORED MOLECULES

In this section, we will identify some of the better characterized GPI-anchored sperm proteins in order to highlight their functional significance.

GPI Molecules that Promote Sperm-Egg Interaction

One of the first known GPI-anchored proteins detected in the posterior head region of guinea pig sperm was the PH-20 implicated in sperm-egg adhesion. This protein was shown by the technique of Fluorescence Recovery After Photobleaching (FRAP) to display a lateral mobility on the sperm surface that was far greater than the other transmembrane proteins. In testicular sperm, the protein is spread over the entire surface but soon becomes localized to the posterior head of the acrosome-intact epididymal spermatozoa. Once the spermatozoa undergo the acrosome reaction, the molecules move to the newly exposed inner acrosomal membrane on the anterior head region. This kind of lateral mobility during different stages of sperm maturation is possible for the GPI anchorage that spans over the outer leaflet of the lipid bi-layer. This may allow the protein to interact with other molecules on the membrane surface. One possible implication of the re-localization into new domains could be the regulation of its function (13).

The N-terminal domain of PH-20 has hyaluronidase activity. The enzyme catalyses the hydrolysis of hyaluronic acid, a large polymer of alternating N-acetylglucosamine and glucuronic acid units found in egg vestments (see Chapter 7). Based on DNA homology between bee venom hyaluronidase and PH-20, it was predicted that mammalian sperm hyaluronidase might be GPI anchored. It is now confirmed that mouse sperm hyaluronidase, a 68 kD protein, remains attached to the surface through a GPI anchor both in the acrosome-intact and acrosome-reacted spermatozoa (14). Since hyaluronic acid is distributed in the matrix of the cumulus complex and the space between the zona pellucida and egg plasma membrane, it has been proposed that sperm membrane hyaluronidase performs the role of both a digestive enzyme and adhesion molecule (13). Perhaps the GPI anchorage increases the mobility of the cell surface protein so that a maximum rate of successful penetration of the egg vestments and fusion of egg-sperm plasma membranes is ensured (14).

Mouse epididymal spermatozoa have a surface-associated decapacitation factor (DF) that can be removed by treatment with phosphatidylinositol-specific phopholipase (PI-PLC). However, exogenous DF cannot re-associate PI-PLC treated spermatozoa suggesting that the GPI-anchored protein is a receptor for the DF. Further characterization has indicated that the DF binds to the GPI receptor through fucose residues.

Several other prominent GPI sperm membrane proteins have been detected by PI-PLC treatment (14). These molecules are present on the surface of acrosome-intact spermatozoa suggesting that they are associated with sperm plasma membrane. In addition, several GPI proteins were detected in acrosome-reacted sperm perhaps due to the exposure of the inner acrosomal membrane. However, the precise role of these molecules in fertilization is not yet known.

GPI Molecules Associated with Sperm Maturation

It is generally accepted that one of the aspects that contributes to sperm maturation is the addition of proteins/glycoproteins to the sperm surface from the surrounding epididymal fluid. In the rat, a ~26 kD glycoprotein was found to attach to the cauda (but not caput) spermatozoa by treatment of radiolabeled cauda spermatozoa with PI-PLC and partitioning the released proteins in Triton X-114; the molecule was confirmed to be GPI-anchored (15). Another major antigen associated with the sperm maturation (CD52) has been reported in humans. The appearance of this protein in the cauda spermatozoa leads one to believe that it is acquired from the epididymis, but the exact mechanism of uptake and binding is not yet known. It remains to be established whether the GPI anchor participates in the transfer of this protein to the spermatozoa. Although the precise function of this protein remains unknown, its persistence in ejaculated spermatozoa suggests that it may be important in fertilization.

GPI Molecules that Provide Immuno-Protection

A number of antigens associated with the immune system like CD52 CD55, CD59 and CD73 have been identified on spermatozoa. These proteins are not present in sperm collected from the testis or caput epididymidis, but have been detected in the cauda spermatozoa and persist in ejaculated cells. A cDNA screening procedure revealed that similar transcripts were present in epithelial cells of the distal epididymis and in blood lymphocytes. Some of these lymphocyte-associated antigens have also been identified in secretory vesicles (prostosomes) of seminal plasma. Thus, it is possible that these antigens are not synthesized by spermatozoa but are acquired from the male genital tract.

Sperm agglutination antigen-1 (SAGA-1), a polymorphic GPI-anchored glycoprotein, has been localized in all domains of the human sperm surface. The core peptide of the antigen has recently been shown to be identical to the sequence of CD52 found in lymphocytes. However, the sperm form of

CD52 exhibits N-linked glycan epitopes which are not expressed on lymphocyte CD52. Hence, it is suggested that these two glycoproteins are glycoforms with the same core peptide but different carbohydrate structures. Perhaps the modification in carbohydrates is effected in the male reproductive tract which is rich in both glycosidase and glycosyltransferases.

It is believed that the cluster of differentiation (CD) proteins, which are GPI anchored, protect sperm against immune attack and complement mediated cell lysis during storage in the cauda epididymidis and transport through the female reproductive tracts en route to fertilizing the egg, much the same way parasites are protected to survive in their host (4).

GPI Molecules that Possess Catalytic Properties

Several GPI-anchored ecto-enzymes have been reported in different membranes. PH20/ Hyaluronidase, discussed earlier, was the first known GPI molecule exhibiting catalytic properties on the sperm surface. Thereafter, 5' nucleotidase acquired post-testicularly was detected on human sperm membranes. More recently, among the glycosidases, it was observed that alpha-mannosidase was released from the boar sperm plasma membrane by PL-PLC treatment (16), supporting a GPI mode of anchorage. Alpha-mannosidase activity is found on the sperm plasma membrane of several species, including man. In the rat, the enzyme is first expressed in testicular germ cells and is incorporated into the spermatozoa at the time of its development in the testis (17). The sperm surface molecule is important not only for its catalytic properties but also as a putative receptor for mannose rich residues present on the egg's extracellular coat, the zona pellucida (see Chapter 15). The properties of these ecto-enzymes could be further regulated by their migration into specific domains of the sperm membrane facilitated by the GPI anchor (13).

The number of GPI proteins associated with the male gamete is increasing. It is likely that additional sperm proteins with GPI anchors will be identified and their functional significance in the fertilization process established.

CONCLUSIONS

From the above discussion, it is obvious that the GPI-anchored molecules on the sperm membrane have received a great deal of attention. The structural architecture of the GPI protein is unique since the lipid anchor and the core carbohydrate moieties are to a large extent conserved, while the protein component is highly variable. The anchor which generally spans the outer leaflet of membranes facilitates sorting and trafficking of proteins to the sperm surface right at the time of their biosynthesis in germ cells during spermatogenesis. It has, however, come to light that sperm also acquire several GPI proteins post-testicularly. From the special traits acquired by the sperm through GPI proteins, it is reasonable to conclude that the anchorage offers special advantage and increases the chances of spermatozoa to fertilize an egg.

ACKNOWLEDGEMENTS

We are grateful to the Council for Scientific and Industrial Research, New Delhi, India for supporting our work on GPI-anchored sperm proteins. The authors are thankful to Mrs. Loreita Little and Ms. Lynne Black for editorial assistance, Mrs. S.M.A Pereira for critically reading the manuscript, and to Dr. Vikas Pruthi for preparation of graphics.

REFERENCES

1. Jones R. Plasma membrane structure and remodeling during sperm maturation in the epididymis. J Reprod Fertil 1998; 53 Suppl:73-84
2. Nosjean O., Briolay A., Roax B. Mammalian GPI proteins: sorting, membrane residence and functions. Biochim Biophys Acta 1997; 1331:153-186
3. Low M.G. "Phospholipases that degrade the glycosylphosphatidyl inositol anchor of membrane proteins." In Lipid modification of proteins-A practical approach. N.M. Hooper, A.J. Turner, eds. Oxford: IRL Press, 1992, pp 117-154
4. Kirchhoff C., Hale G. Cell to Cell transfer of glycosylphosphatidyl inositol-anchored membrane proteins during sperm maturation. Mol Human Reprod 1996; 2:177-184
5. Wirtz K.W.A. Phospholipid transfer proteins revisited. Biochem J 1997; 324:353-360
6. McConville M.J., Ferguson M.A.J. The structure, biosynthesis and function of glycosylated phosphatidyl inositols in the parasitic protozoa and higher eukaryotes. Biochem J 1993; 294:305-324
7. Takeda J., Kinoshita T. GPI anchor biosynthesis. Trends Biochem Sci 1995; 20:367-371
8. Kirchhoff C., Pera I., Derr P., Yeung C.H., Cooper T. The molecular biology of the sperm surface. Post-testicular membrane remodeling. Adv Exp Med Biol 1997; 424:221-232
9. Moran P., Raab H., Kohr W.J., Caras I.W. Glycophospholipid membrane anchor attachment: Molecular analysis of the cleavage/attachment site. J Biol Chem 1991; 266: 1250-1257
10. Simons K., Ikonen E. Functional raft in cell membranes. Nature 1997; 387:569-572
11. Myles D.G. "Sperm cell surface proteins of testicular origin: expression and localization in the testis and beyond." In Cell and Molecular Biology of the Testis, C. Desjardins, L. Ewing, eds. New York: Oxford University Press, 1993: 452-473
12. Kooyman D.L., Byrne G.W., Logan J.S. Glycosylphosphatidylinositol anchor. Exp Nephrol 1998; 6:148-151

13. Myles D.G., Primakoff P. Why did the sperm cross the cumulus? To get to the oocyte. Functions of the sperm surface proteins PH-20 and fertilin in arriving at, and fusing with, the egg. Biol Reprod 1997; 56:320-327

14. Thaller C.D., Cardullo R.A. Biochemical characterization of a glycosyl phosphatidyl inositol linked hyaluronidase on mouse sperm. Biochemistry 1995; 34:7788-7795

15. Moore A., White T.W., Ensrud K.M., Hamilton D.W. The major maturation glycoprotein found on rat cauda epididymal sperm surface is linked to the membrane via phosphatidylinositol. Biochem Biophys Res commun 1989; 160:460-468

16. Kuno M., Yonezawa N., Amari S., Hayashi M., Ono Y., Kiss L., Sonohara K., Nakano M. The presence of a glycosyl phosphatidylinositol-anchored α-mannosidase in boar sperm. IUBMB Life 2000; 49:485-489

17. Pereira B.M.J., Abou-Haila A., Tulsiani D.R.P. Rat sperm surface mannosidase is first expressed on the plasma membrane of testicular germ cells. Biol Reprod 1998; 59:1288-1293

Chapter 8

MALE ACCESSORY GLANDS: MOLECULAR MECHANISM OF DEVELOPMENT

Chhanda Gupta
University of Pittsburgh, Pittsburgh, PA, USA

INTRODUCTION

Over the past decade, knowledge of human sexual differentiation has greatly expanded to understand the processes involved in the development of male accessory glands. Sexual differentiation occurs in three consecutive levels. The chromosomal sex is determined by distribution of sex chromosomes during meiosis. This is followed by gonadal sex determination by the Sry locus on the Y chromosome, leading to development of the testis. The testis induces two hormones, namely, anti-Mullerian hormone (AMH) and testosterone, determining the somatic sex. AMH causes regression of Mullerian ducts whereas testosterone maintains the Wolffian ducts and induces male genital tract and male phenotype formation. In the absence of the sex-determining region of the Y chromosome (SRY), the ovary develops from the indifferent gonad. In the absence of AMH and testosterone, the Mullerian ducts constitutively differentiate and the Wolffian duct regresses resulting in the formation of a female external phenotype. Present knowledge on the role of different factors mediating development of male accessory glands is based on key experiments performed by several investigators (1-5). In 1940, Jost demonstrated a role for AMH using a rabbit embryo, castrated in utero (1). In 1942, Raynaud demonstrated a role of testosterone during embryonic development (2) and in 1967, Ohno identified the Sry locus determining the testis (3). Recently, some new factors were discovered in relation to the sex organ development (4,5). Some of these factors appear to act with the known signaling cascade and others have yet to be determined. In this chapter, I will discuss different steps involved in male genital development. In addition, the role of androgen synthesis and androgen receptor in mediating the process of masculine development will be discussed in detail.

GONADAL DIFFERENTIATION

Gonadal differentiation involves two steps: first, the determination of gonadal sex by the genetic sex; second, the differentiation of the genital apparatus and hence the phenotypic sex. Both male (46,XY) and female (46,XX) embryos possess indifferent, common primordia that have inherent tendencies to feminize unless there is active interference by the Y chromosome-induced masculinizing factors (1). The indifferent embryonic gonad develops into an ovary unless it is diverted by a testicular organizing factor regulated by the Y chromosome, resulting in formation of the testis. The sequential steps involved in testicular differentiation leading to male organogenesis are shown in Figure 1. With the onset of testicular differentiation, the sex-determining gene on the Y chromosome acts autonomously to induce Sertoli cell differentiation which then further mediates testicular differentiation (3,6). An early endocrine function

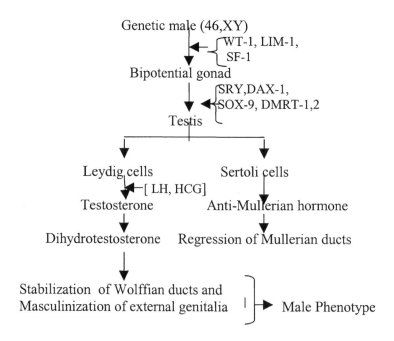

Figure 1: Testicular development leading to hormone secretion and formation of male phenotype

of the fetal testis is the secretion of AMH by the Sertoli cells, which by paracrine diffusion process induces the dissolution of the paired Mullerian ducts (6). The Sertoli cells also secrete inhibin, which causes maturation of the germ cells. Leydig cells of the testis are first found around 60 days of human fetal life and rapidly proliferate during the third month after differentiation of the primitive testicular cords (7). At this time, the interstitial spaces between seminiferous tubules are crowded with Leydig cells. The onset of testosterone biosynthesis by the Leydig cells occurs at about 9 weeks and the synthesis is regulated by two hormones, namely placental human chorionic gonadotropin (HCG) and pituitary luteinizing hormone (LH) (7). Testosterone secreted by Leydig cells causes differentiation of the Wolffian ducts. The hormone acts to stimulate the Wolffian duct inducing development of the epididymis, vas deferens and seminal vesicle (1). Testosterone is also the precursor of the third hormone, dihydrotestosterone, which causes the differentiation of the urogenital sinus, inducing formation of the male urethra, prostate and external genitalia (8).

CHROMOSOMAL DETERMINATION OF GONADAL DIFFERENTIATION:

About 40 years ago, it was first recognized that the Y chromosome is essential for male development (3). Chromosomal errors can arise from faulty replication of the germ cells during spermatogenesis or oogenesis or from faulty mitotic division of cells in the zygote after fertilization. Table 1 describes some of the anomalies associated with chromosomal abnormalities.

Table 1: Human karyotype pertinent to designating sex chromosome abnormalities

Karyotype	Description
46,XX	Normal female
46,XY	Normal male
45X	Monosomy X
47XXY	Karyotype with 47 chromosomes with an extra X chromosome
45,X/46,XY	Mosaic, composed of 45,X and 46,XY cell lines
P	Short arm
Q	Long arm
46,X,del (X) (pter-q21)	Deletion of long arm of X distal to band Xq21
46,X,I(Xq)	Isochromosome of long arm X
46,X,r(X)	Ring X chromosome
46,x,t(y:7) (q11;q36)	Translocation of y chromosome portion to chromosome 7

However, not all patients with anomalies of sex chromosome have abnormal gonads; conversely, congenital defects in gonadal differentiation are not always due to chromosomal errors. The association is so frequent, however, that these topics are inseparable. Exceptions of this association are of special importance in defining the genetic and chromosomal determinations of gonadogenesis.

GENETIC CONTROL OF GONADAL DEVELOPMENT

The development of gonadal, adrenal and urogenital systems is closely linked. Several genes that regulate the process are shown in Figure 1. Abnormal expression of the WT-1 gene (Wilms tumor 1) was found to be associated with failure of gonadal differentiation, nephropathy, development of Wilms' tumor (in Denys-Drash syndrome) and gonadoblastoma (in Frasier syndrome) (9). The role of the WT-1 gene in gonadal differentiation is not sex specific, as gonadal dysgenesis also occurs in 46,XX individuals suggesting that this gene is involved in the formation of bipotential gonad rather than in formation of the testis. Another gene that is involved in the development of the bipotential gonad is the recently cloned LIM-1 gene (10). LIM-1 gene also modulates the formation of the kidney. Homozygous deletion of the LIM-1 gene in mice leads to developmental failure of both gonads and kidneys. New implications of the role of steroidogenic factor 1 (SF-1) in gonadal formation have been recently reported (11). SF-1, a product of the FTZ1-F1 gene, is classified as an orphan receptor. FTZ1-F1 mRNA is present in the urogenital ridge, which forms both gonads and adrenals. The FTZ1-F1 gene expression is also found in the developing brain. Mice lacking SF-1 fail to develop gonads, adrenals and the hypothalamus (11). Recently, a phenotypic female with 46,XY karyotype was identified with adrenal failure. She had normal Mullerian structure but no androgenic response to hCG stimulation. Histology of the gonads revealed poorly differentiated tubules and connective tissue. Within the FTZ1-F1 gene, a heterozygous deletion was characterized which resulted in the absence of binding of SF-1 to its specific binding sites. This finding suggests a role for SF-1 in adrenal and gonadal formation.

Progression from the bipotential gonad towards testicular differentiation requires some gonosomal and autosomal genes. It has been long believed and now proven that a specific testis-determining factor (TDF) was essential for testicular development (3). This encoding gene, located on the Y chromosome, is termed as the sex-determining region of the Y-chromosome (SRY). It is a single-exon gene, which encodes a protein and a DNA-binding motif that acts as a transcription factor; in turn, it regulates the expression of other genes. It has been suggested that SRY binds to the promoter region of the AMH gene and regulates expression of steroidogenic enzymes (6). It may be possible that SRY probably induces expression of AMH to prevent formation of the Mullerian ducts. The evidence that SRY is the TDF was obtained from the experiments performed by Koopman, who demonstrated normal male phenotype formation with introduction of the SRY gene into female mouse embryo (12).

Another autosomal gene related to SRY involved in gonadal development is SOX (SRY-box-related) 9 (13). This gene is especially transcribed following SRY expression in male gonadal structures. The SOX 9 gene is also involved in formation of the extracellular matrix of cartilage. Therefore, defects in SOX 9 lead to sex reversal and skeletal malformations in 46, XY individuals known as campomelic dysplasia (13). DAX 1 is another gene that is involved in adrenal, ovarian and testicular development (13). The gene is located in the X chromosome and is expressed during ovarian development, but it is suspended during testicular development. Interestingly, DAX 1 is repressed by SRY during testicular development. Mutation of the DAX gene leads to lack of adrenal formation and to hypogonadism in congenital adrenal hyperplasia (14). Several other factors may play an important role in male sex determinaton. Deletions in 9p and 10q were shown to be associated with sex reversal in the 46, XY individual (15). Gonadal dysgenesis appears to be linked to the combined hemizygosity of DMRT1 and DMRT2 (15). Defects in developmental genes responsible for gonadal differentiation lead to a complete or partial gonadal dysgenesis; this, in turn, results in complete failure of testicular functions. These individuals have abnormalities in both external and internal genital structures. Associated malformations of the adrenal, urogenital, skeletal and central nervous systems indicate that these genes are involved in multiple developing processes. However, in the majority of partial gonadal dysgenesis subjects, no genetic defects have been identified. Several other genes, as yet unknown, have been implicated. The characterization of these genes will be fundamental to the diagnosis, treatment and counsel of the patients with sexual disorder.

DEVELOPMENT OF INTERNAL GENITALIA

Anatomical Organization

At 7 weeks of gestation, the human fetus is at the indifferent stage of differentiation (7); i.e., it has the anlagens of both female and male reproductive tracts (Figure 2a). The fetus has both Mullerian (paramesonephric) and Wolffian (mesonephric) ducts. The Mullerian ducts and the Wolffian ducts are the anlagen of the female and male reproductive tracts, respectively (7). At the lower end, the Wolffian ducts reach the lateral wall of the cloaca and form the primitive urogenital sinus (Figure 2a). The frontal part of the urogenital sinus produces the vesco-urethral primordium, whereas the caudal part gives rise to the definitive urogenital sinus (Figure 2b). The frontal part of the Mullerian ducts becomes the fallopian tubes and the lower part forms the uterovaginal duct (Figure 2C). The upper portion of the Mullerian duct is located externally to the Wolffian ducts, and the lower portion of the Mullerian duct crosses over the Wolffian duct to position itself internally to the Wolffian ducts (Figure 2C). At the terminal portions, the Mullerian ducts fuse and form the uterovaginal duct. The uterovaginal duct at the lower end makes contact with the posterior wall of urogenital sinus forming the Mullerian tubercle.

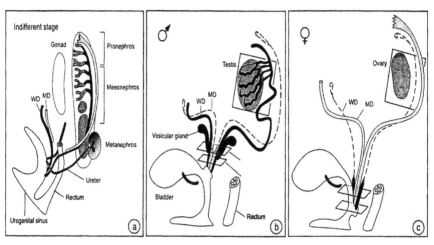

Figure 2: Developmental stages of internal genitalia. a: indifferent stage of development, demonstrating presence of both Wolffian (WD) and Mullerian ducts (MD) at the primitive stage of development; b: development of a male internal genital tract, demonstrating stabilization and differentiation of the Wolffian duct and regression of the Mullerian duct; c: development of a female internal genital tract, demonstrating regression of the Wolffian duct and stabilization of the Mullerian duct. The figure was reproduced with the author's permission from Drews U: Gesxhlechtsspezifische Entwicklung, in wulf KH. Chmidt, H-Mathiesen (Hrsg): Klink der Frauenheikunde and geburtshilfe. Band 1, pp 3-33 (urban and Schwarzenberg. Munchen, 2000)

Biochemical Processes Involved in Internal Genital Tract Differentiation

With the onset of testicular hormonal secretion, the undifferentiated internal genital tract undergoes differentiation. In the XY fetus, the testis differentiates by the end of the 7th week. The Sertoli cells of the testis start to secrete AMH resulting in regression of the Mullerian ducts (6). AMH binds to the receptors present in the mesenchymal cells of the Mullerian ducts and inhibits the formation of the uterine analogues (16). High local concentration of AMH appears to be essential for effective inhibition of Mullerian duct development. Studies of AMH binding have shown that AMH receptor has high binding affinity (16). To ensure a proper pattern of sex determination, AMH must be expressed in the testis but not in the ovary during early fetal life. Thus, the testis–determining factor SRY is the first essential factor for AMH expression (6). It has been suggested that SRY, acting through its ability to bind DNA, could allow the interaction of other transcription factors that bind at sites adjacent to the AMH promoter. SF1 has been considered to be one such factor, capable of binding to a 20bp motif highly conserved in human AMH genes (17). In males, AMH expression is maintained until puberty when it is down-regulated by the negative action of androgens.

By the 8th week, Leydig cells appear in the differentiating testis and secrete androgens causing masculinization of the genitalia. In the first trimester of fetal life, hCG regulates the release of androgenic steroids by the fetal gonads. The hypothalamic-pituitary vascular connection responsible for LH stimulation by gonadotropin releasing hormone GnRH is established at 11.5 weeks of gestation (18). Impaired pituitary gonadotropic function in utero does not result in sexual ambiguity suggesting that pituitary gonadotropins are not essential early in life when the most important steps of sexual differentiation take place. However, most frequently, these patients are born with cryptorchidism and micropenis as a consequence of low testosterone production, suggesting that testicular production of androgens is under the control of pituitary LH production only at the last trimester of fetal life (18). Driven by androgen action, each Wolffian duct forms the epididymis at its distal ends, near the testes (7). The major part of the Wolffian duct is converted into the vas deferens. The vas deferens end with the formation of the seminal vesicle at its proximal end near the urethra. Finally, the segment of the duct lying between the seminal vesicle and urethra forms the ejaculary duct.

DEVELOPMENT OF EXTERNAL GENITALIA

During the first 2 months of fetal life no difference is noticed between sexes in the development of external genitalia (labioscrotal swelling and genital tubercle formation). Masculinization starts approximately at day 65. The anogenital distance lenthens and the labioscrotal swellings gradually fuse. The rims of the urethral groove fuse to form the penis. The process of masculinization is completed in the 14-week old fetuses.The development of male external genitalia depends upon dihydrotestosterone (DHT) action, requiring normal production of testosterone and DHT (8). The defects in androgen production results in impairment of both internal and external genital development whereas reduction in 5-alpha reductase activity causes reduction in DHT production and leads to the impairment of external genital development; the epididymis and vas deferens develop normally (8).   In the early part of the differentiation, gonadal primordia are located in an abdominal position in both male and female fetuses during the indifferent stage of sexual development. The testis is anchored to the inguinal region by the caudal ligament or gubernaculum, while the cranial ligament which holds the urogenital tract progressively regresses. The gubernaculum thickens and shortens, retaining the testis over the inguinal ring. The testes reach the internal inguinal ring by week 24 in the human fetus (19). The second phase of testicular descent takes place in the last 2 months of the intrauterine life. The testes pass through the inguinal canal to reach the scrotum. Androgens are absolutely essential during the inguinoscrotal migration of the gubernaculum across the pubic region to the scrotum (19). Calcitonin-related-peptide, a neurotransmitter, has also been suggested to be involved in the stimulation of gubernaculum via cAMP.

REGULATION OF TESTICULAR ANDROGEN SYNTHESIS AND ITS ROLE IN MALE DEVELOPMENT

Testicular androgen synthesis plays a major role in male development. As shown in Figure 3, androgenic hormones are synthesized from cholesterol within the mitochondria. The acute stimulation of the synthesis of these steroids is mediated by the steroidogenic regulatory protein (StAR), an active transporter of cholesterol. Mutations within StAR lead to a severe lack of adrenal steroidogenesis, as well as lack of virilization in 46 XY individuals in congenital adrenal hyperplasia (20). A p450 enzyme (p450scc) in the target organ cleaves the cholestrol transported in the organ and produces pregnenolone. Pregnenolone gives rise two products, namely, progesterone and 17α-hydroxy (17α-OH)-pregnenolone by the action of 3β-hydroxy steroid dehydrogenase (3β-HSD) and 17α-hydroxylase, respectively. Progesterone and 17-OH pregnenolone are subsequently converted into 17-OH progesterone and dehydroepiandrosterone, respectively by 17α-hydroxylase and 17/20-lyase activities present in a p450c17 enzyme (21). The third important enzyme, that plays a major role in androgen biosynthesis is 3β-HSD; it catalyzes the formation of androstenedione, the major precursor of testosterone (22). Androstenedione is also produced from 17-OH progesterone by 17/20 lyase activity. The next key enzyme for androgen synthesis is 17β-hydroxysteroid dehydrogenase (17β- HSD), which converts andronstenedione to testosterone within the testes (22). Another enzyme that acts on peripheral target cells is 5α reductase which converts testosterone to dihydrotestosterone (8). At least five different isoenzymes of 17β-HSD exist, but only mutations in the type 3 enzyme have been shown to cause defective sexual differentiation. This disorder is characterized by a severe virilization defect in 46 XY individuals who conversely show strong virilization during puberty with marked phallic enlargement (22). This enzyme, expressed in the testes, is compatible with its important role in testicular androgen formation.

For 5α-reductase, two isoenzymes have been detected in diverse tissues. In genital structures, type 2 is more abundant. The type 1 enzyme is necessary for the conversion of testosterone to 5α-dihydrotestosterone (DHT). In 5α-reductase deficiency, DHT formation is severely diminished. However, testosterone levels are normal or even elevated. Affected individuals are usually born with ambiguous genitalia, but the differentiation of Wolffian structures, largely dependent on testosterone, is not obstructed. At the time of puberty, virilization may occur due to high endogenous testosterone (8).

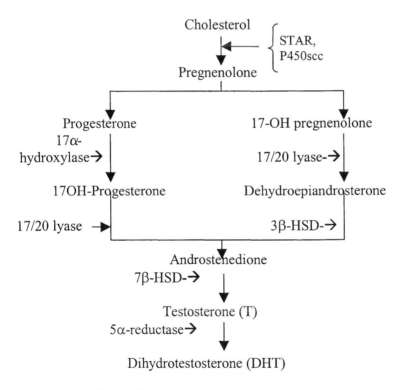

Figure 3: Testicular synthesis of androgens

136

ROLE OF ANDROGEN RECEPTOR IN MALE DEVELOPMENT

Normal androgen action is another essential process for male development (Figure 4). Androgen action on target tissues is dependent on normal expression of a functionally intact androgen receptor (AR). The AR is a hormone-activated DNA-binding transcription factor.

Transcriptional regulation of target genes by steroid receptors is a complex process involving hormone binding, receptor phosphorylation, dissociation of heat-shock proteins, dimerization, intracellular trafficking, nuclear translocation, DNA binding, and transcription activation (23). The predominant form of AR is a 110 kD protein, but in various genital and nongenital tissues, an 87 kD isoform has been detected. The 110 kD and 87 kD isoforms are termed as ARα and ARβ, respectively (24). AR is

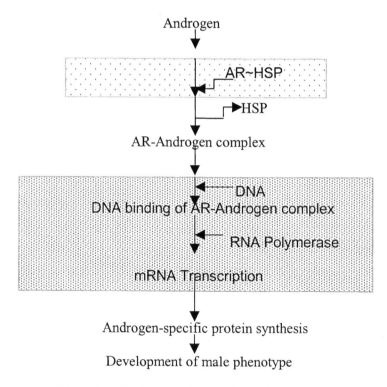

Figure. 4: Mechanism of androgen action in male development. HSP, heat shock protein; ⬚ cytoplasm ; ▨ nucleus.

divided into three major functional domains: a large N-terminal domain, a DNA-binding domain and a C-terminal hormone-binding region (24). Upon entering the target cells, androgens bind specifically to the AR located in the cytoplasm (Figure 4). The cytoplasmic AR is bound to different proteins (e.g. heat shock protein) and has no DNA binding activity. Binding of androgens to AR causes the release of these proteins, resulting in activation and nuclear translocation of the AR (Figure 4) (25). An important step in the AR transactivation cascade consists of homodimerization of two AR proteins. The AR homodimer consequently binds to hormone responsive element (HRE) region of the DNA made up of two half-site sequences within the promoter region of androgen regulated target genes. At this stage, a complex interaction of the receptor with yet unknown factors results in activation of transcription machinery. Specific RNA polymerase is activated resulting in specific gene transcription. Consequently, proteins which are responsible for specific androgen effects, are synthesized (25). Inhibition of androgen action due to the lack of AR is the essential pathological mechanism of androgen insensitivity syndrom (AIS), termed as pseudohermaphroditism (26). The clinical symptoms of these patients result from defective androgen action despite normal or elevated testicular androgen secretion, and normal ability of the fetal testes to produce AMH. Thus, these patients show defective external genitalia and Wolffian duct development in conjunction with usually absent Mullerian ducts.

ROLE OF GROWTH FACTORS IN MALE DEVELOPMENT

Recently, new factors were reported which cause sex reversal or intersex development. The relation of these factors to sex organ development is not yet clear. Along these lines, a number of growth factors were implicated in mediating androgen effects from mesenchyme to epithelium (5). Epidemeral growth factor (EGF) and keratin growth factor (KGF) were shown to mimic the effects of testosterone on the development of the vesicular gland, prostate and Wolffian duct (4,5). It has been reported that EGF enhances AR transcriptional activity at the androgen response element site and thereby mimics the androgen effects (27).

CONCLUSIONS

This chapter focuses on developmental aspects of male accessory glands. Since the major part of the structural organization of these glands is completed before birth, the report primarily highlights the fetal period of differentiation. I have particularly focused on the biochemical and molecular events involved in the development of male organs. Additionally, I have included relevant information on different diseases that are associated with abnormal genetic information leading to the blockade of normal processes of development.

REFERENCES

1. Jost A: Embryonic sexual differentiation (morphology, physiology, abnormalities). In: Jones JW Jr. and Scott WW (eds): Hermaphroditism genital abnormalities and related disorders. Williams and Wilkins: Baltimore, 1971.
2. Raynaud A: Modification experimentale de la differentiation sexuelle des embryons des souris par action des hormones androgens et oestrogens. Herman: Paris, 1942.
3. Ohno S. Sex Chromosomes and sex-linked genes. In: Labhart A., Mann T, Samuels LT., Zander J (eds). Monographs in Endocrinology (1) Springer-Verlag: Berlin, 1967.
4. Cunha GR, Alarid ET, Turner T, Nonjacour AA, Boutin GL and Foster BA. Normal and abnormal development of the male urogenital tract: Role of androgens, mesenchymal-epithelial interactions, and growth factors. J. Androl 1992; 13: 465-475.
5. Gupta C, Siegel S, and Ellis D   The role of EGF in testosterone-induced reproductive tract differentiation. Develop Biol 1991; 146: 106-16.
6. Haqq CM, King Cy, Ukiyama E. Falsafi S. Haqq TN, Donahue PK and Weiss MA: Molecular basis of mammalian sexual differentiation: Activation of Mullerian inhibiting substance gene expression by SRY. Science 1994; 266: 1494-1500.
7. Rey R and Picard J-Y Embryology and endocrinology of genital development Balliere's Clin Endocrinol Metabol 1998; 12: 17-33.
8. Griffin JE, Wilson JD. The androgen resistance syndromes: 5α reductase deficiency, testicular feminization and related disorders. In: Scriver CR, Blaudet AL, Sly WS eds, The molecular basis of inherited disease, 6th ed, McGraw Hill: NY, 1989.
9. Barbaux S, Niaudet P, Gubler MC, Grunfeld JP, Jaubert F. Kuttenn F. Fekete CN, Souleyyreau-Therville N, Thibaud E, Fellous M and Mcelreavey K. Doner: Splice-site mutations in WT1 are responsible for Frasier Syndrome. Nat Genet 1997; 17: 467-470.
10. Lim HN and Hawkins JR. Genetic control of gonadal differentiation. Balliere's Clin Endocrinol Metabol 1998;12: 121-161.
11. Achermann JC, Ito M. Hindmarsh PC and Jameson JL. A mutation in the gene encoding steroidogenic factor1 causes XY sex reversal and adrenal failure in humans. Nat Genet 1999; 22: 125-126.
12. Koopman P. Gubbay J. Vivian N, Goodfellow P and Lovell-Badge R. Male development of chromosomally female mice transgenic for SRY. Nature 1991; 351: 117-121.
13. Kwok C, Waller PA, Guioli S., Foster JW, Mansour S. Zuffardi O, Punnett HH et al Mutations in SOX-9, the gene responsible for campomelic dysplasia and autosomal sex reversal. Am J Hum Genet 1995; 57: 1028-1036.

14. Goodfellow PN, Camerino G Dax-1 an anti-testis gene. Cell.Mol Life Science 1999; 55: 857-863.
15. Raymond CS, Parker ED, Kettlewell JR, Brown LG, Page DC, Kusz K., Jaruzelska J, Reinberg Y, Flejter WL, Bardwell VJ, Hirsch B and Zarkower D. A region of human chromosome 9p required for testis development contains two genes related to known sexual regulators. Hum Mol Genet 1999; 8: 989-996.
16. Imbeaud S, Faure E, Lamarre I et al Insensitivity to anti-Mullerian hormone due to a spontanuous mutation in the human anti-Mullerian hormone receptor. Nat Genet 1995; 11: 382-388.
17. Giuili G, Shen WH and Ingraham HA The nuclear receptor SF1 mediates sexually dimorphic expression of mullerian inhibiting substance in vivo. Devel 1997; 124: 1799-12807.
18. Thliveris JA and Currie RW Observations on the hypothalamo-hypophyseal portal vasculature in the developing human fetus. Am J Anat 1980; 157: 441-444.
19. Hutson JM and Donahue PK The hormonal control of testicular descent. Endo Rev 1986; 7:270-283.
20. Bose HS, Sugawara T, Strauss JF lll and Miller WL The pathophysiology and genetics of congenital lipoid adrenal hyperplasia. New Eng J Med 1996; 335: 1870-1878.
21. Yanase T 17α-Hydroxylase/17-20-lyase defects. J Steroid Biochem Mol Biol 1995; 53: 1-6.
22. Geissler WM, Davis DL, Wu L, Bradshaw kd, Patel SM. Pseudohermaphroditism caused by mutations of testicular 17-β-hydroxysteroid dehydrogenase 3. Nat Genet 1994; 7:34-39.
23. Evans RM The steroid and thyroid receptor superfamily. Science 1998; 240:889-895.
24. Brinkmann AO, Faber PW, Rooji HCJ, Kulper GGJM, Ris C, Klaassen P, Vander korput JAGM, Voorhorst MM, van Laar JH, Mulder E and Trapman J The human androgen receptor: domain, structure, genomic organization and regulation of expression. J Steroid Biochem 1989; 34: 307-310.
25. Beato M, Sebastian C and Truss M Transcriptional regulation by steroid hormones. Steroids 1996; 61: 240-251.
26. Keenan BS, Meyer WJ, Hadjian AJ, Jones HW and Migeon CL Syndrome of androgen insensitivity in man: absence of 5α-dihydrotestosterone binding protein in skin fibroblasts. J Clin Endocrinol Metabol 1974; 38: 1143-1146.
27. Gupta C: Modulation of androgen receptor (AR)-mediated transcriptional activity by EGF in the developing mouse reproductive tract primary cells. Mol Cell Endocrinol 1999; 152: 169-178.

Chapter 9

FORMATION AND STRUCTURE OF MAMMALIAN OVARIES

Yoshihiko Araki
Yamagata University School of Medicine, Yamagata-City, Japan

INTRODUCTION

It is generally accepted that the ability to reproduce is one of the most essential properties of living organisms. Various forms of life are capable of reproducing themselves from one generation to the next under appropriate conditions. Some species reproduce only once during their life span, whereas others, such as mammals, have reproductive cycles that are hormonally regulated; the female ovary is central to the process of reproduction and produces germ cells, as well as gonadal hormones.

The precise origin of primordial germ cells, from which ova are derived, remains unclear. However, germ cells that reach the genital ridge in the female differentiate into oogonia. Although primordial germ cells in the genital ridge can differentiate into either male or female gametes, ovarian somatic cells affect the differentiation of primordial germ cells into oogonia (1). The oogonia undergo a number of mitotic divisions and become arranged in clusters that are surrounded by a layer of flat epithelial cells (the primordial follicle) by the end of the third month of gestation in humans. In primates, the number of germ cells reaches a maximum (estimated at approximately 6 million per ovary) by the fifth month of gestation (2, 3). However, 50-70% of the oogonia disappear before birth, mostly due to apoptosis. The surviving oogonia proliferate and differentiate into primary oocytes. In mammals, basic oogenesis, i.e., mitotic proliferation of oogonia and the meiotic prophase of the first meiotic division in primary oocytes, is completed before birth. These cells subsequently enter the dictyotene stage, a resting stage between prophase and metaphase (see below). It is important to note that primary mammalian oocytes do not complete their first meiotic division before puberty. With the onset of puberty, primordial follicles start to mature under the hormonal control of each ovarian cycle. After puberty, the ovary has a relatively thick cortex that contains the ovarian follicles, corpora lutea, and the medulla, which is composed of connective tissues with multiple elastic fibers, smooth muscle cells and numerous blood vessels.

This chapter focuses on the process of mammalian oocyte maturation, including folliculogenesis and atresia, and emphasizes the morphological characteristics of oogenesis in the ovary.

OOCYTE MEIOSIS AND MATURATION

In most mammals such as small rodents, cattle and primates, the proliferation of oogonia is completed during a specific stage of embryonal development. Since mammalian reproduction is initiated by the fusion of spermatozoa and oocytes during fertilization, each gamete is a haploid cell and contains half the DNA of the parent before fertilization. In order to generate haploid chromosomes, the primitive germ cells undergo two meiotic divisions, whereby the number of chromosomes is reduced to half the original number.

The meiotic divisions are classified as meiosis I and II. Meiosis I can be subdivided into prophase, metaphase, anaphase, and telophase. The

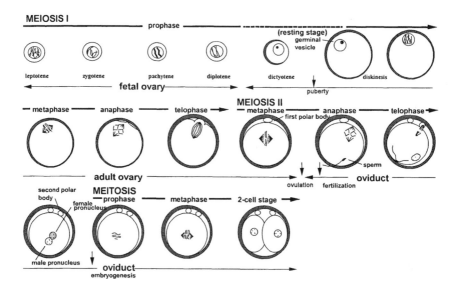

Figure 1. Process of mammalian meiosis during oogenesis The primordial germ cells migrate to the genital ridges, which develop into ovaries. After a period of mitotic proliferation, the germ cells begin meiosis and differentiate into mature oocytes. Most of the oocytes are in the dictyotene stage at birth, and re-start meiosis after puberty, under the control of sex hormones.

prophase is further classified into leptotene, zygotene, pachytene, diplotene, and diskinesis (Fig. 1). The primary oocytes in the ovary replicate their DNA just before the start of the first meiotic division (meiosis I) (4). Therefore, at the beginning of the meiotic division, a primary oocyte contains double the normal number of chromosomes (4n). The first characteristic of meiosis I is the pairing of homologous chromosomes. An interchange of chromatid segments then occurs between two pairs of homologous chromosomes. During the separation of the homologous chromosomes, the points of interchange remain temporarily joined, and the chromosomal structure has a cross-like shape (chiasma). The separated chromosomes subsequently enter diskinesis, at which point they are coiled and incompletely separated. After the first meiotic division, each daughter cell has the double-chromosome composition (2n). No DNA synthesis occurs before the second meiotic division (meiosis II). The 2n chromosomes separate at the centromere; each daughter cell contains half of the normal amount of DNA in a somatic cell (n: haploid). The first and second mitotic divisions in the ovary are shown schematically in Figure 2. As a result of meiotic division, four haploid cells derived from a primary spermatocyte become four mature spermatozoa undergoing spermatogenesis (see Chapter 1); however, haploid cells from a primary oocyte develop into one large cell and three small cells as a result of unequal division of the cytoplasm. The large cell becomes a mature oocyte through the process of oogenesis, whereas the small cells become polar bodies.

At birth, the majority of mammalian oocytes are in the resting stage, called "dictyotene" (Fig. 1). This stage is believed to occur between the prophase and metaphase of meiosis I. The oocytes in dictyotene have relatively large nuclei and are called "germinal vesicles". The dictyotene stage continues until just before ovulation. After sexual maturation, gonadotropin released from the anterior pituitary stimulates resting oocytes to reinitiate meiosis I, resulting in the appearance of the first polar body in the perivitelline space of the ovarian oocyte. Oocyte meiosis II is finally completed in the oviduct in an event that is triggered by fertilization (Fig. 1).

OVARIAN FOLLICULOGENESIS

A primordial follicle consists of a layer of flattened epithelial cells surrounding the primary oocyte (Fig. 3). Although several hundred thousand primary follicles are present in the ovary at birth, their numbers decrease steadily throughout life. Most of them degenerate and die in a process known as "atresia" (see below). As maturation of the follicle proceeds after the onset of puberty, the primary oocyte grows in size, whereas the follicular epithelial cells become cuboidal (Fig. 3). At this stage, the follicular cells start to synthesize the zona pellucida (ZP), the extracellular glycocalyx that surrounds the mammalian oocyte. The ZP mediates species-specificity of sperm binding and penetration, blocks polyspermy, induces the sperm-acrosome reaction, and protects the growing embryo during mammalian fertilization and implantation (5). Although the site of ZP biosynthesis within the follicle remains controversial in several mammalian species (6-11), at least in the mouse, it is believed to originate in the oocyte (12-14). The primary follicles develop into secondary follicles under the influence of

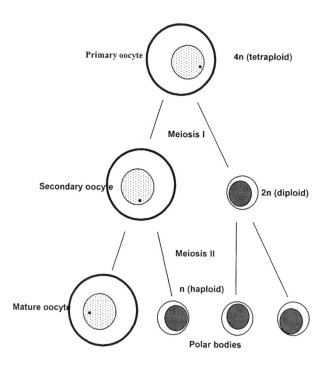

Figure 2. The process of mature oocyte production. Haploid cells from a primary oocyte, unlike spermatids, differentiate into mature oocytes and polar bodies.

follicle-stimulating hormone (FSH) released from the anterior pituitary. As follicular maturation progresses, the follicular cells (granulosa cells) form an increasingly thick layer ("membrana granulosa") around the oocyte. The follicular cells continue to proliferate via mitosis, fluid accumulates, and the follicular antrum develops through coalescence of the intercellular spaces.

At maturity, the follicular cells surrounding the oocyte remain intact and form the cumulus oophorus. Mature follicles are called 'Graafian follicles' (Fig. 3). The Graafian follicle is surrounded by a cellular layer, the theca interna, which is rich in blood vessels and is a major source of estrogen (15),

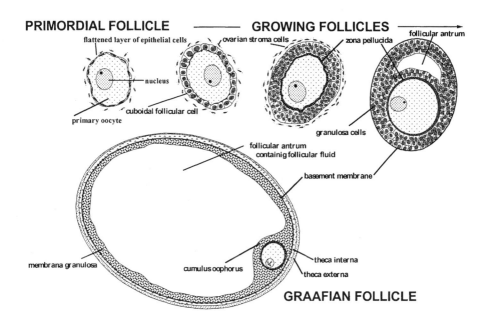

Figure 3. Folliculogenesis in the ovary. A primary oocyte surrounded by a single layer of flat epithelial cells forms a primordial follicle. These epithelial cells become cuboidal in shape when folliculogenesis begins after puberty (primary follicles). In the growing follicles, the follicular cells surrounding the oocyte further thicken (follicles with 2-3 layers of follicular cells are called "secondary follicles"). As follicular maturation proceeds, the follicular cells synthesize the zona pellucida, which is deposited between the membrana granulosa and the surface of the oocyte. With further maturation (the tertiary stage), follicular cells continue to proliferate, and the follicular antrum, called the "antral follicle", is observed in the follicle. A cavity containing follicular liquid appears between the cells and separates them into two compartments. The outer layer forms the membrana granulosa, and the inner set forms the cumulus oophorus of the matured follicle (the Graafian follicle). The stromal cells of the ovarian cortex form a sheath, composed of two layers: the theca interna, which surrounds the basement membrane of the follicle and is permeated by a capillary plexus, and the theca externa, which is a fibrous outer layer.

and an outer fibrous layer, the theca externa. At this stage, the oocyte reaches its maximal diameter. Mature oocytes in the Graafian follicle are exceedingly large (117-142 µm in humans, 70-85 µm in rodents, and 120-140 µm in cattle) as compared with other cells of the body. The follicular cells secrete estrogen which is essential for the growth, development and function of the female reproductive organs. In humans, although a number of follicles begin to develop at each ovarian cycle, normally only one reaches full maturity. The remaining follicles degenerate and become atretic, and the degenerated oocyte and follicular cells are replaced by connective tissue (see below). The majority of follicles degenerate without ever reaching full maturity. During the reproductive years of the average woman, only 400-500 follicles reach maturity and release their oocytes.

Ovulation

Just before ovulation, the volume of a Graafian follicle rapidly increases, due to the production of follicular fluid (Fig. 3). In the final stages of follicular maturation, the ovum and surrounding granulosa cells become detached from the cumulus oophorus and float free in the follicular fluid. This is termed "preovulatory swelling". The ripening follicle enlarges, ruptures and finally projects an ovum surrounded by cumulus cells from the ovarian surface. The collective name for the follicular cells that adhere to

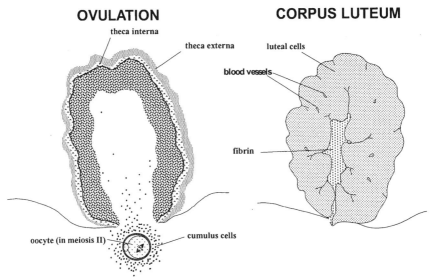

Figure 4. Ovulation and formation of the corpus luteum in the ovary. At the time of ovulation, an oocyte (secondary oocyte in meiosis II), surrounded by the cumulus oophorus cells, is ovulated from the ovary. The inner-most layer of the cumulus cells form the corona radiata. The ovulated oocyte normally passes into the oviduct, the site of fertilization. After ovulation, the cells of the membrana granulosa become greatly enlarged and a yellowish carotinoid pigment is deposited in the cytoplasm. The so called luteal cells form a major part of the corpus luteum. The remaining cavity portion of the follicle (formerly the follicular antrum) is filled with fibrin.

the liberated oocyte is the "corona radiata" (Fig. 4). At the time of ovualtion, The second meiotic division begins in the oocyte (see Fig. 1). In most cases, there is some bleeding into the cavity of the follicle after ovulation. Generally, the number of ovulated follicles depends on the mammalian species, and the processes leading to ovulation are controlled by pituitary gonadotropins. FSH is essential for the secondary stage of follicular growth and retains activity during the tertiary stage, when the Graafian follicles are formed. Luteinizing hormone (LH) levels increase at the mid-point of the ovarian cycle, and cause the FSH-primed follicles to rupture and discharge their ova.

The rupture point in the ovarian wall is sealed by aggregates of follicular fluid and blood into which stromal and epithelial cells grow. The ruptured follicle is transformed into a temporary glandular structure, the "corpus luteum", producing progesterone and other steroid hormones (Fig. 4). After ovulation, the follicular granulosa cells differentiate into relatively large, granulosa lutein cells, which form a thick, folded layer around the remains of the follicular cavity. Cells of the theca interna, which also increase in size prior to ovulation, form theca lutein cells. The two cell types become virtually indistinguishable, although the theca lutein cells remain in distinct groups in the outer folds of the granulosa wall, and subsequent changes in the granulosa lutein cells result in morphological differences. The theca externa retains its regular ovoid outline, and its cells do not undergo transformation. If the discharged ovum is not fertilized, the corpus luteum attains maximal development after ovulation and then begins to degenerate. The lutein cells accumulate lipids and ultimately degenerate. The connective tissue between lutein cells expands and becomes hyalinized, and gradually the corpus luteum transforms into a white scar, the "corpus albicans". Hormone-production capability is maximized only during pregnancy (the lutein cells of nonpregnant human females show degenerative changes approximately 10 days after ovulation). If ovulation is followed by fertilization, the corpus luteum increases in size and becomes the "corpus luteum of pregnancy". The cells continue to grow in size and persist until the mid-term of gestation, then gradually decline in the advancement to full-term.

FOLLICULAR ATRESIA

As stated above, the mammalian female possesses far more gametes than will ever be ovulated. At birth, vast numbers of oogonia are observed in the ovaries (for example, 20,000-30,000 in mice and rats, 60,000 in pigs, and 60,000-100,000 in cows). The majority of these oocytes are lost when the follicles in which they are enclosed undergo a degenerative process known as follicular atresia. In mammals, it is generally accepted that atresia occurs in both prepubertal and adult individuals, and that a follicle can become atretic at any stage during its growth and development. Apoptosis is an underlying mechanism of ovarian follicular degeneration during atresia (16, 17), and appears to be mediated via Fas (expressed in oocytes), and FasL genes (expressed on granulosa cells) (18-24).

One of the earliest morphological signs of atresia seen by light microscopy is the presence of darkly stained pyknotic nuclei in the granulosa cells, although a small percentage of granulosa cells with pyknotic nuclei are also present in the putative dominant follicles (25). Apoptotic granulosa cells in electron micrographs are characterized by structural features, such as condensed chromatin granules in the neoplasm and increased electron density in the cytoplasm (Fig. 5).

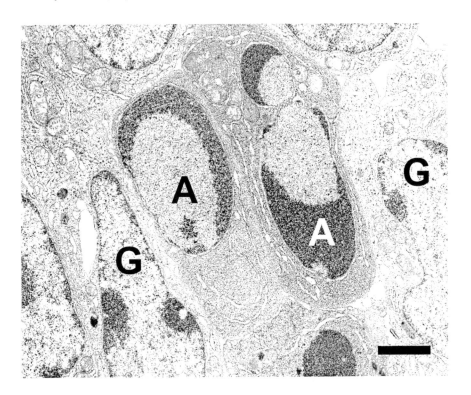

Figure 5. Apoptotic granulosa cells in an atretic follicle (B6C3 mice, aged 32-35 days). Normal granulosa cells (G) and apoptotic granulosa cells (A). Apoptotic cells are characterized by condensed chromatin granules in the nucleoplasm and increased cytoplasmic density. Bar = 1 μm

Atresia is also associated with alterations in the gonadotropin receptors of granulosa cells. The number of LH-binding sites on sheep granulosa cells decreases with the severity of atresia (26). In rats, the granulosa cells of atretic follicles do not exhibit any LH-binding sites (27, 28). There are reportedly less LH- and FSH- binding sites on hamster atretic follicles than on healthy antral follicles (29). However, controversy persists as to whether gonadotropin receptors decrease as a consequence of atresia, or if atresia is the result of never having acquired a sufficient number of gonadotropin receptors.

The composition of atretic follicular fluid is also known to differ among healthy follicles. In bovine, healthy follicles have high concentrations of estrogen and relatively low levels of progesterone and chondroitin sulfate, whereas atretic follicles have low amounts of estrogen and elevated levels of progesterone and chondroitin sulfate (30). Current evidence suggests that the regulation and development of follicular atresia is a complex process that involves interactions between endocrine factors (gonadotropins) and intra-ovarian regulators (sex steroids, growth factors and cytokines, etc.) leading to the control of follicular proliferation, differentiation and programmed cell death (31). Several lines of experimental evidence suggest that macrophages eliminate apoptotic granulosa cells by phagocytosis of degenerated cells in the atretic follicles of mice (32), rats (33, 34), humans (35), rabbits (36) and guinea pigs (37). On the other hand, experimental evidence based on electron microscopic studies indicates that the granulosa cells in pre-ovulatory mature follicles, as well as those in atretic follicles, have the ability to phagocytose since phagocytic "granulosa" cells possess characteristic gap junctions (38-40). Apoptotic granulosa cells or their cytoplasmic remnants are engulfed by granulosa cells of normal appearance that have gap junctions and frequently contain internalized annular gap junctions in the cytoplasm (Fig. 6).

CONCLUSIONS

This chapter has focused on the basic processes and morphological characteristics of ovaries during oogenesis. One of the key functions of the mammalian ovary is the production of haploid gametes, oocytes, in a cyclical, hormone-mediated manner for fertilization and subsequent development into embryos. From this point of view, the female ovary is equivalent to the male testis as a site of gamete production. Indeed, early embryonal development of the reproductive systems is similar in both sexes. However, the process of oocyte formation is quite distinct from that of spermatogenesis. It should be noted that, unlike the events of

150

spermatogenesis, ovarian meiosis is initiated during fetal development, is paused in the middle of the first meiotic division at the time of birth, and reinitiates after puberty. In contrast, male meiosis initiates after puberty and continues throughout life due to the persistence of mitotically-active spermatogonia. Furthermore, the majority of ovarian follicles undergo atresia, except for the few that are destined to reach full maturity. The oocyte is the largest cell in the human body and has a large amount of cytoplasm. The differences in size, number, and formative processes between female and male gametes would suggest that these cells are not

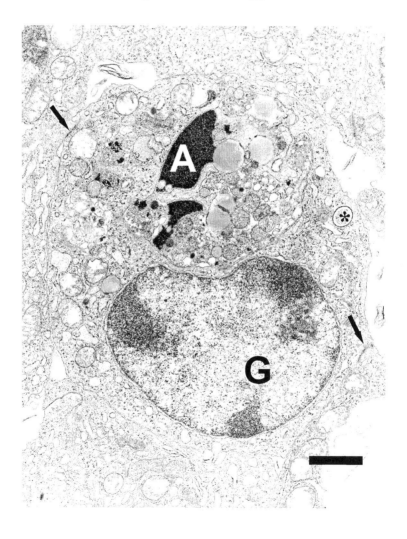

Figure 6. An apoptotic cell is phagocytosed by a normal-looking granulosa cell (B6C3 mice, aged 32-35 days). The granulosa cell (G) has formed gap junctions (arrows) and contains an annular gap junction (asterisk). Bar = 1 μm.

always equivalent as gametes, even though they are the same "haploid" karyotype. The mammalian oocyte has unique gametic characteristics and is highly specialized to function effectively in reproduction.

ACKNOWLEDGMENTS

The author is deeply indebted to Dr. Hiroshi Watanabe (Yamagata University School of Medicine) for helpful discussions and technical support.

REFERENCES

1. Upadhyay S., Zamboni L. Ectopic germ cells: natural model for the study of germ cell sexual differentiation. Proc Natl Acad Sci USA 1982; 79:6584-6587.
2. Baker T.G. A quantitative and cytological study of germ cells in human ovaries. Proc Roy Soc B 1963; 158: 417-433.
3. Baker T.G. A quantitative and cytological study of oogenesis in the rhesus monkey. J Anat 1966; 100: 761-776.
4. Alberts B., Bray D., Lewis J., Raff M., Roberts K., Watson J.D. Germ cells and fertilization. In: Molecular Biology of the Cell (3rd ed.), New York, Garland Publishing; 1994; 1011-1035.
5. Yanagimachi R. Mammalian fertilization. In: Knobil E, Neill JD (eds.) The Physiology of Reproduction, vol.1. New York, Raven Press; 1994: 189-317.
6. Tesoriero J.V. Formation of the chorion (zona pellucida) in the teleost, Oryzias latipes. II. Polysaccharide cytochemistry of early oogenesis. J Histochem Cytochem 1977; 25: 1376-1380.
7. Guraya SS. Recent advances in the morphology, histochemistry, and biochemistry of the developing mammalian ovary. Int Rev Cytol 1977; 51:49-131.
8. Takagi J., Araki Y., Dobashi M., Imai Y., Hiroi M., Tonosaki A., Sendo F. The development of porcine zona pellucida using monoclonal antibodies: I. Immunochemistry and light microscopy. Biol Reprod 1989; 40: 1095-1102.
9. Takagi J., Dobashi M., Araki Y., Imai Y., Hiroi M., Tonosaki A., Sendo F. The development of porcine zona pellucida using monoclonal antibodies: II. Electron microscopy. Biol Reprod 1989; 40: 1103-1108.
10. Akatsuka K., Yoshida-Komiya H., Tulsiani D.R.P., Orgebin-Crist M.-C., Hiroi M., Araki Y. Rat zona pellucida glycoproteins: Molecular cloning and characterization of the three major components. Mol Reprod Dev 1998; 51: 454-467.
11. Haddad A., Nagai M.E. Radioautographic study of glycoprotein biosynthesis and renewal in the ovarian follicles of mice and the origin of the zona pellucida. Cell Tissue Res 1977; 177: 347-369.
12. Bleil J.D., Wassarman P.M. Structure and function of the zona pellucida: identification and characterization of the proteins of the mouse oocyte's zona pellucida. Dev Biol 1980; 76:185-202.
13. Greve J.M., Salzmann G.S., Roller R.J., Wassarman P.M. Biosynthesis of the major zona pellucida glycoprotein secreted by oocytes during mammalian oogenesis. Cell 1982; 31:749-759.

152

14. Zamboni L., Upadhyay S. Germ cell differentiation in mouse adrenal glands. J Exp Zool 1983; 228:173-193.

15. Liu Y.X., Hsueh A.J. Synergism between granulosa and theca-interstitial cells in estrogen synthesis by gonadotropin-treated rat ovaries: studies on the two-cell, two-gonadotropin hypothesis using steroid antisera. Biol Reprod 1986; 35:27-36.

16. Hughes F.M., Gorospe W.C. Biochemical identification of apoptosis (programmed cell death) in granulosa cells: evidence for a potential mechanism underlying follicular atresia. Endocrinology 1991; 129:2415-2422.

17. Tilly J.L., Kowalski K.I., Johnson A.L., Hsueh A.J. Involvement of apoptosis in ovarian follicular atresia and postovulatory regression. Endocrinology 1991;129:2799-2801.

18. Yonehara S., Ishii A., Yonehara M. A cell-killing monoclonal antibody (anti-Fas) to a cell surface antigen co-downregulated with the receptor of tumor necrosis factor. J Exp Med 1989; 169:1747-1756.

19. Suda T., Takahashi T., Golstein P., Nagata S. Molecular cloning and expression of the Fas ligand, a novel member of the tumor necrosis factor family. Cell 1993;75:1169-1178.

20. Guo M.W., Mori E., Xu J.P., Mori T. Identification of Fas antigen associated with apoptotic cell death in murine ovary. Biochem Biophys Res Commun 1994;203:1438-1446.

21. Hakuno N., Koji T., Yano T., Kobayashi N., Tsutsumi O., Taketani Y., Nakane P.K. Fas/APO-1/CD95 system as a mediator of granulosa cell apoptosis in ovarian follicle atresia. Endocrinology 1996;137:1938-1948.

22. Guo M.W., Xu J.P., Mori E., Sato E., Saito S., Mori T. Expression of Fas ligand in murine ovary. Am J Reprod Immunol 1997;37:391-398.

23. Mori T., Xu J.P., Mori E., Sato E., Saito S., Guo M.W. Expression of Fas-Fas ligand system associated with atresia through apoptosis in murine ovary. Hormone Res 1997;48 Suppl 3:11-19.

24. Xu J.P., Li X., Mori E., Sato E., Saito S., Guo M.W., Mori T. Expression of Fas-Fas ligand system associated with atresia in murine ovary. Zygote 1997; 5:321-327.

25. Koering M.J., Goodman A.L., Williams R.F., Hodgen G.D. Granulosa cell pyknosis in the dominant follicle of monkeys. Fertil Steril 1982; 37:837-844.

26. Carson R.S., Findlay J.K., Burger H.G., Trounson A.O. Gonadotropin receptors of the ovine ovarian follicle during follicular growth and atresia. Biol Reprod 1979; 21:75-87.

27. Bortolussi M., Marini G., Reolon M.L. A histochemical study of the binding of 125I-HCG to the rat ovary throughout the estrous cycle. Cell Tissue Res 1979; 197:213-226.

28. Uilenbroek J.T., Richards J.S. Ovarian follicular development during the rat estrous cycle: gonadotropin receptors and follicular responsiveness. Biol Reprod 1979; 20:1159-1165.

29. Oxberry B.A., Greenwald G.S. An autoradiographic study of the binding of [125] I-labeled follicle-stimulating hormone, human chorionic gonadotropin and prolactin to the hamster ovary throughout the estrous cycle. Biol Reprod 1982; 27:505-516.

30. Bellin M.E., Ax RL. Chondroitin sulfate: an indicator of atresia in bovine follicles. Endocrinology 1984; 114:428-434.

31. Asselin E., Xiao C.W., Wang Y.F., Tsang B.K. Mammalian follicular development and atresia: role of apoptosis. Biol Signals Receptors. 2000; 9:87-95.

32. Byskov A.G. Cell kinetic studies of follicular atresia in the mouse ovary. J Reprod Fertil 1974; 37:277-285.

33. Peluso J.J., England-Charlesworth C., Bolender D.L., Steger R.W. Ultrastructural alterations associated with the initiation of follicular atresia. Cell Tissue Res 1980; 211:105-115.

34. Bukovsky A., Chen T.T., Wimalasena J., Caudle M.R. Cellular localization of luteinizing hormone receptor immunoreactivity in the ovaries of immature, gonadotropin-primed and normal cycling rats. Biol Reprod 1993; 48:1367-1382.

35. Bukovsky A., Caudle M.R., Keenan J.A., Wimalasena J., Foster J.S. Van Meter S.E. Quantitative evaluation of the cell cycle-related retinoblastoma protein and localization of

Thy-1 differentiation protein and macrophages during follicular development and atresia, and in human corpora lutea. Biol Reprod 1995; 52:776-792.

36. Kasuya K. The process of apoptosis in follicular epithelial cells in the rabbit ovary, with special reference to involvement by macrophages. Arc Histol Cytol 1995;58:257-264.

37. Kasuya K. Elimination of apoptotic granulosa cells by intact granulosa cells and macrophages in atretic mature follicles of the guinea pig ovary. Arc Histol Cytol 1997; 60:175-184.

38. Koike K., Watanabe H., Hiroi M., Tonosaki A. Gap junction of stratum granulosum cells of mouse follicles: immunohistochemistry and electron microscopy. J Elect Microscopy 1993; 42:94-106.

39. Watanabe H., Tonosaki A. Gap junction in the apoptosis: TEM observation of membrana-granulosa cells of mouse ovarian follicle. Prog Cell Res 1995; 4: 37-40.

40. Inoue S., Watanabe H., Saito H., Hiroi M., Tonosaki A. Elimination of atretic follicles from the mouse ovary: a TEM and immunohistochemical study in mice. J Anat 2000; 196:103-110.

Chapter 10

THE OVARIAN CYCLE

Firyal S. Khan-Dawood
Morehouse School of Medicine, Atlanta, GA, USA

INTRODUCTION

The ovarian cycle is a recurring expression of the synchronized interaction between the hormones of the hypothalamus-pituitary-ovarian system. The goals of each cycle are to: 1) Mature a follicle containing an ovum, leading to ovulation and fertilization. 2) Cause structural and functional changes in target tissues (uterus, oviducts, and vagina) to facilitate fertilization and implantation. 3) Support early pregnancy. The cycle of ovarian activity is divided into three main phases: the follicular phase during which oocytes are recruited for ovulation; the ovulatory phase in which the ovum is expelled from the ovary and the luteal phase during which, the egg may be fertilized and pregnancy occurs. Thus the ovary has two functions, to produce the female gamete (gametogenesis) and to support the growing gamete by steroidogenesis. The predominant hormones involved in the primate and human female are gonadotropin releasing hormone (GnRH from the hypothalamus), follicle stimulating hormone (FSH), luteinizing hormone (LH) from the anterior pituitary, and estradiol (an estrogen), progesterone (a progestogen) and inhibin from the ovary. Although the duration of each phase, and therefore the length of the cycle, differs in various species (Table 1), the processes occurring within the ovary are basically similar. In the lower animals, the cyclic activity is called the estrus cycle, and in primates and the human female it is called the menstrual cycle. These names refer to the external behavioral and physiological manifestations of estrogen and progesterone. Day one of the estrus cycle is the day on which estrus behavior begins and it is a period that is conditioned by the estrogens secreted by the ovary prior to ovulation. During this period, the female becomes receptive to the male. In primates, the menstrual cycle results from the cyclical growth and destruction of the endometrium (the lining of the uterine cavity) each month. Day one of the cycle is the first day of shedding of the endometrial lining, which occurs in response to low levels of progesterone secreted by the regressing corpus luteum of the ovary. In the following section, the ovarian cycle of the human female will be discussed as an example of ovarian cyclicity. Ovarian cycles of lower species are extensively discussed elsewhere (1,2).

Table I. Ovarian Cycle Length in Various Species

Species	Length of Cycle (days)	Follicular Phase (days)	Luteal Phase (days)
Human	24-32	10-14	12-15
Cow	20-21	2-3	18-19
Pig	19-21	5-6	15-17
Sheep	16-17	1-2	14-15
Horse	20-22	5-6	15-16
Mouse/rat	4-5	2	2-3
Rabbit	1-2	1-2	0

Reproduced and modified with permission Johnson M. H., Everitt B. J. In *"Adult Ovarian Function"*. In *"Essential Reproduction"* 5th Edition 2000; p. 79

PREPUBERTAL OVARY

Development of the Ovary

To understand the cyclical events taking place in the pubertal adult ovary, it is important to understand the development of the germ cells prior to sexual maturity. During fetal life the development of the ovary goes through four main stages. The developmental stages are classified by the morphological appearance of the germ cells. The four main stages of germ cell development are: 1) The indifferent stage of the gonad. 2) The stage of differentiation. 3) The period of oogonial multiplication and follicle formation. 4) The final stage of follicle formation (3).

Stage of the Indifferent Gonad

Genetic sex is determined at conception. Within 4-5 weeks of conception the rudiments of the gonad appear as a thickening of the coelomic epithelium on the medial aspect of the mesonephros. The gonad evolves from four different components: 1). The coelomic epithelium. 2). The mesenchyme of the mesonephric ridge. 3). Cells from the mesonephric tubules, which eventually become the intraovarian rete (a network of blood vessels and nerves). 4). Primordial germ cells (precursors of an egg or sperm), which arise from a small number of stem cells (1000-2000) outside the genital ridge from the dorsal endoderm of the yolk sac (4). The coelomic epithelial cells differentiate into cells of the cortex of the ovary and give rise to granulosa cells of the follicle. The mesenchymal cells differentiate into the ovarian stroma and the theca-interstial cells surrounding the granulosa cells. Germ cells, mainly by amoeboid movement, chemotaxis and adhesive peptides such as the cadherins (5) migrate to the genital ridge and occupy the cortex of the ovary by four weeks of gestation. The cadherins are a large family of homotypic, calcium dependent, and intercellular adhesion molecules. These primitive germ cells are the only precursors of adult germ cells. They cannot survive outside of the genital ridge and are necessary for the development of the gonads. At this stage, the appearance of the gonads is similar in both the male and female fetus. Division of germ cells by mitosis begins during the migration and by the 6th week of gestation, at the end of the indifferent stage of the gonads, the primordial germ cells now called the oogonia, total 10,000. The development of the ovary occurs in the absence of the Y chromosome. The Sex-Determining gene (SRY), which is located on the short arm of the Y chromosome, (sex-determining region-Y chromosome) is responsible for the development of the male gonad. A number of other genes are also critical in gonadal development, including steroidogenic factor 1 (SF-1), Wilms' tumor suppressor 1 (WT1), DAX-1, (dosage-sensitive sex reversal, adrenal hypoplasia congenita, X-linked) SOX 9 (6) and SOX 3. The SRY gene codes for a DNA-binding protein 'testis-determining factor' (TDF). This is a transcription factor that regulates the expression of other genes or interacts with other transcription factors. SRY regulates the expression of genes coding for P450 aromatase, an enzyme required for estrogen synthesis. DAX-1 is on the short arm of the X chromosome and known as the dosage-sensitive sex (DSS) reversal locus. It represses the expression of StAR, a protein that transports cholesterol to the inner mitochondrial membrane where the first step in steroid synthesis occurs.

The germ cells in the female contain two X (XX) chromosomes, but during the migration one X chromosome is inactivated. The indifferent stage of gonadal development lasts about 7-10 days. During this period, the indifferent gonad contains rapidly dividing germ cells.

Stage of the Differentiation of the Gonads

Gonadal sex differentiation occurs between 6-9 weeks of gestation. By the 8th week of gestation, continuous mitotic division of the oogonia results in 600,000 oogonia. At this time, some of the oogonia begin to enter meiotic division [two successive nuclear divisions in which a single diploid (2N) cell forms haploid gametes]. In the human, oogonia enter the leptotene stage at the 12th week of gestation (7), zygotene and (N) pachytene at 14-19 weeks of

158

gestation (8), and diplotene between 14-28 weeks of gestation (9). The number of oogonia present becomes dependent on three different interacting mechanisms: mitotic divisions of the oogonia, meiosis, and oogonial cell death by atresia (Figure 1). By 16-20 weeks of gestation, 6-7 million oogonia are present in the two human ovaries. Mitosis occurs until mid-

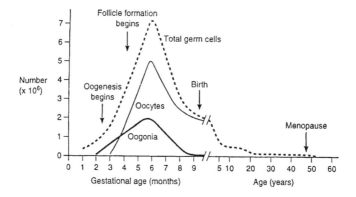

Figure 1. The number of germ cells correlated with age. Approximately 2000 stem cells from the dorsal endoderm migrate to the genital ridge by four weeks of gestation. Peak numbers of cells (oocytes, oogonia) are formed at five to six months of gestation (7 million). At birth 2 million primary oocytes remain and by puberty 400,000 oocytes are present. This occurs because follicles are continously recruited to grow but undergo atresia. Of the 400,000, 400 will be recruited under the influence of pituitary gonadotropin FSH during 400 cycles until the ovary is depleted at menopause. (Reproduced with permission "Clinical Reproductive Medicine Ed. Cowan B. D. and Seiter D. B. p 146, Lippincott-Raven, 1997)

gestation when a peak of approximately 7 million oocytes are present. The termination of mitosis and entry into meiosis is evoked by a meiosis-initiating factor derived from the cells of the in-growing mesonephric tissue. A consequence of the termination of mitosis is that the number of oocytes that a female will be endowed with is now complete.

As some of the oogonia enter meiotic prophase I, they become primary oocytes well before actual follicle formation. At the time of peak germ cell number, about 60% of the total germ cells are intra-meiotic primary oocytes, and the remaining are oogonia (Figure 1). Entry into meiosis initiates the formation of the primordial follicle by the condensation of surrounding ovarian mesenchymal cells derived from invading mesonephric cords. With the formation of the primordial follicles, the oocytes undergo a second major change. They abruptly arrest their progress through first meiotic prophase at diplotene and the chromosomes become enclosed in a nucleus called the germinal vesicle. The primordial follicle may stay arrested for up to 50 years in women, with the oocyte metabolically active and waiting for the signals(s) to resume development. Some oocytes may become atretic while others are recruited (leave the follicle pool and begin to develop) into the growth phase. Although a few follicles may sporadically and incompletely resume development, during fetal and neonatal life, regular recruitment of primordial follicles into a pool of growing follicles occurs only at puberty (attainment of reproductive ability).

The primordial follicle is the basic functional unit of the ovary. Most follicles are small (25mm) non-growing that consist of a small oocyte arrested in meiosis surrounded by a basal lamina and a single layer of pre-granulosa cells (Figures 2 and 3). When a primordial follicle begins to grow, it progresses through several stages in which it is called a primary follicle

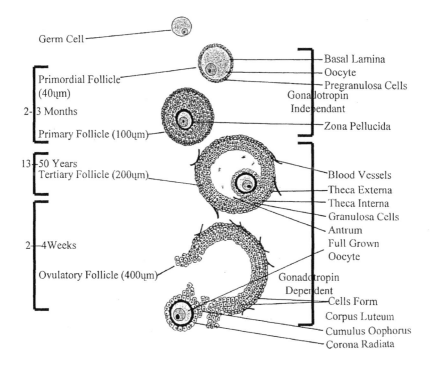

Figure 2. Schematic representation of folliculogenesis. Development of the follicle from the primordial follicle to the pre-antral or secondary follicle is a gonadotropin independent continous process occurring in the foetal ovary to the prepubertal ovary with all follicles becoming atretic. Gonadotropin dependant growth occurs during puberty of a cohort of follicles (10-12) of which one becomes the dominant follicle and ovulates. The oocyte remains arrested in the diplotene stage in the foetal ovary but increases in size and becomes surrounded by the supporting theca and granulosa cells.

(preantral follicle), a secondary follicle (small and large antral follicles) and finally a tertiary, preovulatory or Graafian follicle. In the final stage of folliculogenesis (the process by which follicles containing oocytes are recruited for development, maturation and ovulation in each post-pubertal

160

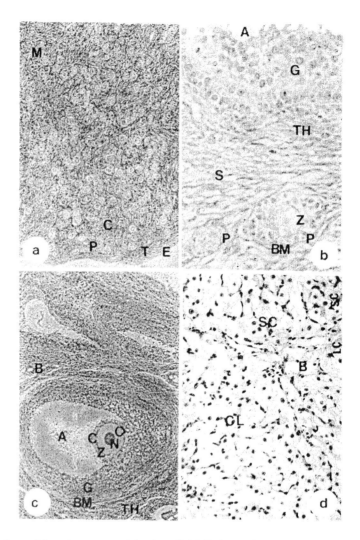

Figure 3. Histology of the primate ovary (a) primordial follicles (P) lying in the cortex of the ovary surrounded by the tunica albuginea (T) and a single layer epithelium (E) in an ovary from a two year old female baboon.(b) Follicles in an adult pubertal ovary with an ovulated follicle showing the antrum (A), the remaining granulosa (G) and theca (TH) cells. Also shown are two primary follicles one with a single layer of granulosa cells, the second with two-three layers of granulosa cells. The zona pellucida (Z) and the basement membrane (BM) are indicated. (c) A preovulatory follicle with an oocyte containing a nucleus, (N) surrounded by the zona pellucida (Z) and the cumulus oophorus (C). Also shown is a blood vessel (B) in the theca layers. (d) A section of a corpus luteum showing "small" (SC) and "large (LC) cells and a blood vessel (B). Magnification (b), (c) and (d) x40 and (a) x10.

cycle) a dominant follicle, one that will ovulate is selected. The secondary follicle houses an enlarging oocyte whose diameter has increased from 25mm to a maximum of 80mm. The granulosa cells proliferate into several layers of cells surrounding the enlarged oocyte. Outside the basal lamina, the pre-theca cells are organized concentrically to form the theca layer. This layer of cells is networked by lymph and blood vessels that do not transverse the basal lamina. Accumulation of cytoplasm and differentiation into steroid producing secretory cells separate the theca interna cells from the outer theca externa cells. The tertiary follicle arises from increased proliferation of granulosa cells in the secondary follicle and is accompanied by localized accumulation of fluid among rosettes of granulosa cells called, Call-Exner bodies. Coalescence of these bodies produces a central fluid-filled cavity called the antrum. Antrum formation forces the oocyte into a more eccentric position surrounded by several layers of granulosa cells that constitute the cumulus oophorus (Figures 2 and Figure 3).

The process of oogonia transformation to primary oocyte occurs at 11-12 weeks of gestation, and may be under the influence of secretions of the rete ovarii. At this time, the oocytes are found in clusters or nests in the ovarian cortex. The progression of meiosis to the diplotene stage occurs throughout birth. The arrest of meiosis at this time is maintained by an inhibiting substance called oocyte maturation inhibitor (OMI) produced by the granulosa cells (10). A decline in the number of germ cells follows a peak at 20 weeks of gestation and is partly due to a progressive decline in the rate of oogonial mitosis, which terminates at 28 weeks of gestation. Balancing this is the increased rate of oogonial atresia which peaks at 5 months of gestation caused by programmed cell death, (11, 12).

Stage of Follicle Formation and Oogonial Multiplication

The formation of the follicles begins at 18-20 weeks of gestation. This process involves the separation of the oocytes from the surrounding stroma, the association of granulosa cells around the surface of each oocyte, and the formation of a basement membrane around this unit (Figures 2 and 3). The follicle formation and growth always begin in the inner most part of the ovarian cortex where oocytes come into contact with the rete ovarii. There is a paucity of information on factors involved in this process. Experimental evidence suggests the existence of an oocyte-granulosa cell regulatory loop, which orchestrates normal follicular differentiation, and the maturation of an oocyte competent to undergo fertilization. The concept of the oocyte being actively involved in its growth and maturation is recent (13). The granulosa cells and the oocyte actively communicate by several mechanisms. These include passage of low molecular weight (<1000 Daltons) substances through gap junctions (junctions between adjacent cells that allow the passage of low molecular weight materials between two cells), paracrine (intercellular communication which involves the action of substances from one cell or an adjacent cell), and autocrine (intercellular communication which occurs when the action of a substance produced by a cell acts within the same cell) factors. The gap junctions are essential for follicle development since targeted mutagenesis of connexin-37, a protein that forms these junctions, impairs follicle development. These junctions allow passage of sugars, amino acids, lipid precursors and nucleotides between the two cell types. Granulosa cells also produce a variety of cytokines and growth factors. Stem Cell Factor (SCF), also known as kit-ligand (KL), is secreted by the granulosa cells and binds to a tyrosine kinase receptor, c-kit, on the oocyte surface (14). Spontaneous mutations of the respective gene (Steel, W) cause germ cell deficiency and arrested development of follicular oocytes.

The oocyte also influences follicle development. It secretes proliferating factors, such as Growth Differentiation Factor-9 (GDF-9) and Bone Morphogenetic Protein (BMP-15), also known as GDF-9β. These are members of the transforming growth factor β (TGFβ) family and are expressed in human, ovine, bovine and mouse oocytes (15, 16). These are detected in the oocyte from primary, one layer granulosa cell follicle stage, until after ovulation. They are not expressed in non-growing follicles (15). Targeted deletion to produce GDF-/- mice results in an infertile phenotype in which follicle development does not go beyond the one-layer stage and the oocytes are enlarged.

Gonadotropins influence the growth of the oocyte indirectly since the cells do not express FSH receptor. The FSH receptor is constitutively expressed by granulosa cells at the primordial stage (17). If the FSH receptor is deleted, the follicles arrest at the pre-antral stage (18). Further formation of the follicle involves penetration of vascular channels from the medulla of the ovary. The process of primordial follicle development continues until all oocytes in the diplotene stage develop into follicles shortly before birth. Some primary follicles can undergo further development to the pre-antral follicle stage in the fetal ovary. With the proliferation of the granulosa cells to several layers and differentiation of a theca layer of cells from the surrounding mesenchyme, the follicle becomes a secondary or pre-antral follicle. Once the initiation of follicular growth has begun, the entry of the follicle into the growth phase is a continuous process throughout fetal, neonatal and adult life. By the fourth to sixth month post-natally all stages of follicle development up to the point of antrum formation can be found in the ovary. The number of oogonia at birth is approximately 2 million (19). The decrease in number is associated with apoptotic cell death (programmed cell death) of both the follicular granulosa cells and the oocyte. The reason for this occurrence is not clear. Although all stages of follicle growth can be observed in the prepubertal ovary, the final maturation of the oocyte, culminating in ovulation, does not normally occur before sexual maturity.

NEONATAL OVARY

At birth, the number of germ cells dramatically declines to 1-2 million as a result of prenatal oocyte depletion over the previous 20 weeks.

Compartmentalization of the gonad into the cortex and medulla is achieved. In the cortex, almost all the oocytes are involved in primordial follicle units. Varying degrees of maturation in some units is seen Table 2.

Table 2. Stages of Human Ovary Development.

Stage and Time	Morphology and Function
1. Pregonadal Day 4-5	Primordial germ cells differentiate
2. Indifferent Days 7-10 Up to day 20 Weeks 5-6	Gonadal ridge differentiates Primordial germ cells migrate to gonad. Indifferent gonad formed
3. Primary Sex Differentiation From Week 8 Weeks 9-10 Weeks 11-12 Neonate Neonate to puberty	Ovarian morphological differentiation Primordial germ cells develop into oogonia Meiosis begins up to diplotene of prophase of 1st meiotic division Resting pool/initial recruitment/atresia
4. Secondary Sex Differentiation Puberty	No gonadal factors produced.
5. Age 12-35 Years	Ovarian cyclic activity produces one egg per cycle. All stages of folliculogenesis present. Cyclic recruitment of follicles.
6. Age 35-51 years	Perimenopause, erratic cycles
7. **Age 51-and older**	Menopause-few follicles

Reproduced and Modified with Permission *from "Adolescent Endocrinology"* Ed. R. Stanhope. Bio Scientifica 1998; p 34

OVARY IN CHILDHOOD

The ovary is not quiescent in childhood. It is a very active gland containing many gonadotropin sensitive antral follicles with a capacity to synthesize estrogens and androgens. Follicles begin to grow at all times and frequently reach the antral stage. Lack of gonadotropin support prevents full follicular development and function. Ovarian function is not necessary until puberty, but the oocytes continue to be active.

ADULT OVARY

Structure

The ovary is covered by a single cell epithelium and a thin layer of connective tissue, the tunica albuginea. These two layers surround the outer cortex and inner medulla. The cortex displays various stages of follicle development and regressing corpora lutea. The medulla contains connective tissue supporting blood vessels, lymph vessels and nerves entering the ovary through the hilum (Figure 4).

The Menstrual Cycle

At puberty, the hypothalamic-pituitary-ovarian system becomes completely operational. This is necessary for the initiation of ovarian cyclic activity. The complex interaction of these components allows the release of an ovum episodically, following recruitment of a dominant follicle. At the time of puberty, there is an average of 400,000 follicles remaining in the two ovaries.

During reproductive life, continuous growth of primordial follicles leads to a gradual decrease in the follicle pool. Death of primordial follicles by apoptosis also contributes to the decrease in the pool. Ten years before menopause, concomitant with subtle increases in serum FSH and decrease in circulating inhibins, increasing percentages of follicles are lost from the resting pool. As a result of ovarian follicle exhaustion, menopause occurs at about 51 years of age.

During the first few months of the onset of puberty, the hypothalamus-pituitary-ovarian axis is not completely synchronized resulting in anovulatory cycles with irregular lengths. Irregular cycles are also the hallmark of the perimenopausal transition. The duration of the menstrual cycle can vary from 24 to 32 days with an average of 28-29 days in the active reproductive years. By convention, the first day of menstrual bleeding is the first day of the menstrual cycle. The cycle days are also numbered relative to the LH peak. The LH peak is considered as day 0, days in the follicular phase as −1 to −14 and luteal phase as +1 to +14. Variability in length is due to the time it takes for the maturation of the dominant follicle in the follicular phase. Ovulation occurs on day 14 of the cycle. This is followed by a luteal phase, which lasts 14 days. In parallel with the ovarian cycle and influenced by the hormones secreted by the ovary is the endometrial cycle or the menstrual cycle. The first 3-5 days are the menstrual flow days. Days 5-14 are the proliferative phase days during which the endometrium increases in thickness as a result of the stimulatory activity of estrogens produced by the ovarian follicle. Mitotic activity is increased in both the lamina propria and the epithelium. Epithelial cells of the glandular epithelium become highly active. After day 14, when ovulation occurs, the endometrium becomes secretory. This secretory phase lasts 14 days. If pregnancy does occur, then day 22 is the day on which nidation (implantation of the fertilized ovum) would occur.

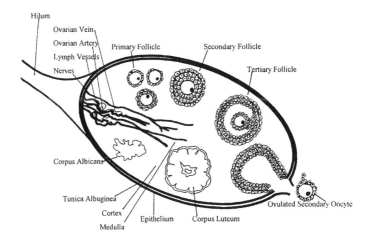

Figure 4. Diagram illustrating morphological features of a cross section through the adult pubertal ovary. It is covered by a simple squamous epithelium and an inner connective tissue layer called the tunica albuginea. Follicles at various stages of preantral development are present in the cortex connective tissue, housing a primary oocyte at the end of meiotic prophase.

Follicular Phase

Recruitment and Selection of the Dominant Follicle

Throughout prepuberty, the resting pools of primordial follicles are continuously being recruited, a few a day, into the growing follicle pool (20). However, in the absence of gonadotropic support, most become atretic. This form of recruitment is termed initial recruitment (21). At puberty, in the presence of rising levels of FSH, however, those primordial follicles that reach the antral follicle stage are selected into a cohort, which will provide a single dominant follicle. To reach the ovulatory stage a primordial follicle goes through several stages of development. These include primary or preantral follicles. Secondary or antral follicle, graafian or the ovulatory follicle (Figure 2).

It takes about 150 days for a primordial follicle to develop into a primary follicle and another 120 days to become a secondary preantral follicle. Thus a primary follicle goes through 9 menstrual cycles before it becomes part of the selectable cohort. A further 15 days are required for the selected follicle to become a dominant ovulatory follicle. This time coincides with the length of the follicular phase. The selection of the follicle takes place at about the time of menstruation of the preceding cycle with the rise in FSH (22) (Figure 5).

At the end of a menstrual cycle, the level of FSH secreted by the pituitary begins to increase as a result of the decreasing levels of estrogen, progesterone and inhibin secreted by the regressing corpus luteum. This increase occurs during the premenstrual period several days before menstruation. The consequence of the FSH increase is that a critical threshold concentration is reached within the microenvironment of the fastest growing follicle. A cohort of antral follicles escape the route to apoptosis due to the survival action of FSH. Among this group of about 10-12 antral follicles, one of the follicles grows faster than the rest of the group and produces higher levels of estradiol and inhibin. This antral follicle, measuring 2-5mm in diameter, develops into the dominant follicle.

The hormones secreted by the developing dominant follicle, estradiol and inhibin-B, suppress pituitary FSH released during the midfollicular phase (Figure5). As a result, the remaining growing antral follicles are deprived of adequate FSH stimulation required for survival. FSH withdrawal is involved in the massive apoptosis of the granulosa cells of the non-dominant follicles. The first indication that a dominant follicle has been selected is that the granulosa cells in the chosen follicle continue to proliferate at a fast rate, while the rate of proliferation slows in non-dominant follicles. The FSH receptor signaling plays a fundamental role in the growth and differentiation of the dominant follicle because of its ability to promote follicular fluid formation, cell proliferation, estradiol production, LH-receptor expression, and inhibin production. The temporal pattern and level of expression of these FSH-dependent genes are crucial for gene expression and normal physiological functions associated with the dominant follicle. The FSH stimulation of LH receptors in the granulosa cells is required for LH to induce ovulation and luteinization of the remaining follicle cells. The LH receptor expression induction occurs in the dominant follicle immediately prior to ovulation. The LH-dependent signaling pathways in the theca interstitial cells elicit changes in gene expression that are critical for estradiol synthesis.

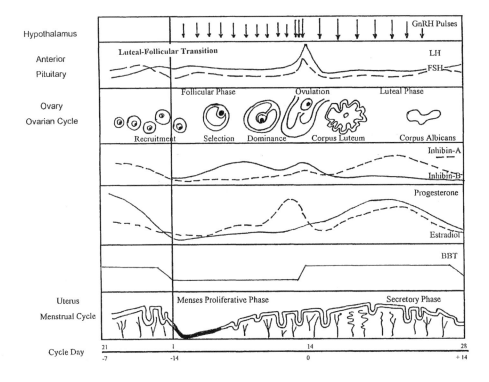

Figure 5. Interrelationships between hypothalamic-pituitary and ovarian hormones during a cycle showing the luteal-follicular transition and a complete ovarian-endometrial (menstrual) cycle. Days of the cycle are shown with the conventional numbering from day 1-28 and when the LH peak is considered as day 0. GnRH pulses during the follicular phase are one per hour and luteal phase one per ninety minutes. The ovarian cycle shows the recruitment of the follicles, ovulation and the corpus luteum. Changes in basal body temperature (BBT) induced by progesterone are also shown. An increase in body temperature of 0.5^0 occurs following ovulation.

Activation of the LH receptors in the theca cells leads directly to the stimulation of high levels of androstenedione production, which is the substrate required for estradiol synthesis by the granulosa cells. Preovulatory granulosa cells do not express the enzyme 17α hydroxylase needed for androgen production. Because theca interna cells produce the androgen precursors for granulosa estrogen production both cell types are needed for optimal estrogen synthesis. This is the two gonadotropin-two cell concept of dominant follicle estrogen biosynthesis. Thus estradiol is a useful marker for monitoring dominant follicle development.

Ovulation

Just prior to ovulation, usually day 14 of the menstrual cycle, the selected dominant follicle grows rapidly. A positive (stimulatory) feedback effect of estradiol on gonadotropin secretion occurs. This results in a LH and FSH surge from the pituitary, meiotic maturation of the oocyte and ovulation or release of the ova into the abdominal cavity. The remaining tissue forms the corpus luteum. The length of time from selection of the dominant follicle to ovulation is variable. The estrogen-FSH levels initially promote the increase in FSH receptors, followed by a stimulation of LH receptors on the granulosa cells. The LH receptors in the granulosa cells stimulate pre-ovulatory synthesis of progesterone and 17α hydroxyprogesterone. Thus, just before ovulation, a rise in plasma levels of these hormones occurs. The pre-ovulatory rise in estradiol and progesterone promote the pituitary gonadotropes to increase their responsiveness to GnRH (Figure 5).

A mid cycle gonadotropin surge results from the activation of a positive estradiol feedback at the level of the pituitary and hypothalamus. For the initiation of the LH and FSH surge, the levels of estradiol produced by the growing follicle need to be above a threshold of 300-500pg/ml for approximately 48 hours. The mid-cycle LH surge with a duration of 48 hours triggers oocyte extrusion, stimulates resumption of meiosis in the oocyte, initiates luteinization of the granulosa cells, as well as synthesis of progesterone and prostaglandins. The process of ovulation lasts 36-40 hours after the onset of the LH surge.

During this time the oocyte completes prophase I of the first meiotic division as it passes from the diplotene stage into diakinesis. The second meiotic division generally begins 30-34 hours after the onset of the LH surge. The oocyte reaches the haploid stage (metaphase 2 oocyte) approximately 32-36 hours from the onset of the LH surge and 1-2 hours before ovulation. The mature oocyte is now capable of being fertilized.

Meiosis resumption of the oocytes is mediated by LH. The initial stimulation in LH receptors together with the onset of the LH surge is followed by a refractory stage where the LH receptors are down regulated. The LH surge is associated with a decrease in both the number and the sensitivity of the LH receptors leading to decreased levels of cAMP and oocyte maturation inhibitor (OMI), decreased estradiol production and disruption of the gap junctions which connect the oocyte to its surrounding granulosa cells. The LH surge also initiates luteinization (process of forming a corpus luteum) of the granulosa cells and shifts steroidogenesis in favor of progesterone secretion through the induction of the enzyme 3β-hydroxysteroid dehydrogenase.

The FSH surge stimulates the production of a plasminogen activator that aids in the detachment of the oocyte/cumulus mass from the follicle wall. The hormone also plays a role in the conversion of granulosa cells to functional luteal cells by facilitating LH receptor induction.

The rupture of the follicle occurs 38-42 hours after the onset of the LH surge and is evoked by LH mediated stimulation of prostaglandin synthesis [mainly Prostaglandin F2α (PGF2α) and prostaglandin E2 (PGE2)] and production of proteolytic enzymes such as collagenase. The follicle wall becomes thin, the LH and prostaglandins decrease follicular blood flow and stimulate the release of granulosa cell plasminogen activator. This converts plasminogen to plasmin. Collagenase and plasmin (proteolytic enzymes) break down the follicle wall. The oocyte-cumulus complex is extruded from the follicle at a weakened point of the ovary wall known as the "stigma." This mass is picked up by the fimbriated end of the fallopian tube. Ciliated cells of the fallopian tube aid in the movement of the oocyte-cumulus complex through the ampullary portion of the tube where fertilization generally occurs.

Luteal Phase

The effect of the LH surge at ovulation is to alter the expression of genes of the enzymes necessary for steroidogenesis in the cells remaining in the ovulated follicle (23). This is initiated about 24 hours before ovulation. The cells of the follicle become luteinized and begin to predominantly synthesize progesterone. Progesterone synthesis continues for 11-14 days. The granulosa and theca cells enlarge, accumulate a yellow carotenoid pigment called lutein. By mid-luteal phase (day 21-22), the corpus luteum is 2.5gms in weight and secretes up to 40 mg of progesterone. The formation of the corpus luteum involves breakdown of the basement membrane of the follicle, reorganization of the granulosa and theca cells. The vascular cells of the preovulatory follicle are rapidly invaded by blood vessels through angiogenesis (growth of capillaries from preexisting blood vessels) (24). Blood accumulates in the antral space forming a transient corpus hemorrhagicum. The fibrin clot is further penetrated by blood vessels, fibroblasts and collagen fibers. Factors that have been implicated in this process are vascular endothelial growth factor (VEGF) and basic fibroblast growth factor (bFGF). Maximal vascular growth occurs at the time of peak circulating levels of plasma progesterone (day 21-22). This facilitates the uptake of low-density lipoprotein cholesterol (LDL) from the blood for the synthesis of progesterone and also for the delivery of progesterone to the endometrium.

The corpus luteum consists of subpopulation of cells. Cells that synthesize and secrete progesterone are derivations of the granulosa and theca cells of the follicle. The granulosa cells give rise to the "large cell" population. These secrete 10-20 times more progesterone than the "small cells" derived from the theca cells. Several interleukins and growth factors have been identified. The large cells also produce oxytocin, inhibin and relaxin. Gap junction communication between the two cell types may be involved in progesterone synthesis, which is facilitated by oxytocin (25, 26). Estrogen and progesterone receptors present on these cells suggest that hormones produced by these cells may have a local function in the regulation of the corpus luteum (27). In addition to the steroidogenic cells several other types of cells are present which include macrophages, immune cells, fibrocytes and endothelial cells. The interaction between these cell types and their possible role in luteal function and regression are under investigation.

The corpus luteum under the influence of LH is functional for 14 days, the length of the luteal phase of the menstrual cycle. This tissue begins to involute about 12 days after ovulation unless fertilization of the ovum occurs and it is rescued from undergoing regression by human chorionic gonadotropin (hCG) produced by the developing blastocyst (a 64-cell embryo). In the absence of a pregnancy, the corpus luteum degenerates and is replaced by connective tissue forming the corpus albicans. Complete degeneration takes several months.

The mechanisms involved in the loss of activity of the corpus luteum are unclear. It may involve an active luteolytic mechanism. On or around day 24 of the cycle during the last 4-5 days a rapid decrease in plasma levels of estradiol, progesterone and inhibin occur (28) (Figure 5). The enzymes of steroidogenesis particularly the cholesterol side chain cleavage enzymes (P450scc) and 3β-hydroxysteroid dehydrogenase involved in progesterone synthesis begin to decline. The decrease in the circulating levels of these hormones removes the suppressive effects of the high mid-luteal phase concentration at the level of the pituitary. Levels of FSH increase dramatically due to the low levels of inhibin, which selectively suppress FSH. The decrease in estradiol enables the pituitary to respond to GnRH signals from the hypothalamus. The increased pulse frequency of GnRH secretion causes a predominance of FSH secretion. The increasing FSH levels stimulate the recruitment of a new cohort of follicles.

It is also probable that the estradiol may act locally in combination with prostaglandins to effect luteal demise. The presence of lymphocytes and macrophages towards the end of the luteal phase also suggest that secretions from these cells may contribute to luteal demise (29). Regression of the corpus luteum happens in two phases. First, functional luteolysis occurs which is a loss of progesterone synthesizing capacity. This occurs before morphological changes in the luteal cells is apparent and can be reversed by hCG. However, once structural changes have occurred, the tissue cannot be rescued. Structural changes occur well after the initial decline in steroid output. Evidence suggests that apoptosis involving the luteolysin prostaglandin $F_{2\alpha}$ ($PGF_{2\alpha}$) may regulate the reactive oxygen species in the corpus luteum which impairs the ability to synthesize progesterone (30). Other possible mechanisms of luteolysis have been suggested involving pathways, which change bcl-2 and c-myc expression of apoptosis-related proteins (29, 30).

Corpus Luteum of Pregnancy

In the event of a pregnancy, hCG secreted by the gestational trophoblast rescues the corpus luteum and progesterone synthesis is revived. This function is maintained until the luteo-placental shift 8-10 weeks after the establishment of the pregnancy (33). The corpus luteum increases in size due to hypertrophy of the steroid producing cells (the large and small cells) proliferation of connective tissue and growth of vascular tissue. However, following the luteo-placental shift the tissue regresses and at term the size is half that seen during the mid-luteal phase.

HORMONES OF THE MENSTRUAL CYCLE

In the complex and intimate interaction of the hypothalamus-pituitary-ovary and the uterus, the ovarian signals (hormones) play a key role of synchronizing the system. Estradiol, progesterone and inhibin A and B are secreted by the developing dominant follicle and the corpus luteum Testosterone and androstenedione synthesized by the theca cells are also secreted by the ovary. The hypothalamus secretes GnRH and in response to this signal the pituitary secretes FSH and LH (Figure 5).

Estrogens

Estradiol synthesis by the theca cells in cooperation with the granulosa cells by the dominant follicle is crucial for development of the follicle-oocyte unit prior to ovulation and for the regulation of the secretions of the hypothalamus and the pituitary (Figure 6).

In the endometrium, estradiol induces proliferation of the endometrial glands. Estradiol has a dual function in regulating FSH and LH secretion. Low circulating levels of estradiol in the first half of the follicular phase act to suppress gonadotropin secretion thus having a negative feedback effect. At high plasma concentrations (>200 pg/ml) levels in the mid-follicular phase which are maintained for 48 hours LH and FSH secretion is increased such that a surge of LH and FSH occurs. This is a positive feedback effect of estradiol on the pituitary. The local actions of estradiol include induction of FSH and LH receptors, proliferation of the granulosa cells, and proliferation and secretion of follicular thecal cells. Prior to ovulation, levels of estradiol rise to 300-400 pg/mL (Table 3).

Progesterone

Progesterone is secreted by the luteinized follicle cells forming the corpus luteum under the LH receptor mediated effect of LH. The high plasma concentration has two effects at the pituitary level. Firstly, the negative feedback effects of estradiol is enhanced, thus suppression of FSH and LH occurs and the positive feedback effect of estradiol is blocked. At the ovarian level, progesterone stimulates the release of proteolytic enzymes from the theca cells during ovulation. It stimulates angiogenesis and prostaglandin secretion in the follicle. In the endometrium, progesterone induces swelling and stimulates the secretory activity of this tissue in preparation for implantation.

Inhibins A and B

This family of nonsteroidal hormones is defined by the property of suppression of gonadotropin secretion (26) although they have local effects within the follicle and the corpus luteum. Inhibin B is synthesized by the granulosa cells of the dominant follicle under the influence of FSH. It regulates the normal development of the follicle and has a negative feedback effect on FSH secretion. Inhibin A is mainly secreted by the corpus luteum and acts by paracrine/autocrine mechanisms to maintain this tissue. It also has a negative feedback action on FSH secretion.

Activin

Activin is also synthesized by granulosa cells; it enhances FSH action in the induction of LH receptors. It is also active at the hypothalamic-pituitary axis.

Follicle Stimulating Hormone

Secreted by the gonadotropes in the anterior pituitary, this peptide hormone is essential for follicle recruitment and development. The secretion of this hormone is high in the first week of the follicular phase and exerts a negative feedback on hypothalamic GnRH secretion. Plasma levels peak at midcycle.

FSH stimulates estrogen secretion in the follicle by activating P450scc and aromatase enzymes. It also induces the proliferation of granulosa cells and expression of LH receptors.

Luteinizing Hormone

LH is also secreted by the cells of the anterior pituitary gland. It is essential for the growth and maturation of the ovum in the preovulatory follicle, ovulation and luteinization of follicle cells. It initiates the first meiotic division in the oocyte in preparation for fertilization. In the follicle,

LH stimulates androgen synthesis by the theca cells, stimulates differentiation and proliferation of the theca cells and increases LH receptors on the granulosa cells. Following ovulation, LH is essential in the maintenance of the corpus luteum.

Gonadotropin Releasing Hormone

GnRH, a deca-peptide, is secreted by the cells of the hypothalamus and plays a critical role in the ovarian cycle. It is carried via the portal vascular system to the pituitary. It has a half-life of 2-4 minutes and is secreted in pulsatile bursts throughout the menstrual cycle. The frequency of pulses in the early follicular phase is once every 60 minutes with low ampli tudes increases at the time of ovulation and decreases in rate with increased amplitude during the luteal phase once every 90 minutes. The secretion of pituitary FSH and LH follow the pulsatile pattern of GnRH release. GnRH regulates the synthesis and storage of the gonadotropins, activation or movement of gonadotropins from reserve to a pool ready for secretion and immediate release.

Interaction of the Hormones of the Menstrual Cycle

The GnRH pulse generator discharges at the rate of one pulse per hour in the late follicular phase. In response to this, the pituitary secretes hourly pulses of LH and FSH. The ovarian response is folliculogenesis and estradiol synthesis. During the follicular phase, gonadotropin secretion is controlled by a negative feedback action of estradiol on the pituitary gonadotropins and GnRH. Progesterone does not have a significant role in the follicular phase neuroendocrine function. In response to the gonadotropins and as folliculogenesis proceeds, inhibin B is produced by the granulosa cells, the time course paralleling that of estradiol. The mid cycle gonadotropin surge

174

is initiated by the positive feedback action of estradiol, acting at the level of the pituitary gonadotrophs, when plasma estradiol levels exceed a threshold of approximately 300-500 pg/mL for 48 hours. Progesterone acting at the pituitary level facilitates the release of gonadotropins (Figure 5).

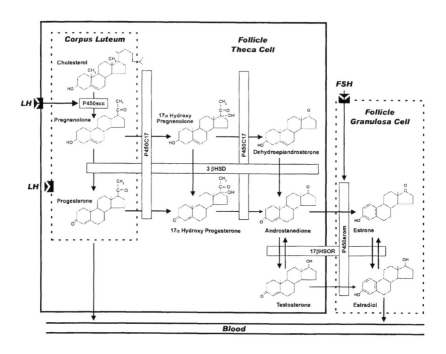

Figure 6. Biosynthetic pathways of estradiol and progesterone in the dominant follicle and the corpus luteum. All the cells have the enzymes necessary for steroid synthesis (P450scc, P450c17, 3βHSD, 17βHSOR and P450$_{arom}$) however, the expression of each enzyme and the availability of precursors determines the main route of synthesis. In the follicle, the two cell-two-gonadotropin interaction is predominant (two right boxes). Thus, precursors from the theca cells (androstenedione and testosterone) are used for estradiol and estrone synthesis. The main secretion of the corpus luteum is progesterone (left box).

In the luteal phase, the GnRH pulse frequency decreases (one pulse per 90 minutes). The action of progesterone on the hypothalamic pulse generator is mediated by endogenous opioids as well as neurotransmitters such as epinephrine/norepinephrine which increase GnRH release, dopamine and serotonin inhibit GnRH release. The lowering levels of luteal estrogen, progesterone and inhibin A relax the inhibition (negative effect) on FSH secretion and the FSH levels rise initiating a new cycle of folliculogenesis (Figures 5 and 7).

Table 3. Plasma Hormone Levels in the Human Female

HORMONE	SI UNITS	CONVENTIONAL
Estradiol		
Basal	70-220 pmol/L	20-60 pg/mL
Ovulatory Surge	>740 pmol/L	>200 pg/mL
Progesterone		
Luteal Phase	6-64 nmol/L	2-20 ng/mL
Follicular Phase	>nmol/L	<2 ng/mL
Gonadotropins Plasma		
Basal FSH	5-20 IU/L	5-20 mIU/mL
Basal LH	5-25 IU/L	5-25 mIU/mL
Ovulatory FSH	12-30 IU/L	12-30 mIU/mL
LH	25-100 IU/L	25-100 mIU/mL
Inhibin A		
Follicular Phase	<20 pg/ml	
Mid-Luteal Phase	60 pg/ml	
Inhibin B		
	85.2 ± 9.6 pg/ml	
Early Follicular Phase	133.6 ± 31.2 pg/ml	
Midcycle Luteal Phase	<20 pg/ml	

MENOPAUSAL OVARY

The menopause is a unique phenomenon of the female homosapiens. It is the complete cessation of spontaneous menstrual cycles and generally occurs between 45 and 55 years (34). The determining factor for the onset of menopause is the finite number of ova laid in the ovary in the fetus. Since they are not replaced and are continuously recruited throughout life, a 50-year life span exhausts the supply of follicles (17). In the absence of the follicles, very little estradiol (90% decline) or inhibin is produced and therefore, a negative feedback control of pituitary secretion of FSH and LH does not occur. The ovulating levels of these hormones become chronically elevated which is a hallmark of the menopause (35). The chemical definition of the menopause is an elevation in the FSH level together with low or absent levels of estradiol. Because a significant loss in thecal-stromal cells in the ovary during the menopause does not occur a large fall in androgen production is not seen. Androgen levels may decrease by 20%.

The transition from premenopause and menopause is gradual lasting 4-5 years. This period is called the perimenopause. In the absence of estradiol, vaginal and urethral atrophy, bone loss, decreased skin thickness and increased risk of cardiovascular disease occurs.

STEROIDOGENESIS

The menstrual cycle is determined by the hormones produced by the developing follicle and following the expulsion of the oocyte by the corpus luteum. The key hormones produced by the ovary are estradiol and progesterone (Figure 6).

These two hormones are primarily derived from plasma low-density lipoprotein (LDL) cholesterol (36). LDL cholesterol enters the cell via a receptor mechanism. Pituitary LH stimulates adenyl cyclase causing an increase in cAMP production. This acts as a second messenger to increase LDL receptor mRNA and the binding and uptake of LDL cholesterol. cAMP also activates the steroidogenic acute regulatory (St AR) protein, which translocates cholesterol from the outer to the inner mitochondrial membrane, the rate-limiting step in steroid hormone synthesis (37).

The conversion of cholesterol to estradiol and progesterone requires five key enzymes (Table 4): Cytochrome P450 side chain cleavage enzyme (P450scc), 3β-hydroxysteroid dehydrogenase (3β-HSD) 17α-hydroxylase cytochrome P450 (P45017α), aromatase P450 (P450arom) and 17β hydroxy steroid dehydrogenase (17β-HSD), type 1 (38). These enzymes convert cholesterol to pregnenolone, pregnenolone to progesterone, progesterone to androgens, androgens to estrogens and estrone to estradiol (Figure 6).

The conversion of pregnenolone to the androgens and estrogens can take two pathways in the ovary. Either via Δ5-3β-hydroxy-steroids (Δ5 pathway) which involves conversion of pregnenolone to dehydroepiandrosterone (DHEA) or the Δ4-3- ketone pathway (Δ4 pathway) which goes through progesterone and 17α-hydroxy progesterone.

The theca, granulosa cells, and the cells of the corpus luteum express all the enzymes needed for steroidogenesis. In the follicle, the predominant pathway is the Δ5 pathway and the Δ4 pathway is more active in the corpus luteum. The rate and type of steroid produced by the ovarian cells is dependent on the amount, activity and expression of enzymes involved in steroidogenesis and the number of gonadotropin receptors present on the cell. Mechanisms controlling these factors are not well understood.

The synthesis of estradiol in the follicle involves intimate interaction between the granulosa and the theca interna cells *(Figure 6)*. This is based on the evidence derived by immunocytochemistry and RNA expression studies indicating that P450scc mRNA is present in both the granulosa and theca cells at the antral stage of follicle development. There is little expression of 3β HSD mRNA in the follicle and P450 17α is only present in the theca interna and is completely absent in the granulosa cells. P450arom is only present on the granulosa cells of the mature follicle. 17β HSD type I is present only in the granulosa cells (38). In addition, FSH receptors are present only on the granulosa cell membranes and are induced by FSH. LH receptors are present on the theca cells and initially absent on the granulosa cells but with follicular growth, FSH induces the appearance of LH receptors on the granulosa cells. FSH also induces P450arom activity in the granulosa cells. These observations have given rise to the concept of the two-cell/two gonadotropin hypothesis. The LH-dependent androgens (androstenedione and testosterone) are derived from cholesterol in the theca cells. These are transferred across the basement membrane of the follicle to the granulosa cells where they are converted to the estrogens (estrone and estradiol) by the FSH-inducible granulosa cells (39). Progesterone synthesis in the follicle occurs predominantly in the granulosa cells.

The pituitary gonadotropins FSH and LH are required for estrogen synthesis and mainly LH for progesterone synthesis. As the follicle grows, the increased secretion of estradiol stimulates the expression of estradiol receptors, which stimulate granulosa cell proliferation as well as endometrial cell proliferation. Estradiol also affects the pituitary and hypothalamus. In the mature follicle, FSH together with estradiol stimulates an increase in LH receptors on granulosa cells. LH stimulates progesterone synthesis and secretion by the granulosa cells. This then stimulates FSH release at ovulation.

Table 4. Nomenclature of Enzymes and Genes Involved in Ovarian Steroidogenesis

Trivial name	Enzyme	Gene
Cholesterol side chain Cleavage enzyme	P450scc	CYP 11A1
17α-hydroxylase	P45017α	CYP 17
3β-Hydroxysteroid Dehydrogenase	3β-HSD	
17-20 lyase	P450 17α	CYP 17
Aromatase	P450arom	CYP 19

The LH surge at ovulation is critical to the synthesis of progesterone by the cells of the corpus luteum. The expression of steroid synthesizing enzymes changes as the cells luteinize. The enzymes, P450scc and 3β-HSD, increase in both granulosa and theca derived cells. P45017α is only expressed in the theca-derived cells and P450arom in the granulosa derived cells.

The synthesis of estradiol by the growing follicle is regulated both by endocrine (LH and FSH) and paracrine/autocrine factors. Thecal insulin-like growth factor-II (IGF-II) enhances LH stimulation of androgen synthesis in the theca cells as well as FSH-mediated aromatization in the granulosa cells. Theca transforming growth factor promotes the growth of granulosa cells and FSH induction of LH receptors. FSH stimulates the production of inhibin and activin in the granulosa cells and activin in turn increases by stimulating FSH receptor expression. Inhibin increases LH stimulation of androgen synthesis in the theca to serve as substrate for estrogen production in the granulosa cells whereas activin suppresses androgen synthesis. This paracrine effect of the inhibin and activin occurs mainly by changes in the expression of the steroidogenic enzymes.

Transport of Steroids

Estradiol and progesterone, secreted by the ovary into the blood, are transported mainly bound to proteins. Steroids are sparingly soluble in plasma. Sixty-five percent of the estradiol is bound to sex hormone-binding globulin (SHBG), 20% is bound to albumin, and the remaining 20% is found as free estradiol. Progesterone is bound mainly to transcortin and albumin.

Metabolism of Steroids

Estradiol and progesterone are inactivated in the liver, the metabolites are conjugated to sulfate or glucuronide and mainly excreted in the urine. The main metabolites of estradiol are estrone, estriol and catecholestrogens. The major metabolite of progesterone is pregnanediol. It is conjugated to glucuronic acid and excreted in the urine as glucuronide.

Mechanism of Action of Steroids

Estradiol and progesterone enter cells by diffusion and bind to nuclear receptors. This releases heat shock proteins (HSPs) and causes a conformational change in the receptor. This increases its DNA affinity and the ligand bound receptor associates with the hormone response element (HRE) of the DNA. It acts as a transcription factor to regulate gene expression (40). The actions of estradiol and progesterone on various target tissues are shown in Table 5.

Gonadotropins

Synthesis

FSH and LH are members of the glycoprotein hormone family that also includes thyroid-stimulating hormone (TSH) and chorionic gonadotropin (CG). These hormones are heterodimers consisting of a common α subunit and bound by noncovalent linkages. The α subunit gene is located on chromosome 6. The FSH β subunit protein is encoded by a single gene located close to several homologous CGβ genes and pseudogenes on chromosome 19. The specificity of the hormone receptor interaction is due to the β subunit.

Mechanism of Action

FSH and LH interact with target tissues via high affinity transmembrane receptors. The genes for these receptors have been mapped to chromosome 2p 16-p 21. Both receptors are glycoproteins linked to G proteins and adenylate cyclase. Both cAMP and calcium channels are important in gonadotropin signaling pathways (41).

THE ENDOMETRIAL CYCLE

In parallel with the ovarian cycle, the endometrium in the uterus undergoes cyclical changes in preparation to receive the fertilized ovum. The changes are associated with the action of the hormones secreted during the follicular and luteal phase of the cycle by the growing follicles and the corpus luteum. Thus the key hormones are estradiol and progesterone (42).

The uterine endometrial cycle is divided into three phases, the menstrual phase, the proliferative and secretory phase (43). During the adult reproductive life, a human female will experience this cycle 400 times. The endometrium is divided morphologically into the upper functionalis layer, which undergoes changes cyclically and is shed during menstruation. The lower basales remains constant through the cycle and regenerates the functionalis each cycle. The role of the functionalis is to prepare for implantation of the blastocyst.

180

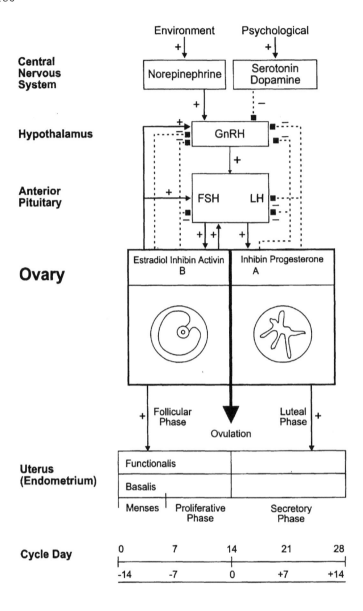

Figure 7. The interrelationship between the hypothalamus-pituitary-ovary components in the regulation of the feedback mechanisms (+positive feedback or - negative feedback). Note, during the early follicular phase, estradiol in low plasma concentrations has a negative feedback to the pituitary and hypothalamus, during the late follicular phase high plasma concentrations of estradiol have a positive feedback on the pituitary and hypothalamus. During the luteal phase high levels of progesterone and estradiol inhibit FSH and LH secretion. Low levels of progesterone and estradiol at the end of the cycle triggers secretion of FSH and stimulates the development of a cohort of follicles.

Table 5. Hormone dependant changes in the target tissues during the menstrual cycle.

Target Tissue	Estrogen	Progesterone
Fallopian Tube	Increase tubal muscle contractility Increases height and number of ciliated cells Stimulates tubal fluid secretion	Decrease tubal muscle contractility Decrease height and number of ciliated cells Atrophy and cell differentiation of epithelium.
Endometrium	Proliferation, increase thickness, size and number of glands Increase progesterone receptors	Secretory, secretion of glycogen into gland lumen Stromal edema
Myometrium	Increase size and number of muscle cells. Rhythmic contractions.	Inhibits contractions
Vagina	Increase epithelial thickness	Increase epithelial thickness
Cervix	Opens os, increase mucus Positive ferning, allows sperm penetration High spinnbarkeit	Close os Mucus scant, thick, no ferning Decreased spinnbarkeit and sperm penetration

The menstrual phase lasts 4-5 days and is the initial phase of the cycle. It occurs following the withdrawal of the ovarian hormones, estradiol and progesterone, as a consequence of corpus luteum regression. Prior to the onset of bleeding, cell death by apoptosis causes endometrial regression and vasoconstriction of the blood vessels. The arteries rupture following ischemia and blood is released into the uterus. The functionalis layer is completely shed. Arterial and venous blood, remains of endometrial stroma and glands, leukocytes and red blood cells are present in menstrual blood. The opposing actions of the vaso-constricting factor, prostaglandin PGF2α from the glandular cells, endothelin-1 from the stromal cells, and vasodilator action of the PGF2α affects the amount of blood loss.

The Proliferative Phase

The proliferative phase (also called the estrogenic phase) is associated with ovarian follicle growth and lasts about 9 days. It begins from the end of menstruation up to the time of ovulation. Increasing levels of estradiol, produced by the maturing follicles, induce growth of the endometrial glands and stromal connective tissue. The endometrial glands become straight and narrow and the spiral arteries become highly networked. The peak of activity occurs on day 8-10 of the cycle corresponding to the peak estradiol levels in the circulation and maximal estrogen receptor concentration in the endometrium. (40). Cytokines, a group of proteins that act by autocrine or paracrine mechanisms [interleukin-1β, interleukin-6 and interleukin-8, tumor necrosis factor (TNFα)] as well as interferons, the regulatory peptides with anti-viral and immunoregulatory activities (1FN-α, 1FN-β and 1FN-y), peptide growth factors (TGFα and β), epidermal growth factor, and insulin-like growth factors, androgenic factors, fibroblast growth factors (FGFs), and vascular endothelial growth factor (VEGF) play important roles in the regeneration of the endometrium during this phase. During this phase, the endometrium regenerates from 0.5mm to 3.5-5.0mm in height.

Secretory Phase

Following ovulation, the endometrium responds to both estrogen and progesterone secreted by the corpus luteum. This phase is also called the progestational phase and is of 13-14 days duration. The endometrial glands become highly secretory and the epithelium accumulates glycogen. The spiral arteries extend into the superficial layer of the endometrium. Peak secretory activity, which includes secretion of glycoproteins, peptides, plasma transudates and immunoglobulin, occurs at the expected time of blastocyst implantation during the mid-luteal phase. If fertilization does not occur by day 23 of the menstrual cycle, the corpus luteum begins to regress with a concomitant decrease in estradiol and progesterone levels. The endometrium involutes and by day 25-26, vasoconstriction of the arteries leads to ischemia, apoptosis of the functionalis, and menstruation follows.

CONCLUSION

The ovarian cycle is a continuous series of changing events, which is orchestrated by the follicle (Figure 8).

The main phenomenon continuously occurring from fetal life to the menopause is follicle atresia. This is controlled by genes regulating programmed cell dealth or apoptosis. FSH is the key anti-apoptotic survival hormone of the growing follicle. It facilitates, in each cycle, terminal growth of a selected follicle during approximately 400 cycles. LH allows the initiation and development of the corpus luteum and supports luteal function. Both the selected dominant follicle and the corpus luteum have a fixed life span, which determines the length of the cycle. It is clear that many areas need further investigation. These areas include: 1) What factors attract the stem cells to migrate to the gonadal ridge? 2) What regulates the continuous growth of primordial follicles to antral follicles until the menopause? 3) What is the mechanism of action of local ovarian modulators of the effects of gonadotropins?

SUMMARY

1) The germ cells that form the pool of recruitable follicles are laid down in the first few weeks of gestation. Cyclic ovarian activity that will recruit about 400 dominant follicles begins at puberty (Figure 8). 2) The follicles
within the ovary orchestrate the ovarian cycle. 3) Folliculogenesis is independent of FSH action before the preantral stage. FSH-dependent maturation only occurs at sexual maturation. Cyclic ovarian activity begins at puberty following the complete maturation of the hypothalamus-pituitary-ovary-axis. 4) As the follicle grows and matures, under the influence of GnRH from the hypothalamus that stimulates pituitary FSH, it secretes estrogens (estradiol) in increasing amounts. 5) Estradiol synthesis requires the close cooperation of the granulosa and theca cells of the follicle (two-cell

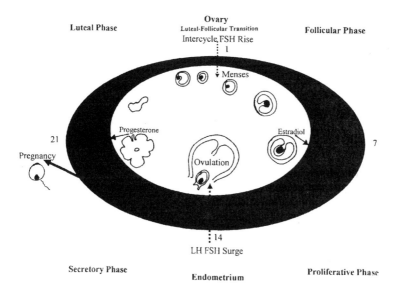

Figure 8. Summary of the interrelationship between the ovarian and menstrual cycle. The ovary functions as the site of gametogenesis and steroidogenesis. The follicle contains the developing ova surrounded by the predominant hormone producing cells (granulosa and theca). Estradiol, produced by the dominant follicle, prepares the follicle for ovulation and target tissue, endometrium, to receive the fertilized ovum at pregnancy. The corpus luteum secretes progesterone and prepares the endometrium for nidation. In the absence of pregnancy, the cycle continues with the decline in luteal hormones and the inter-cycle FSH surge triggering a new cohort of follicles.

theory). 6) The rising levels of estradiol exert a negative feedback on the pituitary to gradually lower the levels of FSH. 7) Estradiol also has a positive feedback on the hypothalamus and GnRH secretion. 8) Prior to ovulation, GnRH from the hypothalamus causes the pituitary to secrete a burst of LH. 9) The LH burst causes the maturing follicle to rupture, the ovum is released and a corpus luteum is formed in the ovary. 10) LH stimulates the corpus luteum to secrete progesterone and estrogen to prepare the uterine endometrium for implantation. 11) The increasing levels of progesterone/estradiol in the blood exert a negative feedback effect on the hypothalamus and pituitary causing LH levels to decrease. 12) The corpus luteum disintegrates as LH decreases. 13) If a pregnancy does not occur, the corpus luteum no longer produces estradiol/progesterone, the endometrium is not maintained and is shed. 14) If a pregnancy occurs, the developing embryo produces hCG, which rescues the corpus luteum. 15) With the low estradiol and progesterone at the end of the cycle, the hypothalamus is activated to secrete GnRH and a new cycle begins.

REFERENCES

1. Yeh J., Adashi E. Y. The Ovarian Cycle; Yen S. S. C., The Human Menstrual Cycle; Strauss J., III Coutifaris C., The Endometrium and Myometrium, Regulation and Dysfunction; In: Reproduction and Endocrinology: Physiology, Pathophysiology and Clinical Management. Ed Yen S. S. C., Jaffe R. B., Barbieri R. L. 4th Ed., Philadelphia: Saunders W. B., 1999.
2. Greenwald A. S., Roy S. K., Espey L. L. Follicular Development and its Control; Espey L. L., Lipner H., Ovulation; Niswender A. D., and Nett T. M. The Corpus Luteum and its Control in Infraprimate Species; Tsafriri A., Adashi E. Y., Local Nonsteroidal Regulation of Ovarian Function: Zeleznik A. J., Benyo D. F., Control of Follicular Development, Corpus Luteum Function and the Recognition of Pregnancy in Higher Primates; Hotchkiss J., Knobil E; The Menstrual Cycle and its Neuroendocrine Control. In: Physiology of Reproduction 2nd Ed., Eds. Knobil., Neill J. D., Greenwald A. S., Market C. L., Pfaff D. W. , N.Y.: Raven Press, 1994.
3. Speroff L., Glass R. H., Kase N.G. The Ovary-Embryology and Development. In: Clinical Gynecologic Endocrinology and Infertility, 6th Ed. Philadelphia: Lippincott Williams and Wilkins, , 1999.
4. Witschi E. Migration of the germ cells of human embryos from the yolk sac to the primitive gonadal folds. Contrib Embryol 1948; 32:69-80.
5. Bendel-Stenzel M.R., Gomperts, M., Anderson, R., Heasman., J., Wylie C. The role of cadherins during primordial germ cell migration and early gonad formation in the mouse. Mech Dev 2000; 91:143-152.
6. Parker K. L., Schedl A., Schimmer B. P. Gene interactions in gonadal development. Annu Rev Physiol 1999; 61:417-433.
7. Manotaya T., Potter E. L. Oocytes in prophase of meiosis from squash preparations of human fetal ovaries. Fertil Steril 1963; 14: 378-392.
8. Speed R. M. The prophase stages in human foetal oocyte studied by light and electron microscopy. Hum Genet 1985; 69: 69-75.
9. Ohno S., Klinger H. P., Atkin W. B. Human oogenesis. Cytogenetics 1962; 1:42-58.

10. Tsafriri A., Dekel N., Bar-Ami S. The role of oocyte maturation inhibitor in follicular regulation of oocyte maturation. J Reprod Fertil 1982; 64 541-551.

11. Gougeon A. , Regulation of ovarian follicular development in primates: Facts and hypothesis. Endocr Rev 1996; 17:121-155.

12. Hsueh A. J. W., Billig H., Tsafriri, A. Ovarian follicle atresia: A hormonally controlled apoptotic process. Endocr Rev 1994; 15:707-724.

13. Eppig J. J. Oocyte control of ovarian follicular development and function in mammals. Reproduction 2001; 122: 829-838.

14. Horie K., Takakura, K., Taii, S., Narimoto, K., Noda, Y., Nishikawa, S., Nakayama, H., Fujita, J., Mori, T. The expression of c-kit protein during oogenesis and early embryonic development. Biol Reprod 1991; 45, 547-552.

15. McGrath, S. A., Esquela A. F., Lee S. J. Oocyte-specific expression of growth differentiation factor–9. Mol Endocrinol 1995; 9:131-136.

16. Dong J., Albertini D. F., Nishimori, K., Kumar T. R., Lu N., Matzuk, M. Growth differentiation factor-9 is required during early ovarian folliculogenesis. Nature 1996; 383:531-535.

17. Oktay K., Briggs D., Gosden R. G. Ontogeny of follicle-stimulating hormone receptor gene expression in isolated human ovarian follicles. J Clin Endocrinol Metab. 1997; 82:3748-3751.

18. Abel M. H., Wootton A. N., Wilkins V., Huhtaniemi I., Knight P. G., Charlton H. M. The effect of a null initiation in the follicle stimulatory hormone receptor gene on mouse reproduction. Endocrinol 2000; 141:1795-1803.

19. Baker T. G., A quantitative and cytological study of germ cells in human ovaries. Proc R. Soc B 1963; 150:417-433.

20. Scaramuzzi R. J., Adams N.R., Baird D. T., Campbell B. T., Downing J.A., et al. A model for follicle selection and the determination of ovulation in the ewe. Reprod Fert Dev. 1993; 5:459-478.

21. McGee E. A., Hsueh A. J. W. Initial and Cyclic Recruitment of Ovarian Follicles. Endocr Rev 2000; 21(2): 200-214.

22. Findlay J. K. Factors affecting follicular development and maturation. In: Hormones and Women's Health. The Reproductive years. Ed. Salamonsen L. A., 2000.

23. Richards J. S. Gonadotropin – regulated Gene Expression in the Ovary. In: The Ovary, Eds. Adashi E. Y., Leung P. C. K. N.Y.: Raven Press Ltd., 1993.

24. Augustin H. G. Vascular Morphogenesis in the Ovary: Introduction and Overview. In: Vascular Morphogenesis in the Female Reproductive System. Eds. Augustin G. H., Iruela- Arise M. L., Rogers P. A. W., Smith S. K., Birkhauser.

25. Khan-Dawood F. S., Yang J., Dawood M. Y., Expression of gap junction protein connexin–43 in the human and baboon (Papio Anubis) corpus luteum. J Clin Endocrinol Metab 1996; 81: 835-842.

26. Khan-Dawood F. S. Oxytocin in intercellular communication in the corpus luteum, Semin in Reprod Endocrinol: Corpus Luteum 1997; 15(4):395-407.

27. Revelli A. Pacchioni D., Cassoni P. In situ hybridization study of messenger RNA for estrogen receptor and immuno histochemical detection of estrogen and progesterone receptors in the human ovary. Gynecol Endocrinol 1996; 10:177-186.

28. Burger H. G., Igarashi M. Inhibin: definition and nomenclature, including related substances. J Clin Endocrinol Metab 1988; 66:885-886.

29. Bukovsky A., Caudle M. R., Keenan J.A. Is corpus luteum an immune-mediated event? Localization of immune system components and luteinizing hormone receptor in human corpora lutea. Biol Reprod 1993; 53:1373-1384.

30. Carlson J. C., Wu S. M., Sawada M. Oxygen radicals and the control of ovarian corpus luteum function. Free Rad. Biol Med 1993; 14:78-84.

31. Fraser H. M., Lynn S. F., Cowen G. M., Illingworth P. J. Induced luteal regression in the primate: evidence for apoptosis and changes in c-myc protein. J Endocrinol 1995; 147:131-137.

32. Rodger F. E., Frazier H. M., Duncan W. C., Illingworth R. J. Immunolocalization of Bcl-2 in the human corpus luteum. Human Reprod 1995; 10:1566-1570.

33. Yoshimi T., Strott C. A., Marshall J. R., Lipsett M. B. Corpus luteum function in early pregnancy. J Clin Endocrinol Metab 1969; 29:225-230.

34. Avis N. E., McKinlay S. M. A longitudinal analysis of women's attitudes towards the menopause: results from the Massachusetts Women's' Health Study. Maturitas 1991; 13: 65-79.

35. McNauguton J., Bangal M., McCloud P., Hee J., Burger H. G. Age related changes in follicle stimulating hormone, luteinizing hormone, estradiol and immuno reactive inhibin in women of reproductive age. Clin Endocrinol 1992; 36:339-345.

36. Carr B. R., MacDonald P. C., Simpson E. R. The role of lipoproteins in the regulation of progesterone secretion by the human corpus luteum. Fertil Steril 1982; 38: 303-311.

37. Stocco D. M. Star protein and the regulation of steroid hormone biosynthesis. Annu Rev. Physiol 2001; 63:193-213.

38. Doody K. J., Lorence M. C., Mason J. I., Simpson E. R. Expression of messenger ribonucleic acid species encoding steroidogenic enzymes in human follicles and corpora lutea throughout the menstrual cycle. J. Clin Endocrinol Metab 1990; 70: 1041-1045.

39. Falck B. Site of production of estrogen in rat ovary as studied by micro-transplants. Acta Physiol Scand Suppl 1959; (163) Suppl 47:1-9.

40. O'Malley B. W., Tsai M-J. Molecular pathways of steroid receptor action. Biol Reprod 1992; 46:163-170.

41. Waterman M. R., Keeney D. S. Signal transduction pathways combining peptide hormones and steroidogenesis. Vitam Horm 1996; 52:129-148.

42. Bergeron C., Ferenczy A., Shyamala G. Distribution of estrogen receptors in various cell types of normal, hyperplastic and neoplastic human endometrial tissue. Lab Invest 1988; 58: 338-345.

43. Noyes R. W., Hertig A. W., Rock J. Dating the endometrial biopsy Fertil Steril 1950; 1:3-25.

Chapter 11

ESTROGEN-ASSOCIATED GLYCOPROTEINS IN OVIDUCT SECRETIONS: STRUCTURE AND EVIDENCE FOR A ROLE IN FERTILIZATION

Gary Killian
The Pennsylvania State University, University Park, Pennsylvania, USA

INTRODUCTION

The oviducts are tubular extensions of the uterus that end in close proximity to the ovaries. The lumen of the oviduct is continuous with that of the uterus and also opens to the peritoneal cavity near the ovary. This structural arrangement makes it possible for ovulated ova to enter the tubular organs of the female reproductive tract. Historically, as its name implies, the oviduct was recognized as a conduit to carry the egg from the ovary to the uterus, although the oviduct is now known to have many functional roles. Included among these are sperm, ovum and early embryo transport, and to provide a physiological environment for events surrounding fertilization and early embryo development. Because the oviduct is the site at which fertilization occurs, its influence can have a profound affect on the success of reproduction.

Anatomically, the oviduct is subdivided into regions that are generally associated with distinct functions. The infundibulum is located next to the ovary and is involved in the capture of freshly ovulated ova and directing them into the ostium or opening of the ampulla. Once in the ampulla, the ovum is transported toward the isthmus region. The isthmus region terminates at the uterus where sperm first enter the oviduct and are prepared or "capacitated" for fertilization. The timing of events associated with sperm and ova entering the oviduct is crucial so that the gametes are at their correct stages of preparation to optimize chances for fertilization. Evidence suggests that for several species sperm may be stored in the isthmus and other locations in the reproductive tract prior to the arrival of the egg and fertilization. For several days post fertilization, the isthmus also

188

provides a nurturing environment for the early embryo as it is transported to the uterus where it will complete development.

The oviduct is a fascinating organ to study because it has diverse functions involving sperm, ova, fertilization and embryos. Several methods have been used to study the oviduct and its secretions in order to better understand its various functions.

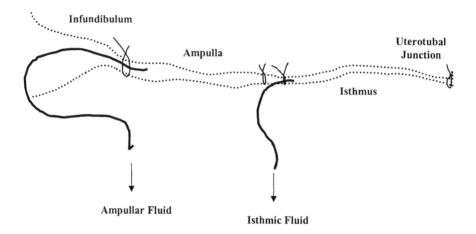

Figure 1. Schematic representation of the placement of catheters in the ampulla and isthmus regions of the cow oviduct for continuous recovery of oviductal fluid throughout the estrous cycle. Ligatures are used to hold the catheters in place and to seal off the ampulla at the ampullary-isthmic junction and the isthmus at the uterotubal junction so that secretions flow into the catheters.

Oviducts collected from animals at slaughter or from women undergoing hysterectomy have provided a source of tissue for study. One approach has been to "flush" the lumen with saline to recover its contents for analysis. Epithelial cell and tissue cultures have also been prepared from the excised oviducts and secretions have been harvested from the culture medium. Tissues collected during different stages of the reproductive cycle have enabled comparisons of the secretions obtained near ovulation with those collected during the luteal phase of the reproductive cycle.

A third approach used extensively by the author of this chapter involves surgical placement of indwelling catheters within the oviducts of live animals. This approach has been used for sheep, horses, cattle, pigs, rabbits and monkeys among other species. Typically the catheter is secured in place in the ostium of the ampulla with ligatures, and a ligature is also placed below the lower isthmus at the uterotubal junction (Fig. 1). This procedure creates a closed system so that as secretions accumulate in the oviduct lumen they pass through the catheter to the collection vials mounted on the outside of the animal. Samples of oviduct fluid collected daily can be correlated with serum hormone concentrations of estrogen and progesterone to establish the stage of the reproductive cycle at which the sample was collected. This approach was further modified to place catheters in both the ampullar and isthmic regions of the same oviduct to enable comparisons of regional oviduct fluid.

The purpose of this chapter is to describe what is known of the functional and biochemical characteristics of oviductal secretions as they relate to gamete preparation, fertilization and early embryo development in the oviduct.

OVIDUCT SECRETIONS

Fluid contained within the oviduct lumen is derived from two sources. The oviduct is well supplied with blood vessels and as blood perfuses the tissue capillaries water and some components from blood serum, ranging in size from ions to large proteins, pass through the tissue and enter the oviduct lumen by a process called transudation (1). Indications are that this process is selective since all components of serum do not find their way into oviduct fluid, or for a given component amounts may differ between the two fluids. Moreover, the protein concentration of oviduct fluid is only 5- 10% of that of serum. Serum contributions are combined with other components that are synthesized and secreted by the oviduct epithelium. As one might expect, many lipids and proteins present in oviduct fluid are also produced by other tissues of the body. However, there are some proteins that are only found in the oviduct. Regardless of the source of the various components, the resulting biological fluid is capable of facilitating a variety of functions involving gametes and embryos.

It is also apparent that the volume of oviduct fluid produced changes during the reproductive cycle. During estrus and ovulation, fluid production is greater than during the luteal phase of the cycle. In cattle for example, the volume of oviduct fluid produced within a 24-hour period near ovulation is typically 1-1.5 ml per oviduct compared with 0.1-0.2 ml during the luteal phase. These changes in fluid production are dependent on changes in circulating levels of serum progesterone and estrogen. Higher concentrations of estrogen and lower concentrations of progesterone favor greater fluid production by the oviduct.

To gain insights into how the environment created in the oviduct lumen is capable of influencing gamete and embryo function the composition of oviduct fluid has been investigated. Although much still remains to be defined, several excellent reviews are available on this topic (2-5). Ions such as calcium, potassium, sodium, magnesium and bicarbonate are present along with pyruvate, lactate and glucose. Concentrations of these components determined in oviduct fluid have served as a basis for formulation of synthetic media used in procedures for *in vitro* fertilization

190

and embryo culture. Cholesterol and several phospholipids are also present in oviduct fluid and evidence suggests that these are both derived from serum and synthesized by the epithelium.

Proteins are the major component of oviduct fluid. Serum-like proteins present include albumins, globulins and high-density lipoprotein. Other proteins include haptoglobin, osteopontin, precollagen and clusterin. Enzymes detected include phosholipase A2, lysozyme, peptidase, diesterase, amylase, β-N-acetylgalactoseaminidase, β-N-glucosaminidase, catalase, along with protease inhibitors TIMP-1 and PAI-1. An array of growth factors have also been detected in the oviduct including EGF, IGF-1 and –2, TGFa and FGF-1 and -2 (5).

Of particular interest is a prominent group of glycoproteins produced specifically by the oviduct called "oviductins" or estrogen-associated glycoproteins. Because maximum production of these proteins occurs during the peri-ovulatory period (Fig. 2) a functional role of the estrogen-associated glycoproteins (EAGs) in events surrounding fertilization and early embryo development is suggested.

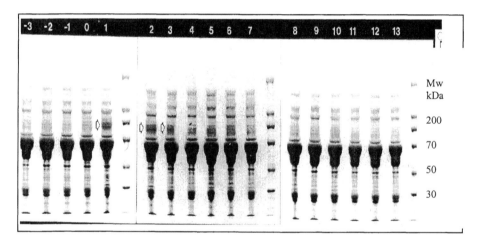

Figure 2. One dimensional protein electrophoresis of cow oviductal fluid collected daily during the estrous cycle. The EAG is most prominent on the day of ovulation (day 1) and 2-3 days thereafter. Adapted from (6).

Influence of Oviduct Fluid on the Gametes

With the development of methods to accomplish fertilization and early embryo development *in vitro* for a variety of species it has been debated whether the oviduct actually plays an essential role in reproduction under natural conditions. Aside from serving as a conduit and a site for the gametes to unite and the embryo to pass to the uterus, do oviduct secretions play an active role in modulating events associated with fertilization and early embryo development? In order to answer this question, studies have been conducted using oviduct fluid, medium conditioned by cultured oviduct tissue and components isolated from these sources to test their affects on the gametes *in vitro*. Selected examples of the results of these studies follow.

In early studies with fluid collected from a single catheter in oviduct ampulla the affect of pooled fluid collected near ovulation with that of the luteal phase on sperm capacitation and motility (7). Capacitation was assessed by the ability of sperm to undergo the acrosome reaction in response to lysophosphotidylcholine. This is a standard bioassay for assessing the readiness of bovine sperm to fertilize an ovum. At concentrations greater than 20%, both types of oviduct fluid capacitated sperm within 4 hours, but at a concentration of 60% oviduct fluid collected near ovulation capacitated sperm in 2 hours, and maintained motility of sperm longer than fluid collected during the luteal phase. Sperm capacitated by either type of oviduct fluid underwent the acrosome reaction when treated with soluble zona pellucida and were capable of penetrating bovine oocytes.

Fluids collected from indwelling catheters in the isthmic and ampullary oviductal regions have enabled studies to explore if functional differences, typically attributed to oviduct regions, were possible to demonstrate with oviduct fluid. Fluid collected daily was pooled according to stage of the cycle (peri-ovulatory or luteal) and region of the oviduct. Ejaculated bull sperm were incubated either in peri-ovulatory isthmic or ampullary fluid, or luteal isthmic or ampullar fluid for up to 6 hours and evaluated for viability, motility, capacitation and fertilizing ability (8). Although no differences in viability were observed for sperm incubated in the different fluid types, the percentage motile sperm in isthmic fluid collected near ovulation was significantly lower than that of the other incubations at 4 and 6h (Fig. 3). These observations suggest that isthmic oviduct fluid inhibits sperm motility and supports the notion that the isthmus functions to store sperm in a quiescent state (8). Incubation in isthmic fluid also increased the number of sperm undergoing capacitation compared with ampullar fluid obtained near ovulation (9). The picture emerging from studies with oviduct fluid is that the isthmic region of the oviduct serves as a sperm reservoir at the time of ovulation and supplies a population of capacitated sperm ready for fertilization. This conclusion is in agreement with previously held beliefs about the function of the oviduct isthmus and demonstrates a regulatory role of oviduct secretions in modifying sperm function.

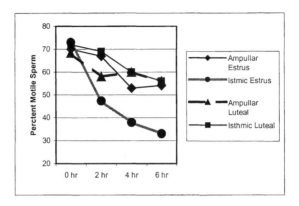

Figure 3. Motility of sperm in fluid recovered from either the isthmic or ampullary regions of the oviduct near ovulation (estrus) or during the luteal phase of the estrous.

Oviduct fluid (ODF) collected near ovulation also affects the ability of spermatozoa to interact with the ovum. The first contact between sperm and the ovum involves binding of sperm to the zona pellucida, the glycoprotein-rich outer most covering of the ovum. Interestingly, more sperm bind to the ovum if they are pre-incubated in isthmic compared to ampullary periovulatory oviduct fluid (Fig. 4). However, spermatozoa

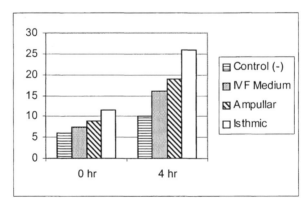

Figure 4. Number of sperm bound per oocyte following capacitation of sperm in either ampullar or isthmic oviductal fluid collected near ovulation or in standard IVF medium. Sperm in control medium were not capacitated. Adapted from (9).

incubated in ampullary fluid have greater fertilization rates than sperm prepared in other treatments (Fig.5). Therefore, sperm appear to be prepared to penetrate the ovum to the greatest degree if they are exposed to both isthmic and ampullary secretions of the oviduct at the time of ovulation.

While these studies clearly demonstrate how sperm are affected by ODF, it is difficult to know whether the outcome would be the same *in vivo* when interactions with the ovum are involved. Ideally, *in vitro* studies assessing the affects of ODF on the interaction of sperm and ovum at fertilization should also be done to evaluate how treatment of the ovum influences the results. Indeed, pre-treatment of sperm *and* ova with ampullary or isthmic ODF collected near ovulation affects sperm-ovum binding and fertilization differently depending on the fluid used (Fig. 6, Table 1). Region of the oviduct from which the fluid is derived influences both sperm-ovum binding and fertilization rates (10). If we assume that during the normal course of events *in vivo*, ova are first exposed to ampullary secretions and sperm are first exposed to isthmic secretions, then the *in vitro* results in Figure 6 and Table 1 are particularly relevant. Concerning sperm-egg binding (Fig. 6), if both gametes are pre-incubated in either isthmic or ampullary nonluteal fluids, binding rates are significantly greater than if sperm are incubated in isthmic and ova in ampullary fluids. In this case, however, more may not be better, since reducing the number of sperm bound to the zona may serve to reduce the incidence of polyspermy, a lethal condition where more than one sperm penetrates the ovum. Physiologically, the consequences of this may be important since it is known that decreased time to first cleavage from insemination for *in vitro*-produced bovine embryos has a major positive influence on the probability of the embryo developing to blastocyst stage (11).

In addition to studies showing that oviduct fluid has the ability to influence fertilization *in vitro*, it is also noteworthy that some of the proteins in oviduct fluid become associated with sperm and ova. This can be demonstrated by labeling proteins in oviduct fluid either by iodination or biotinylation so that oviduct proteins that bind to the gametes can be distinguished from those that are normally present.

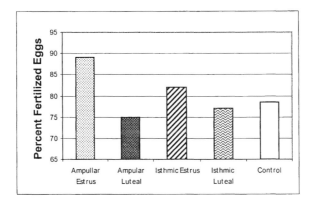

Figure 5. Percentage of ova fertilized when sperm are capacitated in either ampullary or isthmic oviductal fluid obtained near ovulation or during the luteal phase of the estrous cycle. Control represents sperm prepared in standard IVF medium. Adapted from (8).

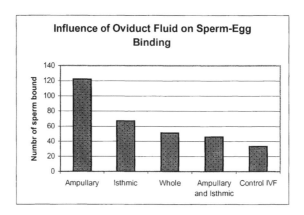

Figure 6. Average number of sperm bound to the zona pellucida when both gametes were incubated in ovulatory oviduct fluid collected either from the ampulla, isthmus or the whole oviduct; or when ova were incubated in ampullary and sperm were incubated in isthmic oviduct fluid. Control incubations were standard IVF condtions. Adapted from (10).

By isolating the membranes from sperm or the zonae pellucidae from ova and visualizing the proteins present using protein electrophoresis and Western blotting techniques it is possible to identify which proteins in oviduct fluid are taken up by the gametes (Figs. 7, 8). For sperm, eight proteins of apparent Mr of 97, 75, 66, 55, 48, 34, 28 and 24 kDa were consistently detected (Fig. 7) while for ova 95, 80, 74, 60, 45 and 30 kDa proteins were detected (Fig. 8).

Table 1. Percentage fertilized bovine oocytes after incubation of both spermatozoa and oocytes in nonluteal oviductal fluid (ODF)*

	Both gametes in:			Spermatozoa in ODF: isthmic oocytes in ampullar ODF	Spermatozoa in heparin: oocytes in medium
Time (h)	Ampullar ODF	Isthmic ODF	Whole ODF		
14	62 ± 10^a	64 ± 7^{ab}	47 ± 8^c	76 ± 8^b	49 ± 10^c
16	66 ± 13^a	72 ± 8^{ab}	57 ± 7^c	75 ± 5^b	51 ± 9^c
18	62 ± 6^a	73 ± 7^{ab}	82 ± 9^b	85 ± 10^b	78 ± 4^b

*Results are expressed as means ± SEM of oocytes fertilized by spermatozoa when both gametes were preincubated in nonluteal ampullary (AODF), isthmic (IODF) or whole (WODF) ODF before insemination. 96-104 oocytes per treatment. Modified from (10).
[a,b,c] Values with different superscripts within rows are significantly different (P<0.05).

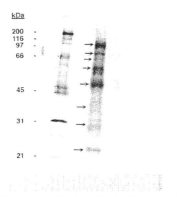
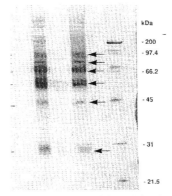

Figure 7. Western blot of biotinylated proteins in bovine oviductal fluid collected near ovulation that associated with the sperm membrane. Modified from (12).

Figure 8. Western blot of biotinylated proteins in bovine oviductal fluid that associated with the zona pellucida of the ovum. Modified from (13).

Estrogen-Associated Glycoproteins

Among the most prominent proteins in oviduct fluid at the time of ovulation are the estrogen-associated glycoproteins (EAG). These or related proteins have been found in all mammalian species studied to date. They are of particular interest because they have been shown to bind to the gametes and early embryo. Partial or complete cDNA sequences obtained for the proteins show that they are well conserved among species and that EAG has sequence similarity to chitinase, an enzyme that catalyzes chitin (14, 15). Chitin is a glycoprotein found in the exoskeleton of invertebrates. Although EAG does not have chitinase activity because it lacks certain key amino acids in its sequence, it has been suggested that the chitinase-like active site of EAG may retain sugar-binding properties that may be important for binding EAG to the gametes or embryo.

EAG also exhibit several properties similar to the super-family of proteins called mucins (16, 17). Mucins are high molecular weight glycoproteins containing over 50% O-linked carbohydrate by weight. The functional significance of the similarity between EAGs and mucins is currently being explored, but possibilities are that they may coat the gametes and embryo or the epithelium creating a negatively charged protective barrier resistant to digestion by proteases or other adverse factors in the environment.

Carbohydrate analyses of EAG have shown the presence of both N-linked and O-linked glycosylation sites, and for several species the presence of several isoforms of EAG with differing degrees of glycosylation.

Galactose, mannose, N-acetylglucosamine, N-acetylgalactosamine, N-acetylneuraminic acid and fucose are among other terminal sugars that have been detected (18, 19). The functional significance of the terminal sugars associated with EAG is yet to be demonstrated, but the importance of carbohydrates and glycosylation in creating molecular surfaces that play a key role in cell-cell recognition and signaling is well known. For example, it is known that mannose receptors on the human sperm surface have been linked with induction of the acrosome reaction and that carbohydrate-binding proteins on the sperm surface appear to be involved with binding sperm to terminal carbohydrates on the zona pellucida of the ovum (20). It is also apparent that glycoconjugates play an important role in binding sperm to the oviduct epithelium to form a sperm reservoir (21).

Functional Studies with EAG

While there is little doubt that oviductal secretions affect gamete function and embryo development, few studies have demonstrated that a specific protein in oviduct fluid influences these events. Because it has been shown that EAG binds to sperm (22, 23), ova (24, 25) and embryos (25, 26) in several species, EAG has received particular attention as a potential molecule in oviduct fluid that may influence gamete or embryo function. Several studies using preparations enriched in EAG have demonstrated a variety of effects on sperm and egg function and embryo development (Table 2).

Bovine oocytes pretreated with EAG have improved fertilization rates (27, 28) even though they bind fewer sperm than controls (27). Similarly, porcine oocytes treated with EAG bind fewer sperm, but embryo production rates are improved (29). Further, evidence from sheep suggests that cleavage rates and developmental competence of oocytes (30, 31) are improved when fertilization occurs in the presence of EAG.

Taken together, these results indicate that EAG is taken up by the ovum and may mediate a sperm selection process and/or a mechanism to reduce the incidence of polyspermy. Direct and indirect affects on the embryo are also apparent as rates of blastocyst development are improved in the presence of EAG.

Bovine EAG has been shown to maintain viability and motility of sperm *in vitro* (23). However, this effect is reduced when EAG is enzymatically deglycosylated (32). This suggests that the motility sustaining effect of EAG is dependent on the type or degree of glycosylation, and perhaps related to the negative charges of the sialic acid moieties (32). This finding is not surprising as there is considerable evidence documenting a significant role of carbohydrates in gamete function and fertilization (see reviews 33-35).

Table 2. Observations and functions affected when sperm, ova or embryos are treated with estrogen-associated glycoproteins.

Sperm	Ovum	Embryo
Binds to sperm surface (22, 23)	Binds to zona pellucida (24, 25) and vitelline membrane	Associates with zona pellucida and blastomeres (24- 26)
Maintains motility and viability (23)	Reduces rates of sperm binding (27, 29)	Facilitates embryo development (30, 31)
Stimulates capacitation (36)	Increases fertilization rates (27-29)	
Increases fertilization rates (36)	Increases development potential (30, 31)	
Improved development (36) of embryos derived from	Increases cleavage (30, 31)	
EAG treated sperm	Increases blastocysts from ova treated with EAG (27-29)	

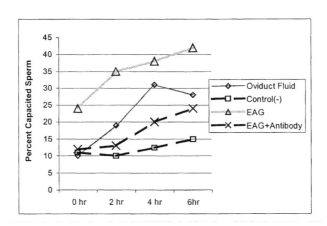

Figure 9. Comparison of rates of bovine sperm capacitation in peri-ovulatory oviductal fluid, purified EAG or in medium containing EAG and antibody to EAG. Modified from (36).

198

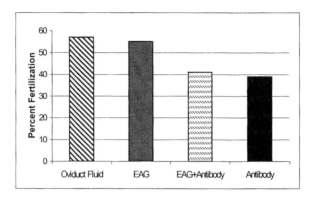

Figure 10. Comparison of rates of fertilization from bovine sperm incubated in peri-ovulatory oviductal fluid, purified EAG or in medium containing EAG and antibody to EAG. Modified from (36).

Because bovine EAG is the most prevalent protein in oviduct fluid at estrus and ovulation, and it becomes associated with sperm it is reasonable to ask if EAG influences sperm capacitation and their ability to penetrate ova. It has been shown that physiological concentrations of EAG do have the ability since treated sperm were functionally capacitated within minutes of the initial treatment to a level not seen in oviduct fluid or standard *in vitro* fertilization conditions for sperm until 3-4 hr incubation. It was also shown experimentally that using an antibody to EAG to "inactivate" EAG in oviduct fluid inhibited the ability of oviduct fluid to stimulate capacitation and fertilization. These studies clearly indicate that EAG facilitates sperm capacitation and fertilizing ability and that its biological activity is similar to that seen for oviduct fluid.

CONCLUSIONS

The oviduct is a dynamic organ that is intimately involved with successful reproduction. Oviduct secretions modulate sperm and ovum functions to influence sperm capacitation, sperm-ovum binding, fertilization and early embryo development. EAG are a unique group of glycoproteins secreted by the oviduct that are able to influence sperm and ovum function and the success of fertilization and early embryo development.

ACKNOWLEDGEMENTS

These studies were supported by the United States Department of Agriculture National Research Initiative Competitive Grants Program. Appreciation is expressed to Linda Killian for editorial assistance in the preparation of this chapter.

REFERENCES

1. Leese, H.J. The formation and function of oviduct fluid. J Reprod Fert 1988; 82:843-856.
2. Hunter, R.H F. ed. The Fallopian Tubes, New York: Springer-Verlag, 1988.
3. Hamner, C.E. "Oviductal Fluid – Composition and Physiology". In *Handbook of Physiology, Vol. 2, Part 2,* Female Reproductive Physiology. Greep, R., Astwood, E.B eds. Baltimore, MD: Waverly Press, Inc., 1973, 141-151.
4. Brackett, B.G., Mastroianni, L., Jr. "Composition of Oviductal Fluid". In *The Oviduct and Its Functions.* Johnson, A.D., Foley, C.W. eds. New York: Academic Press. 1974, 133-160.
5. Buhi, W.C., Alvarez, I.M., Koube, A.J. Secreted proteins of the oviduct. Cells Tiss. Org. 2000; 166:165-179.
6. Gerena, R.L., Killian, G.J. Electrophoretic characterization of proteins in oviduct fluid of cows during the estrous cycle. J. Exp. Zool. 1990; 256:113-120.
7. McNutt, T.L., Olds-Clark, P., Way, A.L., Suarez, S.S., Killian, G.J. Effect of follicular or oviductal fluids on movement characteristics of bovine sperm during capacitation *in vitro*. J. Androl. 1994; 15:328-336.
8. Grippo, A.A., Way, A.L. Killian, G.J. Effect of bovine ampullary and isthmic oviductal fluid on motility, acrosome reaction and fertility of bull spermatozoa. J. Reprod. Fert. 1995; 105:57-64.
9. Topper, E.K., Killian, G.J., Way, A., Engel, B., Woelders, H. Influence of capacitation and fluids from the male and female genital tract on zona binding ability of bovine sperm. J. Reprod. Fertil. 1999; 115:175-183.
10. Way, A.L., Schuler, A.M., Killian, G.J. Influence of bovine ampullary and isthmic oviductal fluid on bovine sperm-egg binding and fertilization *in vitro*. J. Reprod. Fert. 1997; 109:95-101.
11. Lonergan, P., Khatir, H., Piumi, F., Rieger, D., Humbolt, P., Boland, M.P. Effect of time interval from insemination to first cleavage on the development characteristics, sex ratio and pregnancy rate after transfer of bovine embryos. J. Reprod. Fert. 1999; 117:159-167.
12. Rodriguez, C.M., Killian, G.J. Identification of ampullar and isthmic oviductal fluid proteins that associate with the bovine sperm membrane. Anim. Reprod. Sci. 1998; 54:1-12.
13. Staros, A.L., Killian, G.J. Association *in vitro* of six oviductal proteins associated with the bovine zona pellucida *in vitro*. J. Reprod. Fert. 1998; 112:131-137.
14. Sendai, Y., Abe, H., Kikuchi, M., Sattott, T., Hoshi, H. Purification and molecular cloning of bovine oviduct-specific glycoprotein. Biol. Reprod. 1994; 50:927-934.
15. Arias, E.B., Verhage, H.G., Jaffe, R.C. Complementary deoxyibonucleic acid cloning and molecular characterization of an estrogen-dependent human oviductal glycoprotein. Biol. Reprod. 1994; 51:685-694.

16. Malette, B., Paquette, Y., Merlen Y., Bleau, G. Oviductins possess chitinase- mucin-like domains: a lead in the search for the biological function of these oviduct-specific ZP-associating glycoproteins. Molec. Reprod. and Develop. 1995; 41:384-397.

17. Malette, B., Paquette, Y., Bleau, G. Size variations in the mucin-type domain of hamster oviductin: identification of the polypeptide precursors and characterization of their biosynthetic maturation. Biol. Reprod. 1995; 53:1311-1323.

18. Vieira, E.G., Chapman D.A., Killian, G.J. Terminal carbohydrate characterization of multiple forms of estrus-associated protein in bovine oviductal fluid using lectin probes. Biol. Reprod. Suppl. 1. 2000; 62:262..

19. Satoh, T. Abe, H. Sendai, Y. Iwata, H., Hoshi, H. Biochemical characterization of a bovine oviduct specific sialo-glycoprotein that sustains sperm motility *in vitro*. Biochemica Biophysica Acta. 1995; 1266:117-123.

20. Benoff, S., Hurley, I.R., Mandel, F.S., Cooper, G.W., Hershlag, A. Induction of the human sperm acrosome reaction with mannose-containing neoglycoprotein ligands. Molec. Hum. Reprod. 1997; 3:827-837.

21. Suarez, S.S. The oviductal sperm reservoir in mammals: mechanisms of formation. Biol Reprod. 1998; 58:1105-1107.

22. King, R.S., Killian, G.J. Purification of bovine estrus-associated protein and localization of binding on sperm. Biol. Reprod. 1994; 51:34-42.

23. Abe, H., Sendai, Y., Satoh, T., Hoshi, H. Bovine oviduct-specific glycoprotein: a potent factor for maintenance of viability and motility of bovine spermatozoa in vitro. Molec. Reprod. And Devel. 1995; 42:226-232.

24. Wegner, C.C., Killian, G.J. *In vitro* and *in vivo* association of an oviduct estrus associated protein with bovine zona pellucida. Mol. Reprod. Develop. 1991, 29:77-84.

25. Boice, M.L., Mavrogionis, P.A., Murphy, C.N., Prather, R.S., Day, B.N. Immunocytological analysis of the association of bovine oviduct-specific glycoproteins with early embryos. J. Exp. Zool. 1992; 263:225-229.

26. Gandolfi, F., Modina, S., Brevini, T.A.L., Galli, C., Moore, R.M., Lauria, A. Oviduct ampullary epithelium contributes a glycoprotein to the zona pellucida, perivitelline space and blastomeres membrane of sheep embryos. Eur. J. Basic Appl. Histochem. 1991; 35:383-392.

27. Staros, A.L., Killian, G.J. Improvement of *in vitro* fertilization rates following treatment of oocyte with bovine estrus-associated protein (EAP). Biol. Reprod. Suppl. 1. 1997; 56:86.

28. Martus, N.S., Verhage, H.G., Mavrogianis, P.A., Thibodeaux, J.K. Enhancement of bovine oocytes fertilization *in vitro* with a bovine oviductal specific glycoprotein. J. Reprod. Fert. 1998; 113:323-329.

29. Kouba, A.J., Abeydecra, L.R., Alvarez, I.M., Day, B.N., Buhi, W.C. Effects of the porcine oviduct specific glycoprotein on fertilization, polyspermy, and embryonic development *in vitro*. Biol. Reprod. 2000; 63:242-250.

30. Nancarrow, C.D., Hill, J.L. Co-culture, oviduct secretion and the function of oviduct-specific glycoproteins. Cell. Biol. Int., 1994; 18:1105-1114.

31. Hill, J.L., Wade, M.G., Nancarrow, C.D., Kelleher, D.L., Boland, M.P. Influence of ovine oviducal amino acid concentrations and an ovine oestrus-associated glycoprotein on development and viability of bovine embryos. Mol. Reprod. Dev. 1997; 47:164-169.

32. Satoh, T., Abe, H., Sendai, Y., Iwata, H., Hoshi, H. Biochemical characterization of a bovine oviduct specific sialo-glycoprotein that sustains sperm motility *in vitro*. Biochemica Biophysica Acta. 1995; 1266:117-123.

33. Tulsiani, D.R.P., Yoshida-Komiya, H., Araki, Y. Mammalian fertilization, a carbohydrate-mediated event. Biol. Reprod. 1997; 57: 487-494.

34. Zara, J. and Naz, R. The role of carbohydrates in mammalian sperm-egg interactions. How important are carbohydrate epitopes? Front. Biosci. 1998; 3:1028-1038.

35. Topfer-Petersen, E. Carbohydrate-based interactions on the route of a spermatozoon to fertilization. Hum. Reprod. Update 1999; 5:314-329.36.
36. King, R.S., Anderson, S.H., Killian, G.J.   Effect of bovine oviductal estrus-associated protein on the ability of sperm to capacitate and fertilize oocytes.   J. Androl. 1994; 15:468-478.

Chapter 12

STRUCTURE AND FUNCTION OF MAMMALIAN ZONAE PELLUCIDAE

Sarvamangala V. Prasad[1], Gautam Kaul[2], and Bonnie S. Dunbar[1]
[1]Baylor College of Medicine, Houston, Texas USA; [2]National Dairy Research Institute, Karnal, India

INTRODUCTION

The zona pellucida (ZP) is the unique extracellular matrix surrounding the mammalian oocyte. During the fertilization process, sperm must first bind to and penetrate the zona before it comes into contact with its plasma membrane and fuses with the oocyte. While the association of the sperm with the ZP is thought to be a species-specific event in some species, this specificity has been shown to be limited in other mammalian species. After binding, the sperm must penetrate the ZP matrix through limited proteolytic digestion by sperm enzymes. The ZP of some species has also been shown to be modified following fertilization resulting in a block to polyspermy, a process referred to as the "zona reaction". The zona reaction has been attributed to chemical modifications of the ZP presumably due to the release of components from cortical granules initiated by fertilization (1). The ZP matrix remains intact after fertilization in order to support the cleaving cells of the blastocyst. It has also been thought to aid in the movement of embryo in the oviduct, and to ensure proper embryonic development as well as to prevent embryo fusion in the oviduct.

FORMATION OF ZP DURING FOLLICULOGENESIS

The roles of ZP proteins in fertilization and implantation processes are well established. There is, however, increasing evidence that these unique glycoproteins are involved in ovarian follicular development. In mammals, the recruitment of oocytes from meiotic arrest signals the beginning of folliculogenesis. During early stage of folliculogenesis, the primordial follicle consists of an oocyte surrounded by a single layer of squamous granulosa cells

(GCs) which subsequently differentiate into cuboidal epithelial cells to form the primary follicle.

In a highly complex and regulated process the oocyte grows and matures while the GC multiples and differentiates to become a mature preovulatory follicle or the follicle becomes atretic. At this stage of development, two morphologically distinct populations of granulosa cells can be visualized (2). There are two proposed origins of granulosa cells, the ovarian surface epithelium and the ovarian rete. Although neither cellular source has been unequivocally proven to give rise to GC, studies with neonatal rats suggests that both of these structures may give rise to the GC in ovarian follicle (2).

It is during the early stages of follicular development that the ZP proteins are synthesized and secreted between the oocyte and GCs as the extracellular matrix is elaborated (3-5). The cell plasma membranes fuse with the oocyte plasma membrane to form tight junctional complexes and remain in contact as the ZP proteins are secreted around the GC cellular processes. It is this process that leaves a fenestrated matrix as the GC process retract from the oocyte just prior to ovulation (5).

The oocyte also influences the growth and development of the surrounding GC by paracrine factors thus establishing a close association between the oocyte and GC during follicular development (4-6). The formation of the ZP matrix is developmentally regulated, since the synthesis and secretion of the different ZP proteins occurs at specific stages of oogenesis during follicular differentiation (3-4, 7-8). Additional studies have also been carried out to demonstrate that immunization with specific ZP proteins will alter normal ovarian follicular development (9).

Morphological Properties of ZP

The single most distinctive morphological feature of the ZP of different mammalian species is that of relative size. The thickness of the ZP matrices varies in size from 1-2 microns in the opossum, 5 microns in the mouse, 13 microns in the human to 27 microns in the cow (3, 10). More recently, it has been demonstrated that the thickness of ZP of human eggs is a strong predictor of pregnancy outcome in IVF treatment (11). Early studies using light and electron microscopic methods demonstrated that ZP matrix is a relatively amorphous structure. Scanning electron microscopy has been used to show that the outer layer of the ZP appears to be fenestrated while the inner surface has a smoother and less distinct appearance (7, 10). However, the surface structure from different mammals exhibits heterogeneity. Depending on the stage of the oocyte, this structure may vary from a porous net-like structure to a smooth and compact surface (12).

Histochemical methods have further been used to study the molecular heterogeneity of ZP matrices. Ruthenium red staining reveals two layers in the mouse ZP while in the rat, a slight variation in the outer and inner layer is visualized and no staining is observed in the human (3, 13). Histochemical analyses using plant lectins and monoclonal antibody that discriminate the carbohydrate moieties of the ZP glycoproteins have also revealed distinct staining patterns among different species. These studies demonstrate that there is a uniform binding of some lectins throughout the ZP while other lectins localize within discrete domains of the matrix (3, 13). Using a monoclonal antibody (PS1) specific to lactosaminoglycan type carbohydrate moiety present on the ZP of most species, Dunbar et al (14) have demonstrated that the ZP matrix exhibits distinct concentric rings surrounding the oocyte while the untreated ZP appears to be smooth, amorphous structures as illustrated in Figure 1. The appearance of these

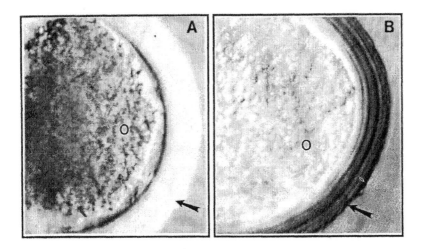

Figure 1. Morphological analysis of pig ZP using PS1mAb. A: Typical smooth and amorphous appearance of pig ZP as observed by modular contrast microscopy. B: Modular contrast microscopic analysis of Pig ZP treated for one hour at 25° C with PS1mAb at 1.0 mg/ml. The binding of the PS1mAb to the ZP gives concentric "rings" or layers in the ZP. O = Oocyte; Arrow Shows ZP. Figure modified from Dunbar et al (14).

rings suggests that there is a specific structural organization of the PS1 determinant within the ZP matrix. The number of rings appears to vary among individual ZP and among species. Regardless of the fixative used, this antibody did not detect the carbohydrate determinant in the oocyte or in the ZP of mice ovaries (Prasad and Dunbar unpublished observations). Collectively, these studies demonstrate that the molecules of the ZP are highly organized within the matrix. This organization is likely the reason for the unique angular trajectory of sperm through the ZP matrix as has been reported for different mammalian species (1).

Physicochemical Properties of ZP

The ZP matrix is permeable to a variety of molecules, including IgG, IgM, and ferritin although these properties vary among species. It is believed that the ability of molecules to pass through the matrix does not depend on the molecular weight but rather on the configuration and/or charge properties of the molecules. Recently, Turner and Horobin (15) have used molecular probes of diverse molecular structure to demonstrate that most biologically active molecules and metabolites can pass through the ZP. Studies in mice have shown that viruses can penetrate the pores in the ZP and manifest its infection. This has serious implications in sexually transmitted diseases (STD) and infertility since STD-associated genital infections may cause permanent damage to the reproductive tract resulting in infertility. Viral sexually transmitted infections are also a major health problem associated with considerable morbidity. Genital herpes has been reported to be the second leading cause of STDs in women (16). Recent evidence suggests that sexually transmitted diseases enhance the transmission of human immunodeficiency virus (HIV) type 1 and therefore prevention of one is dependent on the prevention of the other (17).

A variety of parameters have been shown to influence the solubility of intact ZP as evaluated by light microscopic and biochemical analyses. The structural dissolution of ZP (microscopic evaluation) also varies among species with respect to their susceptibility to treatments such as acid, base, heat, or enzymes including proteolytic enzymes. The ZP of mice and rats are not only morphologically smaller but are more susceptible to chemical and enzymatic dissolution than ZP from other species such as pig, rabbit, human, or cow (7, 10).

Proteins of the ZP

The ZP matrix of all mammalian species studied to date are comprised of three major glycoproteins which exhibit considerable heterogeneity both in charge and molecular weight primarily due to extensive post-translational modifications including glycosylation and sulfation. Because of the extreme variation in these modifications, the nomenclature for classifying ZP proteins of different mammalian species based on their electrophoretic mobility has been confusing and inconsistent (10). While the mouse ZP can easily be resolved into three proteins by non-reducing one-dimensional sodium dodecyl sulfate polyacrylamide gel electrophoresis (SDS-PAGE), the ZP glycoproteins of most other mammalian species cannot be resolved unless high resolution two-dimensional SDS-PAGE (2D-PAGE) is used (7) as shown in Figure 2.

Deglycosylation of these proteins results in the improved electrophoretic resolution of individual proteins into more discrete protein bands (Figure 3; (7,10)).

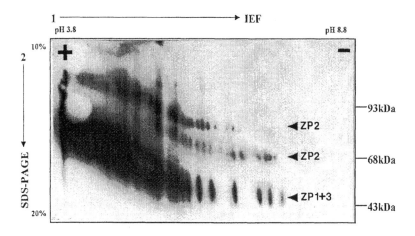

Figure 2. High resolution two-dimensional SDS-PAGE immunoblot analysis of pig ZP. Heat solubilized pig ZP was separated on 2-D SDS-PAGE (1D- charge and 2D- molecular weight) and the proteins were electrophoretically transferred to PVDF membrane. The blot was probed with antibodies to pig ZP and the signal was detected with [125]I Protein A. The immunoblot shows the charge and molecular weight heterogeneity of the ZP proteins. Arrows indicate the migration of the three ZP proteins. Figure modified from Dunbar et al (14).

Molecular Analysis of ZP Proteins

With the isolation and sequencing of a number of ZP cDNA clones from different species, it has now been possible to classify these proteins into three major glycoprotein families, ZP1, ZP2, and ZP3 (mouse ZP nomenclature), based on their amino acid sequence similarities. The proteins of each of these three families range from 50-98% in similarity at the nucleic acid level (2, 7, 19). The genes encoding the mouse and human ZP proteins are regulated by cis-acting sequences at the 5' end and several putative transcription factors (20, 21). Although the relative molecular weights from different species exhibit variation, within each family there is considerable conservation in the number of amino acid residues, in the number and position of cysteine residues, in the number of potential N-linked glycosylation sites and in the intron/exon organization (2, 19).

208

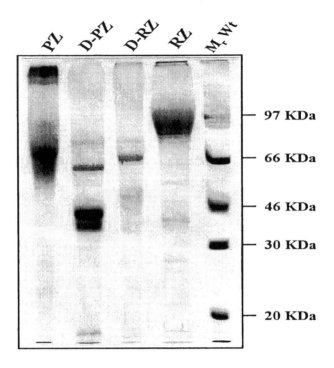

Figure 3. One-dimensional SDS-PAGE immunoblot analysis of pig and rabbit ZP proteins demonstrating heterogeneity of the ZP proteins. Native and deglycosylated pig and rabbit ZP were separated on a 7.5% SDS-PAGE and the proteins were electrophoretically transferred to PVDF membrane. The blot was probed with antibodies to pig ZP and signal was detected with [125]I Protein A. Native pig ZP (PZ) and rabbit ZP (RZ) cannot be resolved into individual components unless they are deglycosylated (chemical deglycosylation) as shown in lanes DPZ and DRZ. MrWt=Molecular Weight Markers.

Recently, it has been suggested that the cysteine residues in the ZP of unfertilized pig eggs are involved in the formation of disulfide linkages which together with specific proteolysis result in the hardening of the zona after fertilization (22). All three ZP protein families share common structural motifs. These include the presence of a ZP module that consists of 260 amino acid sequence with 8 conserved cysteine residues, a N-terminal hydrophobic signal sequence, a potential tetra basic furin processing site and a C-terminal hydrophobic trans-membrane domain 34-47 amino acid down stream of the furin cleavage site (19). The functions of the furin clevage site and trans-membrane domain have not yet been defined for all three proteins.

ZP1 Protein Family

The cDNA encoding the ZP1 protein has been isolated and sequenced from a number of species including human (Table 1). The ZP1 proteins contain an additional domain called the trefoil domain that is not present in the other ZP protein families. This domain is believed to lend resistance to proteolysis, supporting the observation that the pig ZP1 and ZP3 proteins are more resistant to proteolysis than the ZP2 protein (2, 19). Of all the species of ZP1 proteins sequenced, the mouse ZP1 appears to deviate from the consensus features. Mouse ZP1 is only 39% identical to the human ZP1 at the amino acid level while the rabbit and pig ZP1 proteins are 70% identical to human ZP1. The mouse ZP1 gene is 623 amino acids in length (vs 540 in other species) and contains a stretch of sequence in the N-terminus region that is absent in the ZP1 of other species (2, 19). The "mature" mouse ZP1 protein in its native form has been reported to exist as a dimer of two identical subunits (18).

Table 1: Characteristics of ZP1 protein (Mouse nomenclature) of different species[a]

Species	ZP Nomenclature	# Amino acid residues	Relative M_r (kDa)	# N-Glycosyla-tion sites	# Exons
Mouse	ZP1(Orthologue)	623	200	6	12
Rabbit	55kDa/ZP1/ZPB	540	85-95	6	ND
Pig	ZP3α/ ZPB	536	55	5	ND
Bonnet	ZP1/ZPB	539	ND	5	ND
Human	ZP1(Homologue) / ZPB	540	60-65	6	12
Human	ZP1 (Orthologue)	638	ND	4	12

[a] Modified from Prasad et al. (2).

In addition, the trefoil domain in mouse ZP1 is present in a modified form (19). However, using DNA sequence analysis (BLAST search of the database of Expressed Tag Sequences), Hughes and Barratt (23) have shown that the true human orthologue of the mouse ZP1 gene is not human ZPB, but there is a distinct human ZP1 gene that is greater than 90% similar to the mouse ZP1 genomic sequence (67% identical at the amino acid level). This suggests that the true homologue of human, rabbit and pig ZP1 (ZPB) gene exists in the mouse having similar function as the ZP1 (ZPB) genes in other species. In this review the mouse ZP1 and human ZP1 are referred to as the orthologues of the ZP1 family while the ZPB proteins from other species including the human are referred to as homologues of the ZP1 protein family and all data on ZP1 are in reference to the ZPB homologues.

ZP2 Protein Family

Evaluation of the ZP2 proteins from different species (Table 2) shows that this protein is structurally conserved with 60 to 70 % identity at the amino acid level between the mouse, rabbit, pig and human, and 94% identity between the bonnet monkey and human (2, 19). In the mouse, the consensus furin cleavage site (present upstream of the C-terminus transmembrane domain) has been implicated in the secretion of ZP2 (24).

Table 2. Characteristics of ZP2 protein (Mouse Nomenclature) of different species[a]

Species	ZP Nomenclature	#Amino acid residues	Relative M_r (kDa)	#N-Glycosylation sites	# Exons
Mouse	ZP2	713	120	7	18
Rabbit	75kDa / ZP2	684	100-130	7	ND
Pig	ZP1 /ZP2/ ZPA	716	70-100	7	18
Bonnet	ZP2	745	ND	7	ND
Human	ZP2 / ZPA	745	70-100	6	19

[a] Modified from Prasad et al. (2).

ZP3 Protein Family

Of the three ZP protein families, the ZP3 protein family shows a greater degree of structural conservation (Table 3). The deduced amino acid sequences of different mammalian ZP3 proteins are nearly identical in length and have a 22 amino acid signal sequence. The ZP3 proteins of the mouse, rabbit and pig are 70 % identical to the human while the marmoset and bonnet monkey ZP3 are greater than 90% identical to the human (2, 19). As with the other two ZP proteins, ZP3 has a consensus furin cleavage site near the C-terminus and the secretion of nascent mouse ZP3 has been shown to be dependent on the cleavage at the consensus furin cleavage site (24).

It has been suggested that the human ZP3 gene is located in two separate loci and that the second polymorphic locus codes for a truncated protein of only 372 amino acids (25). However, Kipersztok et al. (26) have shown that the truncated gene product is likely due to a partial duplication of the human ZP3 gene (exons 5-8).

Carbohydrate Composition of ZP Proteins

The structural complexity of the mammalian ZP results from the interaction of protein and carbohydrate moieties in a highly organized matrix.

The carbohydrates are thought to greatly influence cell-cell interactions between the oocyte and GCs within the developing follicle as well as those occurring between the fully developed ZP and the fertilizing spermatozoon. Enzymatic and chemical deglycosylation experiments have established that ZP glycoproteins contain both N-linked and O-linked sugars (3, 10, 18). Precise oligosaccharide structures are not yet known but most of the information on the ZP oligosaccharides have come from studies carried out in the mouse and pig (27, 28).

Table 3: Characteristics of ZP3 protein (Mouse ZP nomenclature) of different species[a]

Species	ZP Nomenclature	#Amino acid residues	Relative M_r (kDa)	#N-Glycosyla-tion sites	# Exons
Mouse	ZP3	424	83	6	8
Rabbit	45kDa	419	68-120	5	ND
Pig	ZP3β/ ZPC[4]	421	78-120	4	ND
Marmoset / Bonnet	ZP3	424 / 424	ND	5 / 4	ND
Human	ZP3/ ZPC[4]	424	53-65	4	8

[a] Modified from Prasad et al. (2).

Mouse ZP Glycoproteins

Carbohydrate analysis of the mouse ZP glycoproteins (in particular ZP2 and ZP3 proteins) reveal both N-linked and O-linked glycan units. The presence of high-mannose/hybrid oligosaccharides on ZP2 and ZP3 but not ZP1 has been reported (27). The N-linked sugar chain of ZP3 and ZP2 have been shown to be similar with more than 95% being acidic of which 80% contain sialic acid residues. Tri- and tetra-antennary complex type chains with a fucosyl residue are predominant in the core region with sialic acid residues linked to the core fragments as well as the non-reducing region that contain different numbers of N-acetyl lactosamine residues and other sugar chains (28, 29). Presence of poly-N-acetyllactosaminyl glycans on N-linked sugar chains of mouse ZP2 and ZP3 has also been reported with terminal, nonreducing α-galactosyl residues present only in ZP3 (27, 29, 30). Analysis of O-glycans indicate that majority are core 2-type containing β 1-6-linked branches that terminate with the same sequences as N-glycans (29). The presence of an O-linked trisaccharide has also been reported on mouse ZP2 and ZP3 (27). Further analysis of the sugar residues has involved histochemical analysis using carbohydrate binding specificities of various lectins to decipher the oligosaccharide structure and terminal sugar residues such as α-galactose, β-galactose, N-acetylglucosamine (28). Lectin bnding studies have also revealed differential glycosylation between the inner and outer layers of ZP and in the ZP of fertilized mouse eggs (31). Such differences in lectin binding patterns have been implicated in ZP defects in human infertility (32).

Pig ZP Glycoproteins

More detailed studies on the composition of the carbohydrate molecules in the pig have been possible, due to the development of methods for the large scale isolation of ZP from porcine ovaries obtained from slaughterhouse sources. The porcine ZP consists of 71% protein, 19% neutral hexose, 2.7% sialic acid and 1.3% sulfate (10). Structural analysis of porcine ZP indicates that the N-linked oligasaccharides contain neutral (28%) and acidic (72%) sugar chains. The neutral chains are composed of bi-, tri- and tetra-antennary complex type sugar chains with fucosylated trimannosyl core. In addition, they include linear repeating units of N-acetyllactosamine with N-acetylglucosamine at the nonreducing termini (28). Removal of polylactosaminoglycans by endo-β-galactosidase digestion has been shown to separate the ZP1 and ZP3 proteins (Yurewicz et al., 1987). Analysis of acidic chains has revealed the presence of mono- to –tetraantennary complex type with and without N-acetyllactosamine repeating units and various sulfated and non-sulfated forms of fucosylated structures. Partially sulfated and sialylated N-acetyllactosamines linked to the nonreducing ends of bi-, tri- and tetra-antennary complex type neutral chains forming heterogeneous acidic chains has been reported (28). Tryptic digests of the ZP1 and ZP3 proteins containing N-glycosylation sites has revealed the presence of triantennary and tetraantennary neutral chains on Asn271 of ZP3 and Asn220 of ZP1 while biantennary chains were present on Asn124, Asn146 and Asn271 of ZP3 and Asn 203, Asn220 and Asn 333 of ZP1 proteins (28, 33). Studies on the structure of neutral oligosaccharides associated with porcine ZP O-linked sugar chains have revealed the presence of mostly type-1-core and to a lesser extent type-3-core with oligosaccharides containing terminal α-galactose and β-galactosamine at the nonreducing termini as minor components. The acidic component of the O-linked sugar chains includes sialylation and sulfation with variable structures (28). Analysis of the endo-β-galactosidase digests of N- and O- glycans reveal that both contain similar poly-N-acetyllactosamine chains. However, multiple branching of the core structures is seen in N-linked oligosaccharides resulting in even larger heterogeneity than observed for O-linked sugar chains. The composition of carbohydrates and the structure of N-linked sugar chains of the bovine ZP are similar to that seen in the pig (28).

Developmental Expression and Stage-Specific Synthesis of ZP Proteins

Recent evidence supporting the roles of ZP proteins in early stages of GC differentiation have brought about an increased interest in determining the site of biosynthesis of ZP proteins. To date, however, there has been considerable controversy regarding the site of biosynthesis of ZP proteins (2, 34, 35). In mice, ZP proteins have been reported to be coordinately expressed only in the oocytes (4, 6, 18). However, it has now been shown that all three ZP proteins can be localized in selected populations of GCs of mice (2, 35). The expression of ZP in the oocyte appears to be in the primordial and early primary follicular stage of ovarian development, while in the mature preantral or antral follicles, ZP localization is evident only in the GCs (35). There are also an increasing number of reports demonstrating both ZP mRNA and ZP proteins are expressed in the GCs of developing follicles of other mammalian species as summarized in Table 4. The expression of ZP proteins has been demonstrated in primary cultures of GCs and has further been shown to be dependent on the developmental stage of the follicle. Using *in situ* hybridization, localization of ZP1 and ZP3 mRNA has been demonstrated in the oocyte and GC of porcine, bovine and rabbit ovaries in a stage dependent manner. These studies demonstrate that the ZP proteins are transiently expressed in GC during follicular development. The detection of ZP protein appears to be restricted to distinct populations of GC. Previous studies have demonstrated the presence of two morphologically distinct populations of GC expressing ZP proteins (2, 3, 34). Of significance is that the avian homologue of ZP3 has also been found to be synthesized and secreted by the GCs (36). These studies, therefore, demonstrate that all three ZP proteins are expressed in the oocytes and in GC in a stage-specific manner during follicular development.

Functions of ZP Proteins

Many of the investigations to evaluate the function of ZP proteins have focused on the spermatozoa receptor function of the ZP molecules (Summarized in Table 5; (2, 19, 37)). Most of the data on the function of the individual ZP proteins have come from studies carried out in the mouse. However, data from other mammalian species reveal that the mouse model may not be valid for many mammalian species, including the human. In mouse and hamster, ZP3 has been shown to be the primary spermatozoa receptor with acrosome reaction inducing activity (18). In the pig, the ZP1 protein appears to be involved in the interaction with the sperm.

However, it has been reported that a single pig ZP protein may not function alone as the sperm receptor. It has been shown that pig ZP1 and ZP3 form hetero-complexes that bind with high affinity to boar sperm (38). In line with these observations, Yamasaki et al (39) have demonstrated that both rabbit ZP1 and ZP3 proteins bind to recombinant rabbit spermatozoa antigen Sp17 *in vitro*. Furthermore, rabbit ZP1 protein expressed using the baculovirus expression system binds to rabbit spermatozoa in a dose-dependent manner over the acrosomal region (2, 35, 37).

Table 4. Summary of studies demonstrating expression of ZP proteins / mRNA in the granulosa cells (GCs) of mammalian ovaries[a].

Probe	Detection method (Fixation)	Localization
Anti-human ZP3	Immunohistochemistry (Bouin's)	Primordial follicles and some GCs of growing follicles of rabbit, marmoset, rhesus monkey and human ovaries
Anti-Pig ZP1(ZPB) and ZP3	Immunohistochemistry (Bouin's)	Oocyte and GCs of growing follicles of pig and bovine ovaries
Anti-Pig ZP	Immunohistochemistry (Bouin's)	GCs of secondary follicles of mouse, rat, rabbit, pig, baboon and human ovaries
Bovine ZP3 cDNA	In situ (Frozen; Paraformaldehyde)	Oocyte and GCs of secondary follicles of pig and bovine ovaries
Anti-Rabbit ZP1(ZPB) and ZP3	Immunohistochemistry (Bouin's)	GCs of growing follicles of rabbit ovaries
Rabbit ZP1(ZPB) cDNA	In situ (Paraformaldehyde)	GCs of primary and early secondary follicles but not mature antral follicles
Anti-Rabbit ZP and ZP1(ZPB)	Western Blot Analysis	GCs in culture from primary and early secondary follicles of 6 week old rabbit ovaries.
Rabbit ZP1(ZPB) cDNA	Northern blot	
Anti-recombinant human ZP1(ZPB), ZP3	Immunohistochemistry (Bouin's; formalin)	Developing follicles of mouse, rat and pig with Bouin's fixative but not in formalin fixed ovaries.

[a] See references in (2, 35).

It is interesting to note that mouse ZP1 has been implicated in structural maintenance of the ZP matrix (18). While the true human orthologue of mouse ZP1 has been identified (23) that is 76% similar to the mouse ZP1 at the nucleic acid level, the true homologue of human ZP1/ZPB has not been that the homologue of human ZP1/ZPB should be present in the mouse and would likely be involved in sperm receptor function similar to that seen in the pig and rabbit.

Studies to determine the functional role of the ZP proteins in the human have been hampered by the limited availability of the native ZP. However, solubilized human ZP has been shown to bind and induce the acrosome reaction (40). Therefore, the focus has been to generate sufficient quantities of recombinant ZP proteins that can be used to determine the functional role of individual ZP proteins. Both glycosylated and non-glycosylated recombinant human ZP3 protein has been shown to induce the acrosome reaction *in vitro* (2, 37). The importance of protein backbone in binding to sperm has been demonstrated in human using a 10 amino acid ZP3 peptide that is able to induce acrosome reaction (41). Tsubamoto et al. (42) have demonstrated the binding of recombinant human ZP2 only to acrosome reacted sperm suggesting that the ZP2 has secondary sperm binding function in the human similar to that seen in the mouse and pig. The primary sperm binding function of human ZP1 has not yet been conclusively elucidated. However, binding of recombinant bonnet monkey ZP1 protein to the principal segment of the acrosomal cap of capacitated bonnet monkey sperm has been demonstrated (43). Similarly, in an earlier study VandeVoort et al. (44) had shown inhibition of Cynomolgus monkey sperm from binding to eggs using antibodies generated against recombinant rabbit ZP1. Collectively, these studies suggest that like the pig and rabbit, human ZP1 and ZP3 are involved in sperm-ZP interaction.

Role of Carbohydrates in Sperm -ZP Interaction

It is becoming increasingly apparent that the molecular mechanism of sperm-ZP interaction varies among mammalian species. The initial attachment or adhesion of sperm to ZP is thought to involve carbohydrate moieties that may be involved in some aspects of species-specificity in sperm-ZP interaction. The subsequent tight binding of sperm to ZP, however, appears to be the result of interactions of carbohydrate-carbohydrate, carbohydrate-protein and protein-protein (37, 40, 45). This cascade of molecular interactions results in acrosomal exocytosis and the acrosome reaction. Therefore, the focus of many of the investigations has been to elucidate the glycoprotein molecules involved in sperm-ZP interaction.

Mice: In mice, the ZP ligand primarily responsible for ZP-sperm interaction has been shown to be a class of O-linked oligosaccharides that are covalently linked to ZP3 (18). However, there is controversy regarding the type of terminal sugar residues involved in the interaction. Terminal α-Galactose of Gal$\alpha1\rightarrow$3Gal structure present on mouse ZP3 has been shown to be involved in spermatozoa binding (46). However, humans and

Table 5. Functional role of ZP proteins[a].

ZP Family	Species	Function	Sperm Receptor Activity	
			Native ZP	Recom-binant
ZP1	Mouse (orthologue)	Crosslinks ZP2 and ZP3	-	-
	Rabbit	Sperm receptor	+	+
	Pig	Sperm receptor	+	ND
	Bonnet	Sperm binding	ND	+
	Human (homologue)	ND	ND	ND
	Human (orthologue)	ND	ND	ND
ZP2	Mouse	Secondary sperm receptor	-	-
	Rabbit	ND	ND	ND
	Pig	Secondary sperm receptor	-	-
	Bonnet	ND	ND	ND
	Human	Secondary sperm receptor	ND	+
ZP3	Mouse	Primary sperm receptor and AR activity	+	+
	Rabbit	Sperm binding	+	ND
	Pig	Sperm binding	+	ND
	Bonnet	Sperm binding	ND	+
	Human	Sperm binding and AR activity	ND	+

[a]Modified from (2). See refernces in (2, 35); AR= Acrosome reaction.

monkeys lack this type of oligosaccharide moiety. In addition, mouse, in which the gene encoding this oligosaccharide moiety has been disrupted, develop normally and are fertile (2, 19). In addition, it has been shown that the O-linked oligosaccharides of mZP3 do *not* contain terminal α-galactose residues but contain trisaccharide structures with N-acetylglucosaminyl residue at the non-reducing terminus (27). More recently, Aviles et al. (31) have reported the localization of α-galactose residues to the inner portions of the zona matrix suggesting a role for these residues in sperm penetration. N-acetylglucosamine residues in β-linkage of O-linked oligosaccharides have also been implicated in binding to galactosyl-transferase present on mouse sperm. However, sperm from mice made null for galactosyl transferase by homologous recombination do not bind to solubilized ZP3 but still bind to the zona and achieve low rates of fertilization *in vitro* (47). Using neoglycoproteins, Loeser and Tulsiani (48) have demonstrated that glucosamine-BSA, mannose-BSA, and galactosamine-BSA induce acrosome reaction in mouse sperm while glucose and galactose–BSA had no effect suggesting that sugar residues need to be conjugated to the protein backbone for functional activity. These studies suggest that multiple sugar residues may be involved in sperm recognition.

Pig: In pigs, partially deglycosylated ZP1 was found to be a more effective ligand for spermatozoa than the native protein, suggesting that the terminal oligosaccharides may not be essential for spermatozoa binding. In addition, it has been demonstrated that in the pig the protein backbone is essential for the interaction of the sugar residues with the sperm and that it is the N-linked carbohydrates and *not* the O-linked carbohydrates that are involved in sperm-ZP interaction. Furthermore, tri-or tetra-antennary, neutral, complex-type N-linked carbohydrate chain(s) localized in the N-terminal region (Asn220) of pig ZP1 has been shown to be involved in sperm binding (33).

Human: In human, fucoidan, a polysaccharide, has been shown to inhibit the initial binding of sperm to ZP and it is thought that a "selectin-like" interaction may be involved in mediating gamete interactions (40). Both mnnose and glactose are also implicated in the interaction with sperm and has been shown to be calcium dependent (27, 37, 45). In addition, it has been shown that the removal of sialic acid residues from human ZP protein by mild chemical treatment decreases human sperm binding to ZP by 30% while exhaustive treatment with neuraminidase increased binding by three fold (40). This suggests that not all sialic acid residues may be involved in sperm binding and that excessive removal of sialic acid residues exposes other sperm binding sites on the ZP.

Collectively, these studies suggest that multiple ligands on the ZP interact with multiple sperm proteins that are necessary for the optimal regulation of the sequence of events leading to the successful completion of fertilization.

Structural Organization of the ZP Matrix

Despite the relative abundance of recent information concerning the primary structure of ZP proteins, the specific functions of individual proteins and the way in which they become organized in a three dimensional paracrystalline array is still a subject of intense research. At present, the most developed structure/function model of the ZP is that of the mouse. Mouse ZP has been shown to be composed of ZP2/ZP3 heterodimers located every 140 Å as seen in electron micrographs and organized in filaments that are crosslinked by ZP1 (18). Although support for this structure has come from studies using ZP gene knockout and antisense experiments, this structural model has not yet been proved. Furthermore, from the sequence data it has been reported that the mole ratios of ZP2:ZP3 is close to 1:1 while ZP1 is only 9% of the mole amounts of ZP2 and ZP3 which is far less than the number of ZP1 cross-linking sites (49).

Gene knockout studies have demonstrated that interference with ZP proteins will result in abnormal ovarian follicular development. In the complete absence of ZP matrix, as seen in ZP3 deficient mice, development of ovarian follicles is retarded and females are infertile. In these mice oocytes lacking ZP develop during folliculogenesis but do not form proper cumulus-oocyte complex and granulosa cells are randomly located in the cumulus (6). These studies suggest that ZP formation is necessary for proper follicular development. In contrast, although ZP matrix is formed in ZP1 deficient mice, structural abnormality of the ZP was evident in growing follicles with ectopic clusters of granulosa cells present in the perivitelline space between the ZP matrix and oolemma resulting in abnormal folliculogenesis. The ZP1 null mice were fertile but with significantly reduced number of litters suggesting that ZP1 is necessary for the structural integrity of the ZP matrix to prevent early embryonic loss (6). In ZP2 null mice, an abnormal zona matrix is formed which does not affect the early stages of folliculogenesis but has decreased number of antral follicles resulting in infertility (50). It has been further demonstrated that abnormalities in human ZP3 structure can be detected with an antiserum specific to human ZP3 peptide and the results correlate with fertilization failure during IVF treatment (40, 41). Therefore, the structure / function relationship of ZP proteins in GC differentiation and follicular development needs to be further evaluated.

Immunogenicity and Antigenicity of ZP Proteins

Many studies have demonstrated that the immunogenicity and antigenicity of ZP glycoproteins as well as the physiological effect on ovarian development is extremely complex (7, 9). Antibodies can be identified which recognize amino acid and carbohydrate epitopes as well as conformational or structural epitopes of the ZP. The immune response depends not only on the source of ZP immunogen but also on the species that is immunized. It appears that the immunogenicity of ZP glycoproteins is primarily due to foreign epitopes associated with the ZP of different species, since alloimmunization with ZP protein of the same species does not elicit a significant immune response as determined by the presence of circulating antibodies to 'self' ZP (9, 51).

Carbohydrates appear to influence the immunogenicity of ZP protein. Deglycosylated native or recombinant ZP proteins have been shown to be less immunogenic than glycosylated ZP proteins (2, 7). Bacterially expressed recombinant proteins (non-glycosylated) and ZP3 peptide based immunogens when conjugated to carrier protein have been shown to elicit immune response (44, 52, 53).

Contraceptive Potential of ZP Immunogens

Interest in the immunology of the ZP stems from the potential contraceptive use of antibodies to zona antigens. Immunization with zona pellucida antigens has distinct advantage over other proposed immunological contraceptive methods for the following reasons: (1) It is not abortive but inhibits fertilization. (2) The ZP is directly exposed to antibodies in the follicular fluid of the antral follicle for long periods of time, so only low titres of antibodies are needed to block fertilization. (3) The zona antigens studied to date are tissue specific. (4) The zona of a variety of animal species are immunologically cross-reactive. It has been suggested that binding of the ZP antibodies to the newly elaborated ZP antigens as they are first being secreted in follicles which emerge from meiotic arrest results in the disruption of the junctional complexes between the oocyte and the differentiating granulosa cells leading to oocyte degeneration. Such progressive depletion of oocytes and interference with granulosa cell and theca development result in ovarian damage and infertility (7, 9, 51)

Native ZP as Immunogen

Many of the studies on immunocontraception have used native solubilized pig ZP as immunogen because of the relative ease in obtaining pig ZP and because of its high degree of sequence homology with ZP proteins of other species. These studies have demonstrated that immunization with pig ZP proteins elicits an immune response that is associated with ovarian dysgenesis (7, 9). While this response is desirable for the development of sterilization vaccines for some animals it is not acceptable for the development of a human contraceptive vaccine. It is, therefore, necessary to identify those ZP antigen(s) that elicit antibodies and inhibit fertility without affecting ovarian follicular development. One strategy has been to define such antibodies that bind to the spermatozoa 'receptor' proteins of the ZP of different species and inhibit the fertilization process.

A number of investigations involving native ZP peptides and recombinant ZP sperm receptor molecules from various species have been carried out to determine if these proteins / peptides would elicit antibodies that inhibit fertilization without affecting ovarian follicular development (2, 9, 51).

ZP3 as Immunogen

Although ZP3 has been shown to be involved in sperm binding, immunization of monkeys and rabbits with pig ZP3 results in ovarian dysgenesis (7, 51). Similar observation has been made when marmosets were immunized with recombinant human ZP3 (homologue of mouse ZP3). Significant immune response and infertility was observed that was associated with ovarian pathology (51). In contrast, when mice were immunized with ZP3 expressed using attenuated Salmonella typhimurium or ectromella virus, significant antibody titer was obtained and the females were infertile with no evidence of ovarian oophoritis (54, 55). Mouse ZP3 peptide has also been used as immunogen with varied results. An earlier study using mouse ZP3 peptide showed significant immune response in mice that resulted in infertility without altering ovarian function (4). However, this peptide is not detected immunologically in other species and caused autoimmune oophoritis in certain inbred strains of mice due to cytotoxic T-cell epitopes (52). To circumvent this problem, a chimeric peptide has been designed that contains a promiscuous foreign T-cell peptide capable of eliciting a T-cell response regardless of the MHC haplotype of the inbred mice, and a modified native B-cell peptide of ZP3 that induces antibody to native ZP3. This results in reduction in fertility that is reversible (56). In primate, however, targeting a single or triple ZP3 peptide as an immunogen has not been able to induce a contraceptive effect and no adverse effect on ovarian function was detected (51). In line with these studies, antibodies to various pig ZP3 peptides have been generated that do not individually inhibit boar sperm from binding to pig oocyte but combinations of antisera significantly inhibit sperm-oocyte interaction. In contrast, antisera against bonnet monkey ZP3 peptide inhibited human sperm-oocyte interaction (53). Using transgenic mice in which mouse ZP3 is replaced with human ZP3, it has been demonstrated that administration of antibodies to human ZP3 to these mice results in long-term reversible contraception without affecting ovarian histology (6).

ZP2 as Immunogen

Immunization of monkeys with recombinant rabbit ZP2 protein resulted in ovarian dysgenesis (44) while immunization of mice with a ZP2 peptide containing a T-cell epitope resulted in reduced fertility without ovarian oophoritis (57).

ZP1 as Immunogen

Immunization of Cynomolgus monkey with recombinant rabbit ZP1 expressed in bacteria and conjugated to Protein A resulted in antibodies that inhibited sperm binding to eggs *in vitro* without affecting ovarian follicular development (44). Similar observations have been made when female baboons were immunized with recombinant bonnet monkey ZP1 expressed in E.coli and conjugated to diphtheria toxoid. The antisera inhibited human sperm-oocyte interaction *in vitro* and the immunized animals showed no changes in cyclicity but failed to conceive (58). In contrast, alloimmunization of rabbits with recombinant rabbit ZP1 expressed in the vaccinia virus system or by recombinant myxoma virus resulted in infertility that was associated with follicular degeneration (59).

These studies clearly show the effectiveness of using ZP antigens as immunocontraceptive agents. They also indicate the necessity for elucidating the ZP domains/epitopes involved in sperm binding in order to target these sites for immunocontraceptive effect. Further immunization studies can then be carried out to identify those epitopes / domains that would elicit an immune response without affecting normal ovarian follicular development. This will aid in the development of a safe and effective contraceptive agent for use in the human.

CONCLUSIONS

The studies discussed in this overview demonstrate how much information has been collected concerning the structure and function of the ZP proteins in the fertilization process. Despite this extensive information and observations that ZP proteins are expressed in a stage- specific time frame during ovarian folliculogenesis, we, however, have little information concerning the roles of the ZP proteins with respect to ovarian follicular development and granulosa cell differentiation. Early studies by Dunbar and colleagues (2, 7, 10) demonstrated that extracellular matrix components, including collagen, fibronectin and ZP proteins will dramatically affect the expression of proteins by the granulosa cells cultured from primary ovarian follicles *in vitro*. These studies suggest that the ZP may serve as a unique extracellular matrix and effect the differentiation of granulosa cells in early ovarian follicular development. The interference with normal ovarian follicular development by antibodies to ZP proteins supports these observations (7, 9).

Finally, no studies have been carried out to evaluate the molecular dynamics of the ZP matrix during growth of the oocyte. While the ZP matrix is first elaborated during the early stages of ovarian follicular development when the oocyte mass is small, this matrix must either expand or be remodeled through enzymatic processing during the "maturation" of the ZP matrix. These questions need to be addressed in order for us to fully understand the importance of the ZP proteins in normal ovarian function and ovarian pathologies as well as to develop improved contraceptive methods that target expression of ZP proteins.

222

ACKNOWLEDGEMENTS

The authors gratefully acknowledge Ms. Debra Townley for assistance with the figures. The research reported in this review was supported in part by grants from the CONRAD and by NICHD/NIH through cooperative agreement (U54 (HD07495)) as part of the Specialized Cooperative Centers Program in Reproduction Research.

REFERENCES

1. Yanagimachi R Mammalian fertilization. The physiology of reproduction. Knobil E, Neill JD Editors, Raven Press, New York, 1994.
2. Prasad SV, Skinner SM, Carino C, Wang N, Cartwright J, Dunbar BS. Structure and Function of the Proteins of the Mammalian Zona pellucida.Cells Tissues Organs 2000; 166: 148-164
3. Dunbar BS, Wolgemuth DJ. Structure and function of the mammalian zona pellucida, a unique extracellular matrix. Mod Cell Biol 1984; 3:77-111.
4. Liang LF, Dean J. Oocyte development: molecular biology of the zona pellucida. Vitam Horm 1992; 47: 115-159.
5. Eppig JJ, Chesnel F, Hirao Y, O'Brien MJ, Pendola FL, Watanabe S, Wigglesworth K Oocyte control of granulosa cell development: how and why. Hum. Reprod 1997; 12: 127-132.
6. Rankin T, Soyal S, Dean J. The mouse zona pellucida: folliculogenesis, fertility and pre-implantation development. Mol Cell Endocrinol 2000; 25: 21-5.
7. Dunbar BS, Avery S, Lee V, Prasad SV, Schwahn D, Schwoebel E, Skinner S, Wilkins B. The mammalian zona pellucida: Its biochemistry, immunochemistry, molecular biology and developmental expression. Reprod Fertil Dev 1994; 6: 59-76.
8. Elvin JA, Matzuk MM. Mouse models of ovarian failure. Rev Reprod 1998; 3:183-195.
9. Dunbar BS, Prasad S, Carino C, Skinner SM. The ovary as an immune target. J Soc Gynecol Investig 2001; 8: S43-8.
10. Dunbar BS, Prasad SV, Timmons T. Comparative mammalian zonae pellucidae; in Dunbar, B.S., M. O'Rand (eds): Comparative Overview of Mammalian Fertilization. New York, Plenum Press, 1991.
11. Palmstierna M, Murkes D, Csemiczky G, Andersson O, Wramsby. Zona pellucida thickness variation and occurrence of visible mononucleated blastomers in preembryos are associated with a high pregnancy rate in IVF treatment. J Assist Reprod Genet 1998; 15: 70-75.
12. Magerkurth C, Topfer-Petersen E, Schwartz P, Michelmann HW. Scanning electron microscopy analysis of the human zona pellucida: influence of maturity and fertilization on morphology and sperm binding pattern. Hum Reprod 1998; 14: 1057-1066.
13. Maymon BB, Maymon R, Ben-Nun I, Ghetler Y, Shalgi R, Skutelsky E. Distribution of carbohydrates in the zona pellucida of human oocytes. J Reprod Fertil. 1994; 102: 81-86.
14. Dunbar BS, Timmons TM, Skinner SM, Prasad SV. Molecular analysis of a carbohydrate antigen involved in the structure and function of zona pellucida glycoproteins. Biol Reprod 2001; 65:951-60.
15. Turner K, Horobin RW . Permeability of the mouse zona pellucida: a structure-staining-correlation model using coloured probes. J Reprod Fertil 1997; 111: 259-265.
16. Kawana T. Sexually transmitted diseases of alpha herpes virus in women. Nippon Rinsho. 2000; 58: 883-889.

17. Horgan M, Bersoff-Matcha S. Sexually transmitted diseases and HIV. A female perspective. Dermatol Clin 1998; 16: 847-51.

18. Wassarman PM. Zona pellucida glycoproteins. Annu Rev Biochem 1988; 57: 415-442.

19. McLeskey SB, Dowds C, Carballada R, White RR, Saling PM .(Molecules involved in mammalian sperm-egg interaction. Int Rev Cytol 1998; 177: 57-113.

20. Lira S, Schickler M, Wassarman PM. Cis-Acting DNA elements involved in oocyte-specific expression of mouse sperm receptor gene mZP3 are located close to the gene's transcription start site. Mol. Reprod. Develop 1993; 36: 494-499.

21. Liang L, Soyal SM, Dean J. FIGalpha, a germ cell specific transcription factor involved in the coordinate expression of the zona pellucida genes. Development 1997; 124: 4939-4947.

22. Iwamoto K, Ikeda K, Yonezawa N, Noguchi S, Kudo K, Hamano S, Kuwayama M, Nakano .M Disulfide formation in bovine zona pellucida glycoproteins during fertilization: evidence for the involvement of cystine cross-linkages in hardening of the zona pellucida. J Reprod Fertil l999; 117: 395-402.

23. Hughes DC, Barratt CL. Identification of the true human orthologue of the mouse Zp1 gene: evidence for greater complexity in the mammalian zona pellucida? Biochim Biophys Acta 1999; 1447: 303-306.

24. Litscher ES, Qi H, Wassarman PM. Mouse zona pellucida glycoproteins mZP2 and mZP3 undergo carboxy-terminal proteolytic processing in growing oocytes. Biochemistry 1999; 38: 12280-12287.

25. van Duin M, Polman JE, Suikerbuijk RF, Geurts-van Kessel AH, Olijve W. The human gene for the zona pellucida glycoprotein ZP3 and a second polymorphic locus are located on chromosome 7. Cytogenet Cell Genet 1993; 63: 111-113.

26. Kipersztok S, Osawa GA, Liang LF, Modi WS, Dean J. POM-ZP3, a bipartite transcript derived from human ZP3 and a POM121 homologue. Genomics 1995; 25: 354-359.

27. Tulsiani DR, Yoshida-Komiya H, Araki Y. Mammalian fertilization: a carbohydrate-mediated event. Biol Reprod 1997; 57: 487-494.

28. Takasaki S, Mori E, Mori T. Structures of sugar chains included in mammalian zona pellucida glycoproteins and their potential roles in sperm-egg interaction. Biochim Biophys Acta 1999; 1473: 206-215.

29. Easton RL, Patankar MS, Lattanzio FA, Leaven TH, Morris HR, Clark GF, Dell A. Structural analysis of murine zona pellucida glycans. Evidence for the expression of core 2-type O-glycans and the Sd(a) antigen., J Biol Chem 2000; 275: 7731-7742.

30. Tulsiani DR Structural analysis of the asparagine-linked glycan units of the ZP2 and ZP3 glycoproteins from mouse zona pellucida. Arch Biochem Biophys 2000;382: 275-283.

31. Aviles M, El-Mestrah M, Jaber L, Castells MT, Ballesta J, Kan FW. Cytochemical demonstration of modification of carbohydrates in the mouse zona pellucida during folliculogenesis. Histochem Cell Biol 2000; 113: 207-19.

32. Talevi R, Gualtieri R, Tartaglione G, Fortunato A. Heterogeneity of the zona pellucida carbohydrate distribution in human oocytes failing to fertilize in vitro. Hum Reprod 1997; 12: 2773-2780.

33. Nakano M, Yonezawa N. Localization of sperm ligand carbohydrate chains in pig zona pellucida glycoproteins. Cells Tissues Organs 2001; 168: 65-75.

34. Sinowatz F, Kolle S, Topfer-Petersen E. Biosynthesis and expression of zona pellucida glycoproteins in mammals. Cells Tissues Organs 2001; 168: 24-35.

35. Carino C, Prasad SV, Skinner SM, Dunbar BD, Larrea F, Dunbar BS. Localization of species conserved zona pellucida antigens in mammalian ovaries. Reproductive Biomedicine Online 2002 (in press).

36. Takeuchi Y, Nishimura K, Aoki N, Adachi T, Sato C, Kitajima K, Matsuda T. A 42-kDa glycoprotein from chicken egg-envelope, an avian homolog of the ZPC family glycoproteins in mammalian Zona pellucida. Its first identification, cDNA cloning and granulosa cell-specific expression. Eur J Biochem 1999; 260: 736-742.

37. Prasad SV, Dunbar BS. Human sperm-oocyte recognition and infertility. Semin Reprod Med 2000;18: 141-149.
38. Yurewicz EC, Sacco AG, Gupta SK, Xu N, Gage DA. Hetero-oligomerization-dependent binding of pig oocyte zona pellucida glycoproteins ZPB and ZPC to boar sperm membrane vesicles. J Biol Chem 1998; 273:7488-7494.
39. Yamasaki, N, Richardson RT, O'Rand MG. Expression of the rabbit sperm protein Sp17 in COS cells and interaction of recombinant Sp17 with the rabbit zona pellucida. Mol Reprod Dev 1995; 40: 48-55.
40. Oehinger S. Molecular basis of human sperm-zona pellucida interaction. Cells Tissues Organs 2001; 168: 58-64.
41. Hinsch, KD, Schill WB, Hinsch E. Induction of acrosome reaction through a synthetic ZP3 peptide. J Reprod Fertil Abstr Ser 1998; 22: 17.
42. Tsubamoto H, Hasegawa A, Nakata Y, Naito S, Yamasaki N, Koyama K. Expression of recombinant human zona pellucida protein 2 and its binding capacity to spermatozoa. Biol Reprod 1999; 61: 1649-1654.
43. Govind CK, Gahlay GK, Choudhury S, Gupta SK. Purified and refolded recombinant bonnet monkey (*Macaca radiata*) zona pellucida glycoprotein-B expressed in *Escherichia coli* binds to spermatozoa. Biol Reprod 2001; 64: 1147-52.
44. VandeVoort CA, Schwoebel ED, Dunbar BS. Immunization of monkeys with recombinant cDNA expressed zona pellucida proteins. Fertil Steril 1995; 64: 838-847.
45. Benoff S. Carbohydrates and fertilization: an overview. Mol Human Reprod 1997; 3: 599-637.
46. Litscher ES, Juntunen K, Seppo A, Penttila L, Niemela R, Renkonen O, Wassarman PM. Oligosaccharide constructs with defined structures that inhibit binding of mouse sperm to unfertilized eggs in vitro. Biochemistry 1995; 34: 4662-4669.
47. Nixon B, Lu Q, Wassler MJ, Foote CI, Ensslin MA, Shur BD. Galactosyltransferase function during mammalian fertilization. Cells Tissues Organs 2001; 168: 46-57.
48. Loeser CR, Tulsiani DR. The role of carbohydrates in the induction of the acrosome reaction in mouse spermatozoa. Biol Reprod 1999; 60: 94-101.
49. Green DP. Three-dimensional structure of the zona pellucida. Rev Reprod 1997; 2: 147-56.
50. Rankin T, O'Brien M, Lee E, Wigglesworth K, Eppig J, Dean J. Defective zonae pellucidae in Zp2-null mice disrupt folliculogenesis, fertility and development. Develop 2001;128: 1119-1126.
51. Paterson M, Jennings ZA, van Duin M, Aitken RJ. Immunocontraception with zona pellucida proteins. Cells Tissues Organs 2000; 166: 228-232.
52. Lou Y, Tung KSK. T cell peptide of a self-protein elicits autoantibody to the protein antigen. Implications for specificity and pathogenetic role of antibody in autoimmunity. J Immunol 1993; 151: 5790-5799.
53. Afzalpurkar A, Shibahara H, Hasegawa A, Koyama K, Gupta SK. Immunoreactivity and in-vitro effect on human sperm-egg binding of antibodies against peptides corresponding to bonnet monkey zona pellucida-3 glycoprotein. Hum Reprod 1997; 12: 2664-2670.
54. Zhang X, Lou YH, Koopman M, Doggett T, Tung KS, Curtiss R Antibody responses and infertility in mice following oral immunization with attenuated *Salmonella typhimurium* expressing recombinant murine ZP3. Biol Reprod 1997; 56: 33-41.
55. Jackson RJ, Maguire DJ, Hinds LA, Ramshaw IA. Infertility in mice induced by a recombinant ectromelia virus expressing mouse zona pellucida glycoprotein 3. Biol Reprod 1998; 58: 152-9.
56. Lou Y, Ang J, Thai H, McElveen F, Tung KS. A zona pellucida 3 peptide vaccine induces antibodies and reversible infertility without ovarian pathology. J. Immunol 1995;155: 2715-2720.
57. Sun W, Lou YH, Dean J, Tung KS. A contraceptive peptide vaccine targeting sulfated glycoprotein ZP2 of the mouse zona pellucida. Biol Reprod 1999; 60: 900-907.

58. Govind CK, Gupta SK. Failure of female baboons (*Papio anubis*) to conceive following immunization with recombinant non-human primate zona pellucida glycoprotein-B expressed in *Escherichia coli*. Vaccine 2000; 18: 2970-2978.
59. Kerr PJ, Jackson RJ, Robinson AJ, Swan J, Silvers L, French N, Clarke H, Hall DF, Holland MK. Infertility in female rabbits *(Oryctolagus cuniculus)* alloimmunized with the rabbit zona pellucida protein ZPB either as a purified recombinant protein or expressed by recombinant myxoma virus. Biol Reprod 1999; 61: 606-613.

Chapter 13

TRANSPORT OF SPERMATOZOA IN THE FEMALE GENITAL TRACT

Susan S. Suarez
Cornell University, Ithaca, N.Y., U.S.A.

INTRODUCTION

Millions of spermatozoa are inseminated into the mammalian female genital tract in order to fertilize only one or a few oocytes. The movement of spermatozoa through the female genital tract is actually regulated by the female so that a few reach the oocyte in the oviduct (fallopian tube) and only one of them succeeds in fertilizing it. This is accomplished by placing filters and traps in the path of the spermatozoa and by switching the way in which the tails of the spermatozoa beat. In this chapter, the way in which movement of spermatozoa into and through each female reproductive organ is regulated will be discussed.

The challenge to learning how spermatozoa reach the egg is that there are so many spermatozoa, yet only a few make it to the site of fertilization. How, then, can we determine which spermatozoa will be successful? How can we catch those few spermatozoa at the oocyte and determine how they made it there?

THE PATHWAY OF TRANSPORT OF SPERMATOZOA

Site of Insemination

Spermatozoa are deposited in the vagina at the entrance to the cervix in humans and other primates, as well as in cattle and other ruminants. Some rapidly enter the cervix. In humans, only about 1/1000th of the approximately 300 million spermatozoa deposited in the vagina manage to enter the cervix (1). In cattle, a few billion spermatozoa are normally deposited by the male and only about 3% enter the cervix (2). One reason for rapid entry into the cervix is that the vagina presents a hostile environment to spermatozoa, one that is highly acidic and loaded with immunological defenses against any microbial invaders. Being foreign to the female's immune system, spermatozoa can be treated as infectious organisms and attacked. In fact, insemination is known to trigger an invasion of leukocytes into the vagina, which then proceed to ingest spermatozoa (3). Seminal plasma offers some protection, providing buffers to the acid and inhibitors of immunological responses; however, most spermatozoa leave the vagina and enter the cervix within minutes of deposition (1).

In mice and hamsters, semen is deposited in the vagina but is transported rapidly, along with seminal fluid, into the uterine cavity. In this case, the cervix does not present much of a barrier (4).

In pigs and horses, the semen is deposited directly into the uterus, bypassing the cervix altogether. This places a large volume of seminal plasma into the uterus, about 70 ml in the horse and 250 ml in the pig (5). In contrast, the volume of semen placed in the vagina at the entrance to the cervix in women and cows is only about 4 ml (1,5).

The Cervical Portal

When semen is deposited in the vagina, the cervix can act as a filter to remove some abnormal spermatozoa as well as potentially infectious microbes. Human and bovine cervices produce copious amounts of watery mucus during the fertile phase of the reproductive cycle (the late follicular phase of the human menstrual cycle and the estrous phase of the bovine estrous cycle) that is readily penetrated by normally-shaped, vigorously-motile spermatozoa. Estrogens, which are present at high levels during the fertile phase, are responsible for inducing the production of thin, watery mucus. At other stages of the cycle, when progesterone is the predominant influence, the mucus is less watery and is nearly impenetrable to spermatozoa (1). Essentially, the mucus produced during the nonreceptive phase glues the cervical portal shut, preventing entrance of microbes.

There are grooves in the walls of the bovine cervix that are thought to provide direct passage through to the uterus while sheltering spermatozoa from the outward flow of mucus and uterine debris in the center of the cervical canal (6). Such passageways have not been identified in the human cervix, but it is possible that they exist. In women, there are about 100 crypts invaginating into the wall that appear to trap spermatozoa. The crypts could serve as storage sites. Living spermatozoa have been recovered from the cervix several days after insemination and therefore it has been assumed that the cervix serves to store spermatozoa (1). Nevertheless, it has never been demonstrated that spermatozoa which linger in the cervix eventually make their way up to the site of fertilization. Furthermore, it is not even known if the spermatozoa recovered from the cervix several days after insemination had actually remained there the entire time or had re-entered the cervix with the outflow from the uterus.

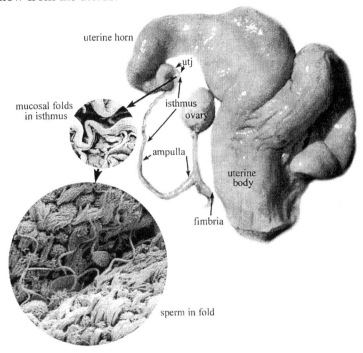

Figure 1. Photograph of a bovine female reproductive tract, showing the reservoir of spermatozoa in the isthmus of the oviduct. The circular inserts show scanning electron micrographs of the wall of the isthmus at the site of the reservoir. Magnification of the small insert is 35 times and the large insert is 1000. Spermatozoa are clearly visible in the larger insert, along with cilia on the surface of oviductal cells, to which they attach. Utj, uterotubal junction

Transport of Spermatozoa Through the Uterus

The uteri of most mammals, except primates, bifurcate into two long tubes called uterine horns (Figure 1). The horns, like other tubular organs, have been observed to undergo coordinated contractions, similar to peristalsis in the gut. Peristaltic activity has been observed during estrus in cows and ewes (7). Even the short, pear-shaped uteri of humans show cranially-directed contractions during the late follicular phase (8). The contractions could serve two purposes: to draw spermatozoa out of the cervix and to propel them towards the oviducts.

Spermatozoa that do not pass on to the oviduct may be flushed out of the uterus through the cervix or phagocytized by leukocytes that enter the uterine cavity after coitus (9).

The Uterotubal Portal

In many mammals, the uterotubal junction between the oviduct and the uterus forms a physiological gate to filter out abnormal spermatozoa. The gate may shut entirely to all spermatozoa at certain times. In the cow, the wall of the uterotubal junction contains a vascular bed that resembles erectile tissue and is thought to enable the closing of the junction (10). There is also a thick layer of smooth muscle around the junction in most species. The female reproductive tract of mice is so small that one can see through the wall with a high-powered compound microscope and it can be seen that the portion of the junction that passes through the uterine wall is completely closed at times (11).

Even when the uterotubal junction is open, the opening is narrow, only about the width of a sperm head in cows. In cows, pigs, rabbits and humans, a mucus has been found within the uterotubal junction and initial segment of the oviduct (12,13). Thus, the junction can act as a filter that allows only gradual passage of vigorous, progressively motile spermatozoa.

Rapid Transport of Spermatozoa

In several species, spermatozoa have been recovered from the ampulla of the oviduct, near the ovary, only minutes after coitus. When Overstreet and Cooper (14) examined this phenomenon of rapid spermatozoal transport more closely in the rabbit, they found that most of the spermatozoa in the ampulla were immotile and damaged. Furthermore, these early arriving spermatozoa were cleared out of the oviduct before the oocytes arrived. They proposed that the waves of contraction stimulated by insemination could incidentally transport some spermatozoa all of the way to the site of fertilization, damaging them by the shearing force of the rapid movement. The function of this rapid transport of spermatozoa is not known, but it has been speculated that the negative pressure created by the waves of contraction aids in pulling spermatozoa into the cervix right after ejaculation, overshooting the target for some spermatozoa. Alternatively, the spermatozoa introduced rapidly into the upper reaches of the reproductive tract, even if they are killed, could serve to somehow signal the tract to prepare for fertilization.

The Oviductal Reservoir of Spermatozoa

As spermatozoa enter the oviduct, they encounter a narrow, labyrinthine lumen that is only about the width of a sperm head and is filled with mucus (13). The inner surface of the wall is arranged in branching folds, creating channels that in some instances end blindly (15). The mucus and the narrow intricate passageway undoubtedly slow down the progress of spermatozoa and increase their contact with the epithelial cells that line the wall. Evidently, when spermatozoa contact the epithelium, they often stick to it. *In vitro*, motile spermatozoa have been observed to bind to oviductal epithelium in cattle (16), humans (17), mice (11), hamsters (18), pigs (19), and horses (20). Spermatozoa bind via the plasma membrane overlying the acrosome in the head. Most often the binding is to cilia on the epithelial surface (16,19), but spermatozoa may also bind to microvilli of nonciliated cells (21,22). As more and more spermatozoa enter the oviduct and become stuck, a reservoir is formed (Figure 1).

Binding of spermatozoa to oviductal epithelium is a specific interaction, mediated by carbohydrates. The first evidence for this came from the hamster, in which the glycoprotein fetuin was found to inhibit binding when it was mixed with spermatozoa infused into oviducts. The component of fetuin that accounted for this effect was found to be the sugar sialic acid, which appears in terminal (external) positions on the oligosaccharides attached to its protein core. Gold-labeled fetuin was observed to bind to hamster spermatozoa over the acrosomal region of the head and to certain proteins extracted from spermatozoa (23).

In the horse, asialofetuin blocked binding of spermatozoa to explant cultures of oviductal epithelium more effectively than fetuin. In this case, galactose, which is the principal sugar remaining at the exposed ends of oligosaccharide chains when sialic acid is removed from fetuin, was found to account for the inhibitory effect (24).

In cattle, binding of spermatozoa to epithelium was specifically blocked by fucoidan and by its constituent fucose (25). Furthermore, fucose in an (α1-4) linkage to N-acetylglucosamine, as in the blood group trisaccharide Lewis a (Le[a]), was more effective at inhibiting binding than any other linkage of fucose (26). It was determined that molecules (glycoproteins or glycolipids) containing fucose are on the surface of the oviductal epithelium and fucose-binding molecules are on the surface of spermatozoa. Specifically, fucose was detected on the surface of bovine oviductal epithelium by the fucose-binding lectins from *Lotus tetragonolobus* (LTL) and *Ulex europaeus* (UEA-I) (25). Pretreatment of epithelium with α-L-fucosidase, an exo-glycohydrolase that cleaves α-linked fucosyl residues from glycoproteins and glycolipids, reduced binding of spermatozoa (25). The fucose-binding molecule was detected on bull spermatozoa using fluorescently tagged fucose- and Le[a]-neoglycoproteins, which labeled living spermatozoa on the plasma membrane over the acrosomal region (26,27).

In each of the three species studied, a different carbohydrate was most effective at inhibiting binding of spermatozoa to epithelium. These species differences may not seem so unusual when one considers that substitution of a single amino acid can change the carbohydrate specificity of some carbohydrate binding proteins (28).

Carbohydrate binding is also implicated in anchoring of germ cells to Sertoli cells during spermatogenesis (29) and in binding of spermatozoa to the zona pellucida during fertilization (30). Different carbohydrate-binding proteins and carbohydrate ligands are involved in these interactions.

While the reservoir of spermatozoa is confined to the region where the uterotubal junction meets the oviductal isthmus, bull spermatozoa bind *in vitro* to epithelium taken from both the isthmus and ampulla. Labeling by LTL and UEA-I lectins indicate that fucose is present on the surface of epithelium throughout the isthmus and ampulla of the cow (25), indicating that binding sites are distributed throughout the oviduct. To examine this apparent discrepancy between the site of the reservoir *in vivo* and spermatozoal binding behavior *in vitro*, spermatozoa were surgically infused into the isthmuses and ampullas of cows. After 10 minutes, the oviducts were removed and opened, unbound spermatozoa were rinsed away, and the tissue was prepared for scanning electron microscopy. In the micrographs, many spermatozoa could be seen associated with epithelium in segments taken from both the isthmus and ampulla. These findings indicate that the reservoir is confined to the isthmus *in vivo*, because it is the first region encountered by spermatozoa when they enter the oviduct (16). In horses and hamsters, spermatozoa will also stick to ampullar epithelium when given the opportunity to do so (20,23).

The protein on the surface of bull spermatozoa that enables them to bind to oviductal epithelium has been isolated and identified as PDC-109 (also called BSP-A1/A2). This is a protein known to be secreted by the seminal vesicles of the bull and to associate with spermatozoa when they become exposed to seminal plasma during ejaculation (31,32). The purified protein blocks binding of bull spermatozoa to oviductal epithelium (33). The identification of the protein as a constituent of seminal plasma explains the observation that seminal plasma blocks the ability of bull spermatozoa to bind fucose (27). The presence of excess PDC-109 in seminal plasma may ensure that spermatozoa do not bind to any fucose-containing molecules in the lower regions of the female reproductive tract.

Because bull spermatozoa do not acquire PDC-109 on their surface until ejaculation, epididymal spermatozoa cannot bind to oviductal epithelium. In contrast, hamster epididymal spermatozoa can bind to oviductal epithelium (23); therefore, the source of the hamster carbohydrate-binding protein may not be the seminal vesicles and is probably the epididymis.

Release of Spermatozoa from the Reservoir

Spermatozoa must be released from the epithelium in order to ascend to the oviductal ampulla and fertilize oocytes. Hormonal changes associated with impending ovulation do not appear to cause significant reduction of binding sites for spermatozoa on the epithelium (16,17,20); instead, it is the state of the spermatozoa that enables them to release. Smith and Yanagimachi (18) reported that hamster spermatozoa which had undergone both capacitation and hyperactivation *in vitro* did not bind to epithelium when infused into hamster oviducts. While examining the behavior of spermatozoa within oviducts of mated mice, DeMott and Suarez (34) noticed that only hyperactivated mouse spermatozoa detached from epithelium. Pacey and colleagues (35) observed that human spermatozoa detaching from cultured oviductal epithelium were hyperactivated. The flagella of hyperactivated spermatozoa beat asymmetrically, producing acute bends in one direction; therefore, hyperactivation could assist spermatozoa in detaching from epithelium by increasing the force they can exert for pulling away from a surface.

When bull spermatozoa were capacitated, fewer bound to oviductal epithelium (36) and to fucosylated BSA (27). Thus, changes in the surface of spermatozoa that occur during the process of capacitation may be responsible for loss of binding affinity of spermatozoa for its oviductal carbohydrate ligand. Loss in binding affinity of bull spermatozoa for fucosylated molecules may be attributed to loss of PDC-109 from the sperm head (33). As binding affinity of bull spermatozoa for fucosylated molecules decreases during capacitation, hyperactivation of flagellar motility probably helps them to pull away from the wall of the oviduct. This process may be gradual, resulting in detachment and reattachment, until spermatozoa finally contact the cumulus surrounding the oocyte (34).

As mentioned above, carbohydrate recognition is also implicated in binding of spermatozoa to the zona pellucida. The carbohydrates involved in zona binding must differ from those involved in binding to oviductal epithelium, because spermatozoa lose binding affinity for the oviductal carbohydrate ligand when they are capacitated and thus ready to bind to the zona. During capacitation, bull spermatozoa lose the capacity to bind fucose and gain the capacity to bind mannose (27), which is present in terminal positions on bovine zona glycoproteins (37).

Although it appears that epithelium does not release spermatozoa by reducing expression of carbohydrate ligand, epithelial cells could bring about release by secreting initiators of capacitation and hyperactivation. Soluble oviductal factors have been shown to enhance capacitation of bull spermatozoa (38,39). Hormones whose levels rise before ovulation could stimulate secretion of initiators of capacitation and hyperactivation to release spermatozoa at the appropriate time for fertilization.

Is There a Reservoir of Spermatozoa in Women?

No evidence has been found for a distinct reservoir of human spermatozoa in the isthmus of the oviduct (fallopian tube). Human spermatozoa appear to move continuously through the oviduct and on out through the fimbriated end into the peritoneal cavity (1). The lumen of the human oviduct increases in complexity from the uterotabal junction to the ampulla, with increased infolding of the inner surface of the wall. Thus, human spermatozoa encounter an increasingly labyrinthine passageway as they ascend towards the ovary. Croxatto (1) has proposed that this arrangement serves to provide a long lasting, continuous flow of spermatozoa through the oviduct that never reaches high enough numbers to risk polyspermic fertilization. Human spermatozoa appear to stick lightly to oviductal epithelium *in vitro*; therefore, their movement through the oviduct in vivo may be slow and intermittent.

As mentioned above, the cervix may serve as a reservoir in humans, storing spermatozoa temporarily in its crypts. Gradual release from cervical crypts could provide the steady stream of low numbers of spermatozoa through the oviduct, if the spermatozoa are not phagocytized by leukocytes in the uterus.

CONCLUSIONS

Although millions of spermatozoa are inseminated into the female in most mammalian species, their movement through the female reproductive tract is regulated to ensure that a few, but not too many, reach the oocytes in the ampulla of the oviduct. In many species, spermatozoa may be stored in the oviduct during the time between receptivity of the female and final maturation and ovulation of the oocyte. While spermatozoa may be swept through the uterine cavity towards the ovary by peristaltic contractions, they may be required to swim through the cervix, uterotubal junction and oviduct under their own power. Abnormal spermatozoa may be prevented from fertilizing by being filtered out at these sites.

REFERENCES

1. Croxatto H.B. Gamete Transport. In Reproductive Endocrinology, Surgery, and Technology, E.Y. Adashi, J.A. Rock, Z. Rosenwaks, eds. Philadelphia: Lippincott-

Raven Publishers, 1996, 385-402

2. Hawk H.W. Transport and fate of spermatozoa after insemination in cattle. J Dairy Sci 1987; 70:1487-1503

3. Phillips D. M., Mahler S. Phagocytosis of spermatozoa by the rabbit vagina. Anat Rec 1977; 189: 61-72

4. Bedford J. M., Yanagimachi R. Initiation of sperm motility after mating in the rat. J. Androl 1992; 13: 444-449

5. Roberts S.J. Infertility in male animals (andrology). In Veterinary Obstetrics and Genital Diseases (Theriogenology), S.J. Roberts, ed. North Pomfret Vt: David and Charles, Inc. 1986

6. Mullins K.J., Saacke R.G. Study of the functional anatomy of bovine cervical mucosa with special reference to mucus secretion and sperm transport. Anat Rec 1989; 226:106-117

7. Hawk H. W. Transport and fate of spermatozoa after insemination of cattle. J. Dairy Sci 1983; 70,:1487-1503

8. Kunz G., Beil D. , Deininger H., Wildt, L.., Leyendecker, G. The dynamics of rapid sperm transport through the female genital tract: evidence from vaginal sonography of uterine peristalsis and hysterosalpingoscintigraphy. Human Reprod 1996; 11: 627-632

9. Harper M. J. K. Gamete and zygote transport. In The Physiology of Reproduction, E. Knobil, J. D. Neill, eds. New York: Raven Press, Ltd, 1994, 123-187.

10. Wrobel, K.-H., Kujat, R., and Fehle, G., The bovine tubouterine junction: general organization and surface morphology. Cell. Tissue Res 1993; 271: 227-239

11. Suarez S.S. Sperm transport and motility in the mouse oviduct: observations in situ. Biol Reprod 1987; 36:203-210

12. Jansen, R. P. S. , Fallopian tube isthmic mucus and ovum transport. Science 1978; 201:349-351

13. Suarez S. S., Brockman K., Lefebvre R. Distribution of mucus and sperm in bovine oviducts after artificial insemination. Biol. Reprod 1997; 56:447-453

14. Overstreet J. W., Cooper G. W. Sperm transport in the reproductive tract of the female rabbit: I. The rapid transit phase of transport. Biol. Reprod 1978; 19:101-114

15. Yaniz J.L., Lopez-Gatius F., Santolaria P., Mullins K.J. Study of the functional anatomy of bovine oviductal mucosa. Anat Rec 2000; 260: 268-278

16. Lefebvre R., Chenoweth P.J., Drost M., LeClear C.T., MacCubbin M., Dutton J.T., Suarez S.S., Characterization of the oviductal sperm reservoir in cattle. Biol Reprod 1995; 53:1066-1074

17. Baillie H.S,. Pacey A.A., Warren M.A., Scudamore I.W., Barratt C.L.R. Greater numbers of human spermatozoa associate with endosalpingeal cells derived from the isthmus compared with those from the ampulla. Human Reprod 1997; 12:1985-1992

18. Smith T.T., Yanagimachi R. Attachment and release of spermatozoa from the caudal isthmus of the hamster oviduct. J Reprod Fertil 1991; 91:567-573

19. Suarez S.S., Redfern K., Raynor P., Martin F., Phillips D.M. Attachment of boar sperm to mucosal explants of oviduct in vitro: possible role in formation of a sperm reservoir. Biol Reprod 1991; 44:998-1004

20. Thomas P.G.A., Ball B.A., Brinsko S.P. Interaction of equine spermatozoa with oviduct epithelial cell explants is affected by estrous cycle and anatomic origin of explant. Biol Reprod 1994; 51:222-228

21. FlŽchon J.-E., Hunter R.H.F. Distribution of spermatozoa in the utero-tubal junction and isthmus of pigs, and their relationship with the luminal epithelium after mating: a scanning electron microscope study. Tissue & Cell 1981; 13:127-139

22. Hunter R.H.F., FlŽchon B., FlŽchon J.E. Distribution, morphology and epithelial interactions of bovine spermatozoa in the oviduct before and after ovulation: a scanning electron microscopy study. Tissue and Cell 1991; 23:641-656

23. DeMott R.P., Lefebvre R., Suarez S.S. Carbohydrates mediate the adherence of hamster sperm to oviductal epithelium. Biol Reprod 1995; 52:1395-1403

24. Lefebvre R., DeMott R.P., Suarez S.S., Samper J.C. Specific inhibition of equine sperm binding to oviductal epithelium. Equine Reproduction VI, Biol Reprod Mono 1995; 1: 689-696

25. Lefebvre R., Lo M.C., Suarez S.S. Bovine sperm binding to oviductal epithelium involves fucose recognition. Biol Reprod 1997; 56:1198-1204

26. Suarez S.S., Revah I., Lo M., Koelle S. Bull sperm binding to oviductal epithelium is mediated by a Ca2+-dependent lectin on sperm that recognizes Lewis-a trisaccharide. Biol Reprod 1998; 59:39-44

27. Revah I., Gadella B.M., Flesch F.M., Colenbrander B., Suarez S.S. The physiological state of bull sperm affects fucose and mannose binding properties. Biol Reprod 2000; 62:1010-1015

28. Kogan T.P., Revelle B.M., Tapp S., Scott D., Beck P.J. A single amino acid residue can determine the ligand specificity of E-selectin. J Biol Chem 1995; 270:14047-14055

29. Raychoudhury S.S., Millette C.F. Multiple fucosyltransferases and their carbohydrate ligands are involved in spermatogenic cell-Sertoli cell adhesion in vitro in rats. Biol Reprod 1997; 56:1268-1273

30. Tulsiani D.R., Yoshida-Komiya H., Araki Y. Mammalian fertilization: a carbohydrate-mediated event. Biol Reprod 1997; 57:487-94

31. Manjunath P., Chandonnet L., LeBlond E., Desnoyers L. Major proteins of bovine seminal vesicles bind to spermatozoa. Biol Reprod 1993; 49:27-37

32. Gasset M., Saiz J., Laynez J., Sanz L., Gentzel M., Toepfer-Petersen E., Calvete J.J. Conformational features and thermal stability of bovine seminal plasma protein PDC-109 oligomers and phosphorylcholin-bound complexes. Eur J Biochem 1997; 250:735-744

33. Ignotz G.G., Lo M., Perez C., Gwathmey T.M., Suarez S.S. Characterization of a fucose-binding protein from bull sperm and seminal plasma responsible for formation of the oviductal sperm reservoir. Biol Reprod 2001; 64:1806-1811

34. DeMott R.P., Suarez S.S. Hyperactivated sperm progress in the mouse oviduct. Biol Reprod 1992; 46:779-785

35. Pacey A.A., Hill C.J., Scudamore I.W., Warren M.A., Barratt C.L.R., Cooke I.D. The interaction in vitro of human spermatozoa with epithelial cells from the human uterine (Fallopian) tube. Human Reprod 1995; 10:360-366

36. Lefebvre R., Suarez S.S. Effect of capacitation on bull sperm binding to homologous oviductal epithelium. Biol Reprod 1996; 54:575-582

37. Katsumata T., Noguchi S., Yonezawa N., Tanokura M., Nakano M. Structural characterization of the N-linked carbohydrate chains of the zona pellucida glycoproteins from bovine ovarian and fertilized eggs. Eur J Biochem 1996; 240:448-453

38. Chian R.I.-C., LaPointe S., Sirard M.A. Capacitation in vitro of bovine spermatozoa by oviduct cell monolayer conditioned medium. Molec Reprod Dev 1995; 42:318-324

39. Mahmoud A.I., Parrish J.J. Oviduct fluid and heparin induce similar surface changes in bovine sperm during capacitation. Molec Reprod Devel 1996; 43:554-560

Chapter 14

CAPACITATION: SIGNALING PATHWAYS INVOLVED IN SPERM ACQUISITION OF FERTILIZING CAPACITY

V. Anne Westbrook, Alan B. Diekman, John C. Herr and Pablo E. Visconti
Center for Research in Contraceptive and Reproductive Health (CRCRH), Department of Cell Biology, University of Virginia, Charlottesville, VA, USA

INTRODUCTION

Mammalian testicular spermatozoa are morphologically differentiated but are neither progressively motile nor able to fertilize an egg. Although the ability to move forward is acquired during maturation in the epididymis, sperm require a finite period of residence in the female reproductive tract before they become fertilization competent. The molecular, biochemical, and physiological changes that occur to sperm while in the female tract are collectively referred to as capacitation. Capacitation is associated with the attainment of a distinct motility pattern, namely hyperactivation. Another endpoint of sperm capacitation is the ability of spermatozoa to undergo the acrosome reaction in response to physiological stimuli such as the zona pellucida (1) and progesterone (2). During capacitation, changes in membrane properties, enzyme activities, and motility render spermatozoa responsive to stimuli that induce the acrosome reaction and prepare spermatozoa for penetration of the egg investments prior to fertilization. These changes are facilitated by the activation of cell signaling cascades in the female reproductive tract *in vivo* or in defined media *in vitro*. The purposes of this chapter are to consider some recent contributions towards our understanding of capacitation, to summarize open questions in this field, and to discuss future avenues of research. Reviews by Florman and Babcock (1), Cohen-Dayag and Eisenbach (3), Yanagimachi (4), Harrison (5),Kopf (6), de Lamirande et al. (7), Visconti et al. (8), and Baldi et al. (9) provide supplementary reading and complement the material of this chapter.

SIGNALING EVENTS FOR THE INITIATION OF CAPACITATION

To facilitate an understanding of capacitation, a discussion of the complex cascade of molecular events that occur may be divided into events that initiate capacitation and events that are a consequence of this process. Molecular events implicated in the initiation of capacitation have been partially defined including removal of cholesterol from the sperm plasma membrane, modifications in plasma membrane phospholipids, fluxes in Ca^{2+} and other intracellular ion concentrations, alterations in sperm membrane potential, and increased tyrosine phosphorylation of proteins involved in the induction of hyperactivation and the acrosome reaction. A working model for initiation of capacitation based on recent work is presented in Figure 1.

Changes in Membrane Lipids and Phospholipids During Sperm Capacitation

Capacitation is associated with significant changes in the membrane properties of spermatozoa, including an efflux of cholesterol from the sperm plasma membrane. Serum albumin, an essential component of *in vitro* capacitation media, is believed to function as a sink for cholesterol by removing it from the sperm plasma membrane (10-17). Based on these *in vitro* findings, it is important to consider what component of the female tract fluid serves as a cholesterol acceptor *in vivo*. Since fluids of the female tract are partially derived from serum, serum-associated sterol acceptors likely serve this function *in vivo*. A recent study reports that seminal fluid phospholipid-binding proteins interacting with follicular fluid high density lipoprotein (HDL) also releases cholesterol from the sperm membrane during capacitation.

Although serum albumin may have other roles during capacitation (18), its ability to facilitate cholesterol efflux is required for capacitation. For example, capacitation is inhibited by the addition of cholesterol and/or cholesterol analogues to the capacitation medium (19). Furthermore, serum albumin can be substituted in *in vitro* capacitation media with cholesterol-binding compounds such as HDL (19-21) and β-cyclodextrins (2, 22-24) to induce capacitation. In addition to its role as a cholesterol sink, albumin may also function as an acceptor for membrane phospholipids including ether phospholipid 1-*O*-alkyl-2-acetyl-*sn*-glyceryl-3-phosphocholine, also known as platelet activating factor (PAF), a potent signaling molecule (25). Incubation of sperm with PAF in albumin-free media appears to increase the percentage of capacitated sperm based on the chlortetracycline (CTC)

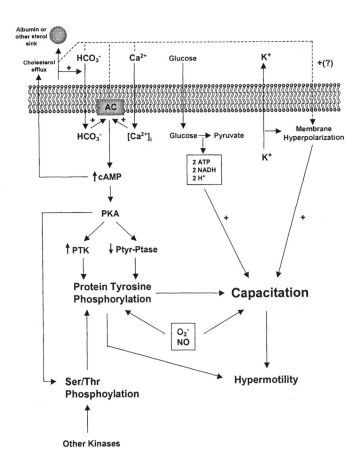

Figure 1. Working model of the transmembrane and intracellular signaling pathways involved in regulating sperm capacitation. This model is based on the work from a number of different laboratories (14, 46, 76, 78). Abbreviations used in this figure: BSA, bovine serum albumin; Chol, cholesterol; 5'AMP, 5' adenosine monophosphate; PTK, protein tyrosine kinase; PTyr-Ptase, phosphotyrosine phosphatase; PDE, cyclic nucleotide phosphodiesterase.

fluorescence assay (25). Recently, release of sperm PAF into the capacitation medium and binding to its receptor on the sperm membrane was shown to induce capacitation *in vitro* by an autocrine loop mechanism (26). The removal of cholesterol, other sterols (e.g. desmosterol) (19), and membrane phospholipids from the plasma membrane is upstream of multiple intracellular signaling events intrinsic to the capacitation process.

In somatic cells, cholesterol removal is thought to disrupt lipid rafts, detergent insoluble micromembrane domains characterized by high content of cholesterol, sphingolipids, gangliosides, and lipid-modified proteins (27). Lipid rafts are important signaling centers in the cell membrane and many proteins involved in cell signaling congregate in these areas including tyrosine kinases, G proteins, inositol phospholipids, GPI-anchored proteins, nitric oxide synthase and/or other molecules (28-30). In the presence of the membrane protein caveolin, these glycolipid rafts can form non-clathrin coated pits termed caveolae. In sperm, cholesterol may likewise be concentrated in specialized plasma membrane microdomains such as lipid rafts and caveolae (28) and removal of cholesterol may activate signaling events during capacitation through the disruption of membrane assemblies (31). This idea is supported by the recent finding that one important component of caveolae, caveolin 1, is present in the plasma membrane overlying the acrosomal region and the flagellum of mouse and guinea pig sperm (32). The hypothesis that lipid rafts and caveolae concentrate signaling complexes in the sperm plasma membrane warrants continued investigation.

Changes in the composition and distribution of membrane phospholipids are another significant attribute of capacitation. The removal of cholesterol from the sperm plasma membrane decreases the cholesterol/phospholipid ratio as assessed by different criteria (12, 33, 34). This could account for the membrane fluidity changes (35, 36) and the redistribution of membrane proteins observed with lectin (37) and immunochemical analysis (38, 39) that occur during capacitation. Although there is little change in the total amount of phospholipids under capacitating conditions, there is an increase in the methylation of phospholipids, in the translocation and concentration of phosphatidylcholine to the inner leaflet, and in the production of fusogenic lysophospholipids during capacitation (4, 40, 41). Furthermore, capacitation-associated alterations in the transbilayer phospholipid behavior resulting in membrane lipid disorder were reported to occur through a cAMP-dependent pathway after exposure of boar sperm to HCO_3^- (42). Therefore, multiple plasma membrane modifications occur during the process of capacitation.

HCO_3^-, Ca^{2+}, and the cAMP Pathway

During transit through the male and female reproductive tracts, sperm are exposed to significant changes in the extracellular millieu, including differences in extracellular ion concentrations that play a role in cell signaling during capacitation. For example, numerous studies have demonstrated that capacitation is a HCO_3^- ion-dependent process (43-47). Little is known about the mechanisms of HCO_3^- transport in sperm.

However, the ability of 4,4'-diidothiocyanatostilbene-2,2'-disulfonic acid (DIDS), a well-known inhibitor of anion transporters, to inhibit the actions of HCO_3^- on various sperm functions suggests that sperm contain functional anion transporters (48-51).

The transmembrane movement of HCO_3^- anions into sperm may be responsible for the known increase in intracellular pH that is observed during capacitation (52, 53) as well as the regulation of sperm cAMP metabolism (50, 54, 55). Adenylyl cyclase, the enzyme responsible for cAMP synthesis, is markedly stimulated by increased levels of HCO_3^- and has been the subject of multiple studies in sperm, but whether this enzymatic activity is represented by one or more proteins remain controversial.

Sperm adenylyl cyclase has several unique biochemical properties (56, 57). For example, unlike somatic cell cyclases, responses of sperm adenylyl cyclase to agents that modulate stimulatory GTP-binding proteins (Gs), such as cholera toxin, AlF_4^- and GTP analogues, are weak or completely absent. Since no cholera toxin-ADP ribosylated substrate has been detected in mammalian sperm, the low sensitivity to G proteins effectors could be due to the lack of stimulatory Gαs protein in these cells (58). Alternatively, sperm adenylyl cyclase may be unable to interact with Gs proteins due to differences in cyclase tertiary structure.

As mentioned above, an important property of sperm adenylyl cyclase is its regulation by HCO_3^- anion (54). Sperm adenylyl cyclase is likely to be a posttranslationally modified form of the testicular soluble adenylyl cyclase (SAC) (59) encoded by a gene distinct from the transmembrane adenylyl cyclase (tmAC) gene family. Two alternatively spliced products of the SAC gene have recently been reported: a truncated SAC mRNA encoding only the catalytic domain and a full-length SAC mRNA. Both forms of testicular SAC protein are regulated by HCO_3^-, as is sperm adenylyl cyclase, and enzymatically active in testicular germ cells (60). Furthermore, antibodies raised against the catalytic domain of testicular SAC recognized two sperm proteins corresponding to the deduced molecular masses of the two forms of the testicular enzyme (61). These data suggest that testicular SAC remains associated with sperm after sperm release from the testis. Interestingly, the sequence from the catalytyic domain of SAC has sequence homology to cyanobacterial adenylyl cyclases which are also HCO_3^--dependent (61). Although testicular SAC has been found in the soluble fraction, it is significant that cyclase activity identified in mammalian sperm remains associated with the particulate membrane fraction. Therefore, testicular SAC would be predicted to have a mechanism allowing for translocation from the cytosol to the membrane at some point during spermatogenesis or epididymal maturation.

Numerous studies have also demonstrated that capacitation is calcium-dependent (46, 62). An increase in the concentration of Ca^{2+} during capacitation has been demonstrated in mammalian spermatozoa by some investigators (63). Mechanisms suggested for regulation of intracellular Ca^{2+} concentration during sperm capacitation include a Ca^{2+}-ATPase extrusion system (64), a Na^{+}/Ca^{2+} antiporter (65), voltage-activated calcium channels (66), and intracellular calcium stores (67). Downstream targets of calcium during capacitation include the calcium-binding protein calmodulin. A recent report indicates that purified calmodulin reverses the blocking effect of calmodulin antagonists on capacitation and on the mannose-BSA-induced acrosome reaction (68).

The initiation and/or regulation of capacitation by Ca^{2+} occur via different targets, some of which are involved with cAMP metabolism. Since Ca^{2+}/calmodulin can activate both the synthesis of cAMP by adenylyl cyclase (69) in sperm and its degradation by cyclic nucleotide phosphodiesterases (70), it is not surprising that Ca^{2+} has both positive and negative actions on capacitation and related signaling events. Ca^{2+} has a positive effect on mouse sperm by inducing capacitation-associated changes in protein tyrosine phosphorylation (46). In contrast, Ca^{2+} has been demonstrated to inhibit protein tyrosine phosphorylation in human sperm during the first two hours of *in vitro* capacitation (71, 72). An increase in intracellular sperm Ca^{2+} during capacitation has been described, whereas others have shown that no changes occur during this maturational event (4). This ambiguity could be due, in part, to the well-demonstrated action of Ca^{2+} on the acrosome reaction and to the inherent difficulties in differentiating between capacitation and the acrosome reaction. However, the action of Ca^{2+} at the level of effector enzymes' involved in sperm signal transduction suggests that this divalent cation is likely to play an important role in capacitation. This idea is further confirmed by the recent finding of a novel Ca^{2+} channel (CatSper) essential for motility in mouse sperm (73). Targeted disruption of the gene results in male sterility in otherwise normal mice. Sperm motility is decreased markedly in CatSper-/- mice and they are unable to fertilize intact eggs.

Ion Fluxes and the Regulation of Sperm Plasma Membrane Potential

Ion fluxes play a fundamental role in sperm physiology of multiple species. Their role in the acrosome reaction has been well established both in invertebrate as well as vertebrate sperm. Movement of ions has also been associated with other aspects of sperm function such as sperm motility (74, 75), regulation of intracellular messengers (76, 77), and mammalian sperm capacitation (66, 78). In all these cases, the influence of the external ion composition and the effect of channel blockers suggest that ion channels actively participate in the regulation of sperm function. Since ion channels can catalyze the flow of millions of ions through the ion-impermeable lipid bilayer, a few of them can cause changes in sperm within milliseconds (79). The ion concentration gradient through the plasma membrane determines the plasma membrane potential via ion-selective channels and controls the extent of channel activity and ion flow.

During epididymal transit, and subsequently in the female tract, sperm are exposed to significant changes in the extracellular millieu including changes in extracellular ion concentrations and osmolality. For example, caudal epididymal sperm are stored in an environment that contains high K^+, low Na^+, and very low HCO_3^- concentrations (4, 80, 81). After ejaculation, in the seminal fluid and then in the female tract, the ionic environment is significantly different; while K^+ concentrations are reduced, there is a significant increase in Na^+ and HCO_3^- concentrations. These dramatic shifts in the external ion concentrations that occur after ejaculation trigger modulations of the sperm intracellular ionic environment and consequently lead to an altered sperm plasma membrane potential.

In the absence of any stimulus, the plasma membrane potential is known as the resting plasma membrane potential. During *in vitro* capacitation, the mammalian sperm resting membrane potential is determined by the relative permeability of the sperm plasma membrane for ions that constitute the capacitation media. The ion composition of capacitation media mimics that of oviductal fluid (4). These media are high in Na^+ (about 130 mM) and Cl^- (about 100 mM), but low in K^+ (about 5.9 mM). Capacitation media also contain Ca^{2+} (about 2.7 mM) and HCO_3^- (10-25 mM). In contrast, intracellular fluids of sperm have a low concentration of Na^+ (about 14 mM) and high concentration of K^+ (about 90 to 120 mM) (78, 82). The free intracellular Ca^{2+} concentration of non-capacitated sperm is approximately 0.1 µM or less, but during the acrosome reaction it may increase to approximately 10 µM (66, 83). To date, the intracellular concentrations of Cl^- and HCO_3^- in sperm have not been determined. Since pH reflects the H^+ ion concentrations, it is also important to note that while capacitation media have usually a pH between 7.2 and 7.4, the sperm cytosolic pH is much lower at approximately 6.5 (53, 82).

Mammalian sperm capacitation is accompanied by the hyperpolarization of the sperm plasma membrane (78). Hyperpolarization is observed as an increase in the intracellular negative charges when compared with the extracellular environment. Although it is not clear how the sperm plasma membrane potential is regulated during capacitation, it appears that membrane hyperpolarization may be partially due to an enhanced K^+ permeability as a result of the decrease in inhibitory modulation of K^+ channels (78). As shown in Fig. 1, extrusion of K^+ ions reduces the intracellular positive charge and hyperpolarizes the sperm plasma membrane. Biochemical and molecular biology data suggest the presence of different K^+ channels in both mammalian testis and sperm (84-86).

However, electrophysiological investigation of ion channels in the plasma membrane of mature sperm has been precluded by the small size of these cells (87). Recently, Muñoz-Garay et al. (88) demonstrated with patch clamp techniques that inward rectifying K^+ channels are active in mouse spermatogenic cells and proposed that these channels may be responsible for the capacitation-associated hyperpolarization. Interestingly, Ba^{2+} blocks these K^+ channels with an IC_{50} similar to that shown to inhibit the capacitation-associated hyperpolarization and the zona pellucida-induced acrosome reaction (88).

The functional role of sperm plasma membrane hyperpolarization during capacitation is at present not well understood. However, one may speculate that since capacitation prepares the sperm for the acrosome reaction, capacitation-associated hyperpolarization may regulate the ability of sperm to generate transient Ca^{2+} elevations during acrosome reaction by physiological agonists (e.g. *zona pellucida* of the egg or progesterone). This hypothesis is consistent with the presence of low voltage-activated (LVA) Ca^{2+} T channels in spermatogenic cells (89, 90) that may also be present in mature sperm. A signature property of LVA Ca^{2+} channels is a low threshold for voltage-dependent inactivation, this means that these Ca^{2+} channels are inactive at high membrane potentials, such as the ones observed before capacitation. In other words, these Ca^{2+} channels are inactivated at holding potentials that are more positive than -80 mV (89, 90). Thus, if LVA Ca^{2+} T channels are involved in the regulation of the acrosome reaction, sperm must maintain a hyperpolarized membrane potential during the early stages of interaction with the egg (66, 91).

When mammalian sperm are studied in a population analysis, the membrane potential overall drops from around –30 to –55 mV. However, when sperm are studied as single cells, Arnoult et al. (66) demonstrated that after 1 h of capacitation, mice sperm were divided into two populations. Approximately 50 % of the sperm remain at a membrane potential close to the uncapacitated population. In the remaining sperm, the membrane potential decreases to approximately –80 mV, a membrane potential compatible with the opening of LVA Ca^{2+} channels. The observation that only a fraction of the sperm population undergoes changes in membrane potential is compatible with the finding that only a subpopulation of sperm are able to acrosome react in the presence of physiological agonists. Moreover, Arnoult and coworkers (66) demonstrated that the zona pellucida-induced acrosome reaction was observed only in the hyperpolarized sperm population.

Capacitation-Associated Changes in Protein Tyrosine Phosphorylation

Capacitation-associated changes in protein tyrosine phosphorylation have been demonstrated in multiple species including the mouse (46), bovine (92), human (2, 93), pig (94), and hamster (51, 95). For example, *in vitro* capacitation of mouse cauda epididymal sperm promotes tyrosine phosphorylation of a subset of proteins between Mr 40,000 - 120,000 (46). The increase in tyrosine phosphorylation is dependent on the presence of serum albumin, Ca^{2+}, and HCO_3 in the medium at concentrations that correlate with those required for capacitation (46). Among different species, the concentration of these media components necessary to promote protein tyrosine phosphorylation varies slightly (51). In addition, it has been demonstrated that glucose, but not lactate or pyruvate, is necessary to promote protein tyrosine phosphorylation during capacitation (96, 97). The absence of any one of these components in the media prevents both tyrosine phosphorylation of sperm proteins and the capacitation of spermatozoa.

The dependence of *in vitro* protein tyrosine phosphorylation on serum albumin, or other cholesterol acceptors, suggests a correlation between cholesterol efflux and tyrosine phosphorylation. Whether cholesterol removal is upstream from or coincident with the action of Ca^{2+} and/or HCO_3^- is not presently known. However, it is hypothesized that cholesterol removal, with a resultant change in sperm plasma membrane fluidity, modulates Ca^{2+} and/or HCO_3^- ion fluxes leading to adenylyl cyclase activation; this hypothesis remains to be tested.

The increase in protein tyrosine phosphorylation by serum albumin, Ca^{2+} and HCO_3^- is regulated by a cAMP-dependent pathway that involves protein kinase A (PKA), a serine/threonine kinase, in spermatozoa of several species including the mouse (76), bull (92), human (2, 93), boar (94) and hamster (51). Inhibitors of PKA activity block tyrosine phosphorylation as well as capacitation. Interestingly, a testis- and sperm-specific transcript encoding a unique catalytic subunit of PKA, designated C_s, has been recently identified in several mammalian species and is localized to the tail of mouse testicular spermatozoa (98, 99).

Since PKA is not able to directly phosphorylate proteins on tyrosine residues, an intermediate tyrosine kinase must be involved during capacitation. Fig. 1 summarizes three possible mechanisms of interaction between PKA and protein tyrosine phosphorylation: 1) Direct or indirect stimulation of a tyrosine kinase by PKA; 2) direct or indirect inhibition of a phosphotyrosine phosphatase; 3) direct or indirect phosphorylation of proteins by PKA on serine or threonine residues that prime these proteins for subsequent phosphorylation by a tyrosine kinase. These inter-related possibilities are currently being explored in several laboratories. The identification of specific enzymes and substrates involved in the PKA/tyrosine phosphorylation pathway(s) will improve our knowledge of the capacitation process. Interestingly, the effect of glucose on protein tyrosine phosphorylation appears to be downstream of PKA/cAMP signaling events and to arise indirectly as a result of metabolism (Travis et al., 2001).

Superoxide anion ($O_2^{\bullet-}$) and nitric oxide (NO) also appear to mediate capacitation by inducing protein tyrosine phosphorylation (100-102) via the cAMP/PKA signaling pathway (103, 104). Reactive oxygen species (ROS), primarily superoxide anion and hydrogen peroxide (H_2O_2), are spontaneously generated by mammalian spermatozoa following their release from the epididymis (105). In addition, NO increases during sperm capacitation (104). The generation of these reactive oxygen species in controlled quantities stimulates oxido-reduction events that appear to be involved in sperm capacitation and the hyperactivation of motility (106).

Other Modulators of Signaling During Sperm Capacitation

Cyclic AMP (cAMP) appears to be a central regulator of several sperm processes, such as motility (107), capacitation (76, 108), and the acrosome reaction (56, 109, 110). Changes in membrane potential also are involved in these sperm functions (66, 74, 78, 79, 111). Presently, it is not known whether these two signaling events interact. However, it is likely that these two processes are related since capacitation is accompanied by both hyperpolarization of the plasma membrane and an increase in cAMP synthesis. Supporting this idea, the presence of a membrane potential-regulated adenylyl cyclase has been reported in non-mammalian species (112). In addition, although PKA is the main downstream effector for cAMP in sperm, the view that PKA mediates all of the effects of cAMP has been amended with the discovery of new types of cyclic nucleotide receptors. These receptors include cyclic-nucleotide-gated channels, exchange factors (113), a cGMP binding cyclic nucleotide phosphodiesterase, and extracellular cAMP receptors (114). Cyclic nucleotide-gated channels were identified in sea urchin (115) and mammalian sperm (116) with specificity for cAMP and cGMP, respectively. Altogether, these data support the idea that changes in membrane potential and changes in the cAMP signaling pathway might interact during capacitation. Alternate possibilities are summarized in Figure 2.

CONSEQUENCES OF CAPACITATION ON SPERM FUNCTION

Although fertilization is the ultimate evidence of sperm capacitation, the ability of sperm to undergo a physiologically-stimulated acrosome reaction (e.g. in response to the zona pellucida or progesterone) can be taken as an earlier hallmark or measure of capacitation (1, 8). Capacitation is also correlated with changes in sperm motility patterns, referred to as hyperactivation or hypermotility, in a number of species (4, 117). When one considers the process of capacitation at the molecular level, events occurring both in the sperm head (i.e acrosome reaction) and in the flagellum (i.e. motility changes) must be taken into account. One may postulate that components of the sperm exocytotic and motility machinery are modified during capacitation. Such alterations may involve changes in protein phosphorylation, changes in protein localization, and/or modification of protein-protein interactions. Experiments leading to the identification and characterization of effector molecules will further increase our understanding of capacitation.

To understand the link between capacitation and the acrosome reaction, an increased knowledge of the mechanisms that regulate this exocytotic event in sperm is necessary. Exocytosis is a tightly regulated, complex process that involves fusion of intracellular vesicles with the overlying plasma membrane and release of vesicular contents. Membrane fusion is apparently governed by a few conserved protein families regardless of whether membrane fusion occurs between intracellular organelles or between trafficking vesicles and the plasma membrane. One such protein family is commonly referred to as SNARE proteins (Soluble N-ethylmaleimide-sensitive attachment protein receptors) (118). Sperm homologues of SNARE proteins as well as SNARE-associated proteins, such as Rab 3A and NSF, have been detected in sea urchin (119, 120) and mammalian sperm (121-123). These observations indicate that the sperm acrosome reaction is regulated in similar ways as exocytotic processes in somatic cells. Since capacitation is necessary for exocytosis in mammalian sperm, elucidation of the mechanisms regulating the acrosome reaction will also increase our understanding of capacitation.

Capacitation is also correlated with moleuclar events that occur in the sperm flagellum. For example, two members of the A Kinase Anchoring Protein (AKAP) family located in the fibrous sheath, a flagellar cytoskeletal structure, become phosphorylated at tyrosine residues during human sperm capacitation (71, 124, 125). AKAPs represent a growing family of scaffolding proteins that function to tether the regulatory subunits of PKA and signalling enzymes, such as calcineurin and protein kinase C, to organelles or cytoskeletal elements. Such proteins regulate signal trans-

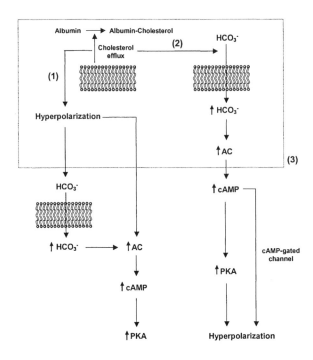

Figure 2. Crosstalk between signaling pathways involved in capacitation. (1) *Hyperpolarization is upstream to the increase in cAMP synthesis.* Cholesterol removal regulates sperm plasma membrane potential through a K^+ channel or through an increase in anionic permeability. Hyperpolarization may directly regulate adenylyl cyclase activity as described in sea urchin and trout sperm (112) or indirectly via an increase in HCO_3^-. (2) *Hyperpolarization is downstream of the increase in cAMP synthesis.* In this model, cholesterol removal regulates HCO_3^- permeability and HCO_3^- stimulates adenylyl cyclase. cAMP could then activate a cyclic nucleotide gated channel directly or indirectly through phosphorylation by PKA leading to plasma membrane hyperpolarization. (3) *Hyperpolarization and cAMP synthesis are independent events associated with capacitation.*

transduction in discrete regions of the cell (126). Tyrosine phosphorylation of AKAPs may modulate the biochemical and biophysical properties of these proteins and the fibrous sheath and thus, contribute to the regulation of events associated with flagellar bending, including observed alterations in tail wave amplitude during hyperactivation.

Although hyperactivation and the ability to undergo the acrosome reaction are well correlated with sperm capacitation, other aspects of sperm function also appear to be affected by capacitation. For example, the increase in intra-acrosomal pH that occurs during capacitation may prepare acrosomal contents for release during the acrosome reaction (127). Furthermore, the sperm of multiple species respond to chemotactic factors secreted by the egg or present in the oviduct; in the human, only capacitated sperm are observed to respond to this chemotaxis (128). Continued investigation of sperm capacitation will undoubtedly identify additional consequences of this maturation process.

CONCLUSIONS

Capacitation is a unique feature of mammalian sperm that occurs in the female tract and is essential for fertilization. During capacitation, sperm acquire hyperactive motility and the ability to undergo a regulated acrosome reaction. Capacitation is accompanied by changes in lipid composition of the plasma membrane as well as posttranslational modifications of proteins regulated by several signaling pathways. Although work emanating from multiple laboratories is leading to a better understanding of capacitation, most of the proteins involved in this process remain to be characterized. The molecular identification of tyrosine kinases, phosphotyrosine phosphatases, tyrosine phosphorylation substrates, and Ca^{2+} channels present in mammalian sperm will certainly increase our understanding of the molecular basis of capacitation.

ACKNOWLEDGEMENTS

This chapter was supported in part by NIH HD38082 (PEV), NIH U54 29099 (JCH) and HD 35523 (ABD and JCH) and by the Andrew W. Mellon Foundation.

REFERENCES

1. Florman HM, Babcock DF. Progress towards understanding the molecular basis of capacitation. In: Chemistry of Fertilization: CRC Uniscience; 1991: 105-132.
2. Osheroff JE, Visconti PE, Valenzuela JP, Travis AJ, Alvarez J, Kopf GS. Regulation of human sperm capacitation by a cholesterol efflux- stimulated signal transduction pathway leading to protein kinase A- mediated up-regulation of protein tyrosine phosphorylation. Mol Hum Reprod 1999; 5: 1017-1026.
3. Cohen-Dayag A, Eisenbach M. Potential assays for sperm capacitation in mammals [published errata appear in Am J Physiol 1995 Mar;268(3 Pt 1):section C following table

of contents and 1995 Jun;268(6 Pt 3):section C following table of contents]. Am J Physiol 1994; 267: C1167-1176.

4. Yanagimachi R. Mammalian fertilization. In: Knobil E, Neill JD (eds.), The Physiology of Reproduction, vol. 1. New York: Raven Press, Ltd.; 1994: 189-317.

5. Harrison RA. Capacitation mechanisms, and the role of capacitation as seen in eutherian mammals. Reprod Fertil Dev 1996; 8: 581-594.

6. Visconti PE, Kopf GS. Regulation of protein phosphorylation during sperm capacitation. Biol Reprod 1998; 59: 1-6.

7. de Lamirande E, Leclerc P, Gagnon C. Capacitation as a regulatory event that primes spermatozoa for the acrosome reaction and fertilization. Mol Hum Reprod 1997; 3: 175-194.

8. Visconti PE, Galantino-Homer H, Moore GD, Bailey JL, Ning X, Fornes M, Kopf GS. The molecular basis of sperm capacitation. J Androl 1998; 19: 242-248.

9. Baldi E, Luconi M, Bonaccorsi L, Muratori M, Forti G. Intracellular events and signaling pathways involved in sperm acquisition of fertilizing capacity and acrosome reaction [In Process Citation]. Front Biosci 2000; 5: E110-123.

10. Davis BK, Byrne R, Hungund B. Studies on the mechanism of capacitation. II. Evidence for lipid transfer between plasma membrane of rat sperm and serum albumin during capacitation in vitro. Biochim Biophys Acta 1979; 558: 257-266.

11. Davis BK, Byrne R, Bedigian K. Studies on the mechanism of capacitation: albumin-mediated changes in plasma membrane lipids during in vitro incubation of rat sperm cells. Proc Natl Acad Sci U S A 1980; 77: 1546-1550.

12. Davis BK. Timing of fertilization in mammals: sperm cholesterol/phospholipid ratio as a determinant of the capacitation interval. Proc Natl Acad Sci U S A 1981; 78: 7560-7564.

13. Go KJ, Wolf DP. Albumin-mediated changes in sperm sterol content during capacitation. Biol Reprod 1985; 32: 145-153.

14. Langlais J, Roberts JD. A molecular membrane model of sperm capacitation and the acrosome reaction of mammalian spermatozoa. Gamete Res. 1985; 12: 183-224.

15. Suzuki F, Yanagimachi R. Changes in the distribution of intramembranous particles and filipin-reactive membrane sterols during in vitro capacitation of golden hamster spermatozoa. Gamete Res 1989; 23: 335-347.

16. Cross NL. Effect of cholesterol and other sterols on human sperm acrosomal responsiveness. Mol Reprod Dev 1996; 45: 212-217.

17. Cross NL. Role of cholesterol in sperm capacitation. Biol Reprod 1998; 59: 7-11.

18. Espinosa F, Lopez-Gonzalez I, Munoz-Garay C, Felix R, De la Vega-Beltran JL, Kopf GS, Visconti PE, Darszon A. Dual regulation of the T-type Ca(2+) current by serum albumin and beta-estradiol in mammalian spermatogenic cells. FEBS Lett 2000; 475: 251-256.

19. Visconti PE, Ning X, Fornes MW, Alvarez JG, Stein P, Connors SA, Kopf GS. Cholesterol efflux-mediated signal transduction in mammalian sperm: cholesterol release signals an increase in protein tyrosine phosphorylation during mouse sperm capacitation. Dev Biol 1999; 214: 429-443.

20. Therien I, Soubeyrand S, Manjunath P. Major proteins of bovine seminal plasma modulate sperm capacitation by high-density lipoprotein. Biol Reprod 1997; 57: 1080-1088.

21. Therien I, Bousquet D, Manjunath P. Effect of seminal phospholipid-binding proteins and follicular fluid on bovine sperm capacitation. Biol Reprod 2001; 65: 41-51.

22. Visconti PE, Galantino-Homer H, Ning X, Moore GD, Valenzuela JP, Jorgez CJ, Alvarez JG, Kopf GS. Cholesterol efflux-mediated signal transduction in mammalian sperm. beta-cyclodextrins initiate transmembrane signaling leading to an increase in protein tyrosine phosphorylation and capacitation. J Biol Chem 1999; 274: 3235-3242.

23. Choi YH, Toyoda Y. Cyclodextrin removes cholesterol from mouse sperm and induces capacitation in a protein-free medium. Biol Reprod 1998; 59: 1328-1333.

24. Cross NL. Effect of methyl-beta-cyclodextrin on the acrosomal responsiveness of human

sperm. Mol Reprod Dev 1999; 53: 92-98.

25. Huo LJ, Yang ZM. Effects of platelet activating factor on capacitation and acrosome reaction in mouse spermatozoa. Mol Reprod Dev 2000; 56: 436-440.

26. Wu C, Stojanov T, Chami O, Ishii S, Shimuzu T, Li A, O'Neill C. Evidence for the autocrine induction of capacitation of mammalian spermatozoa. J Biol Chem 2001; 276: 26962-26968.

27. Anderson RG. The caveolae membrane system. Annu Rev Biochem 1998; 67: 199-225.

28. Brown DA, London E. Functions of lipid rafts in biological membranes. Annu Rev Cell Dev Biol 1998; 14: 111-136.

29. Kabouridis PS, Magee AI, Ley SC. S-acylation of LCK protein tyrosine kinase is essential for its signalling function in T lymphocytes. Embo J 1997; 16: 4983-4998.

30. Roy S, Luetterforst R, Harding A, Apolloni A, Etheridge M, Stang E, Rolls B, Hancock JF, Parton RG. Dominant-negative caveolin inhibits H-Ras function by disrupting cholesterol-rich plasma membrane domains [see comments]. Nat Cell Biol 1999; 1: 98-105.

31. Fielding CJ. Caveolae and signaling. Curr Opin Lipidol 2001; 12: 281-287.

32. Travis AJ, Vargas LA, Merdiushev T, Moss SB, Hunnicutt GR, Kopf GS. Expression and localization of caveolin-1 in mouse and guinea pig spermatozoa. Mol. Biol. Cell 2000; 11: 121a.

33. Hoshi K, Aita T, Yanagida K, Yoshimatsu N, Sato A. Variation in the cholesterol/phospholipid ratio in human spermatozoa and its relationship with capacitation. Hum Reprod 1990; 5: 71-74.

34. Tesarik J, Flechon JE. Distribution of sterols and anionic lipids in human sperm plasma membrane: effects of in vitro capacitation. J Ultrastruct Mol Struct Res 1986; 97: 227-237.

35. Wolf DE, Hagopian SS, Ishijima S. Changes in sperm plasma membrane lipid diffusibility after hyperactivation during in vitro capacitation in the mouse. J Cell Biol 1986; 102: 1372-1377.

36. Wolf DE, Cardullo RA. Physiological properties of the mammalian sperm plasma membrane. In: Baccetti B (ed.) Comparative Spermatology 20 Years After. New York: Raven Press; 1991: 599-604.

37. Cross NL, Overstreet JW. Glycoconjugates of the human sperm surface: distribution and alterations that accompany capacitation in vitro. Gamete Res 1987; 16: 23-35.

38. Rochwerger L, Cuasnicu PS. Redistribution of a rat sperm epididymal glycoprotein after in vitro and in vivo capacitation. Mol Reprod Dev 1992; 31: 34-41.

39. Shalgi R, Matityahu A, Gaunt SJ, Jones R. Antigens on rat spermatozoa with a potential role in fertilization. Mol Reprod Dev 1990; 25: 286-296.

40. Llanos MN, Meizel S. Phospholipid methylation increases during capacitation of golden hamster sperm in vitro. Biol Reprod 1983; 28: 1043-1051.

41. Gadella BM, Miller NG, Colenbrander B, van Golde LM, Harrison RA. Flow cytometric detection of transbilayer movement of fluorescent phospholipid analogues across the boar sperm plasma membrane: elimination of labeling artifacts. Mol Reprod Dev 1999; 53: 108-125.

42. Gadella BM, Harrison RA. The capacitating agent bicarbonate induces protein kinase A-dependent changes in phospholipid transbilayer behavior in the sperm plasma membrane. Development 2000; 127: 2407-2420.

43. Lee MA, Storey BT. Bicarbonate is essential for fertilization of mouse eggs: mouse sperm require it to undergo the acrosome reaction. Biol Reprod 1986; 34: 349-356.

44. Neill JM, Olds-Clarke P. A computer-assisted assay for mouse sperm hyperactivation demonstrates that bicarbonate but not bovine serum albumin is required. Gamete Res 1987; 18: 121-140.

45. Shi QX, Roldan ER. Bicarbonate/CO2 is not required for zona pellucida- or progesterone-induced acrosomal exocytosis of mouse spermatozoa but is essential for capacitation. Biol Reprod 1995; 52: 540-546.

46. Visconti PE, Bailey JL, Moore GD, Pan D, Olds-Clarke P, Kopf GS. Capacitation of mouse spermatozoa. I. Correlation between the capacitation state and protein tyrosine phosphorylation. Development 1995; 121: 1129-1137.

47. Boatman DE, Robbins RS. Bicarbonate: carbon-dioxide regulation of sperm capacitation, hyperactivated motility, and acrosome reactions. Biol Reprod 1991; 44: 806-813.

48. Okamura N, Tajima Y, Sugita Y. Decrease in bicarbonate transport activities during epididymal maturation of porcine sperm. Biochem Biophys Res Commun 1988; 157: 1280-1287.

49. Spira B, Breitbart H. The role of anion channels in the mechanism of acrosome reaction in bull spermatozoa. Biochim Biophys Acta 1992; 1109: 65-73.

50. Visconti PE, Muschietti JP, Flawia MM, Tezon JG. Bicarbonate dependence of cAMP accumulation induced by phorbol esters in hamster spermatozoa. Biochim Biophys Acta 1990; 1054: 231-236.

51. Visconti PE, Stewart-Savage J, Blasco A, Battaglia L, Miranda P, Kopf GS, Tezon JG. Roles of bicarbonate, cAMP, and protein tyrosine phosphorylation on capacitation and the spontaneous acrosome reaction of hamster sperm. Biol Reprod 1999; 61: 76-84.

52. Uguz C, Vredenburgh WL, Parrish JJ. Heparin-induced capacitation but not intracellular alkalinization of bovine sperm is inhibited by Rp-adenosine-3',5'-cyclic monophosphorothioate. Biol Reprod 1994; 51: 1031-1039.

53. Zeng Y, Oberdorf JA, Florman HM. pH regulation in mouse sperm: identification of Na(+)-, Cl(-)-, and HCO3(-)-dependent and arylaminobenzoate-dependent regulatory mechanisms and characterization of their roles in sperm capacitation. Dev Biol 1996; 173: 510-520.

54. Okamura N, Tajima Y, Soejima A, Masuda H, Sugita Y. Sodium bicarbonate in seminal plasma stimulates the motility of mammalian spermatozoa through direct activation of adenylate cyclase. J Biol Chem 1985; 260: 9699-9705.

55. Garty NB, Salomon Y. Stimulation of partially purified adenylate cyclase from bull sperm by bicarbonate. FEBS Lett 1987; 218: 148-152.

56. Leclerc P, Kopf GS. Mouse sperm adenylyl cyclase: general properties and regulation by the zona pellucida. Biol Reprod 1995; 52: 1227-1233.

57. Garbers DL, Kopf GS. The regulation of spermatozoa by calcium and cyclic nucleotides. Adv Cyclic Nucleotide Res 1980; 13: 251-306.

58. Hildebrandt JD, Codina J, Tash JS, Kirchick HJ, Lipschultz L, Sekura RD, Birnbaumer L. The membrane-bound spermatozoal adenylyl cyclase system does not share coupling characteristics with somatic cell adenylyl cyclases. Endocrinology 1985; 116: 1357-1366.

59. Buck J, Sinclair ML, Schapal L, Cann MJ, Levin LR. Cytosolic adenylyl cyclase defines a unique signaling molecule in mammals. Proc Natl Acad Sci U S A 1999; 96: 79-84.

60. Jaiswal BS, Conti M. Identification and functional analysis of splice variants of the germ cell soluble adenylyl cyclase. J Biol Chem 2001; 276: 31698-31708.

61. Chen Y, Cann MJ, Litvin TN, Iourgenko V, Sinclair ML, Levin LR, Buck J. Soluble adenylyl cyclase as an evolutionarily conserved bicarbonate sensor [see comments]. Science 2000; 289: 625-628.

62. DasGupta S, Mills CL, Fraser LR. Ca(2+)-related changes in the capacitation state of human spermatozoa assessed by a chlortetracycline fluorescence assay. J Reprod Fertil 1993; 99: 135-143.

63. Baldi E, Casano R, Falsetti C, Krausz C, Maggi M, Forti G. Intracellular calcium accumulation and responsiveness to progesterone in capacitating human spermatozoa. J Androl 1991; 12: 323-330.

64. Roldan ER, Fleming AD. Is a Ca2+ -ATPase involved in Ca2+ regulation during capacitation and the acrosome reaction of guinea-pig spermatozoa? J Reprod Fertil 1989; 85: 297-308.

65. Ben-Av P, Rubinstein S, Breitbart H. Induction of acrosomal reaction and calcium

uptake in ram spermatozoa by ionophores. Biochim Biophys Acta 1988; 939: 214-222.

66. Arnoult C, Kazam IG, Visconti PE, Kopf GS, Villaz M, Florman HM. Control of the low voltage-activated calcium channel of mouse sperm by egg ZP3 and by membrane hyperpolarization during capacitation. Proc Natl Acad Sci U S A 1999; 96: 6757-6762.

67. O'Toole CM, Arnoult C, Darszon A, Steinhardt RA, Florman HM. Ca(2+) entry through store-operated channels in mouse sperm is initiated by egg ZP3 and drives the acrosome reaction. Mol Biol Cell 2000; 11: 1571-1584.

68. Bendahmane M, Lynch C, 2nd, Tulsiani DR. Calmodulin signals capacitation and triggers the agonist-induced acrosome reaction in mouse spermatozoa. Arch Biochem Biophys 2001; 390: 1-8.

69. Gross MK, Toscano DG, Toscano WA, Jr. Calmodulin-mediated adenylate cyclase from mammalian sperm. J Biol Chem 1987; 262: 8672-8676.

70. Wasco WM, Orr GA. Function of calmodulin in mammalian sperm: presence of a calmodulin-dependent cyclic nucleotide phosphodiesterase associated with demembranated rat caudal epididymal sperm. Biochem Biophys Res Commun 1984; 118: 636-642.

71. Carrera A, Moos J, Ning XP, Gerton GL, Tesarik J, Kopf GS, Moss SB. Regulation of protein tyrosine phosphorylation in human sperm by a calcium/calmodulin-dependent mechanism: identification of A kinase anchor proteins as major substrates for tyrosine phosphorylation. Dev Biol 1996; 180: 284-296.

72. Luconi M, Krausz C, Forti G, Baldi E. Extracellular calcium negatively modulates tyrosine phosphorylation and tyrosine kinase activity during capacitation of human spermatozoa. Biol Reprod 1996; 55: 207-216.

73. Ren D, Navarro B, Perez G, Jackson AC, Hsu S, Shi Q, Tilly JL, Clapham DE. A sperm ion channel required for sperm motility and male fertility. Nature 2001; 413: 603-609.

74. Morisawa M, Suzuki K. Osmolality and potassium ion: their roles in initiation of sperm motility in teleosts. Science 1980; 210: 1145-1147.

75. Gatti JL, Billard R, Christen R. Ionic regulation of the plasma membrane potential of rainbow trout (Salmo gairdneri) spermatozoa: role in the initiation of sperm motility. J Cell Physiol 1990; 143: 546-554.

76. Visconti PE, Moore GD, Bailey JL, Leclerc P, Connors SA, Pan D, Olds-Clarke P, Kopf GS. Capacitation of mouse spermatozoa. II. Protein tyrosine phosphorylation and capacitation are regulated by a cAMP-dependent pathway. Development 1995; 121: 1139-1150.

77. Ward CR, Kopf GS. Molecular events mediating sperm activation. Dev Biol 1993; 158: 9-34.

78. Zeng Y, Clark EN, Florman HM. Sperm membrane potential: hyperpolarization during capacitation regulates zona pellucida-dependent acrosomal secretion. Dev Biol 1995; 171: 554-563.

79. Darszon A, Labarca P, Nishigaki T, Espinosa F. Ion channels in sperm physiology. Physiol Rev 1999; 79: 481-510.

80. Brooks DE. Epididymal functions and their hormonal regulation. Aust J Biol Sci 1983; 36: 205-221.

81. Setchell BP, Maddocks S, Brooks DE. Anatomy, vasculature, innervation, and fluids of the male reproductive tract. In: Knobil E, Neill JD (eds.), The Physiology of Reproduction, vol. 1, second ed. New York: Raven Press; 1994: 1063-1175.

82. Babcock DF. Examination of the intracellular ionic environment and of ionophore action by null point measurements employing the fluorescein chromophore. J Biol Chem 1983; 258: 6380-6389.

83. Bailey JL, Storey BT. Calcium influx into mouse spermatozoa activated by solubilized mouse zona pellucida, monitored with the calcium fluorescent indicator, fluo-3. Inhibition of the influx by three inhibitors of the zona pellucida induced acrosome reaction: tyrphostin A48, pertussis toxin, and 3-quinuclidinyl benzilate. Mol Reprod Dev 1994; 39: 297-308.

84. Salvatore L, D'Adamo MC, Polishchuk R, Salmona M, Pessia M. Localization and age-dependent expression of the inward rectifier K+ channel subunit Kir 5.1 in a mammalian reproductive system. FEBS Lett 1999; 449: 146-152.

85. Schreiber M, Wei A, Yuan A, Gaut J, Saito M, Salkoff L. Slo3, a novel pH-sensitive K+ channel from mammalian spermatocytes. J Biol Chem 1998; 273: 3509-3516.

86. Jacob A, Hurley IR, Goodwin LO, Cooper GW, Benoff S. Molecular characterization of a voltage-gated potassium channel expressed in rat testis. Mol Hum Reprod 2000; 6: 303-313.

87. Espinosa F, de la Vega-Beltran JL, Lopez-Gonzalez I, Delgado R, Labarca P, Darszon A. Mouse sperm patch-clamp recordings reveal single Cl- channels sensitive to niflumic acid, a blocker of the sperm acrosome reaction. FEBS Lett 1998; 426: 47-51.

88. Munoz-Garay C, De la Vega-Beltran JL, Delgado R, Labarca P, Felix R, Darszon A. Inwardly rectifying K(+) channels in spermatogenic cells: functional expression and implication in sperm capacitation. Dev Biol 2001; 234: 261-274.

89. Lievano A, Santi CM, Serrano CJ, Trevino CL, Bellve AR, Hernandez-Cruz A, Darszon A. T-type Ca2+ channels and alpha1E expression in spermatogenic cells, and their possible relevance to the sperm acrosome reaction. FEBS Lett 1996; 388: 150-154.

90. Arnoult C, Cardullo RA, Lemos JR, Florman HM. Activation of mouse sperm T-type Ca2+ channels by adhesion to the egg zona pellucida. Proc Natl Acad Sci U S A 1996; 93: 13004-13009.

91. Florman HM, Arnoult C, Kazam IG, Li C, O'Toole CM. A perspective on the control of mammalian fertilization by egg-activated ion channels in sperm: a tale of two channels. Biol Reprod 1998; 59: 12-16.

92. Galantino-Homer HL, Visconti PE, Kopf GS. Regulation of protein tyrosine phosphorylation during bovine sperm capacitation by a cyclic adenosine 3'5'-monophosphate-dependent pathway. Biol Reprod 1997; 56: 707-719.

93. Leclerc P, de Lamirande E, Gagnon C. Cyclic adenosine 3',5'monophosphate-dependent regulation of protein tyrosine phosphorylation in relation to human sperm capacitation and motility. Biol Reprod 1996; 55: 684-692.

94. Kalab P, Peknicova J, Geussova G, Moos J. Regulation of protein tyrosine phosphorylation in boar sperm through a cAMP-dependent pathway. Mol Reprod Dev 1998; 51: 304-314.

95. Devi KU, Jha K, Shivaji S. Plasma membrane-associated protein tyrosine phosphatase activity in hamster spermatozoa. Mol Reprod Dev 1999; 53: 42-50.

96. Travis AJ, Jorgez CJ, Merdiushev T, Jones BH, Dess DM, Diaz-Cueto L, Storey BT, Kopf GS, Moss SB. Functional relationships between capacitation-dependent cell signaling and compartmentalized metabolic pathways in murine spermatozoa. J Biol Chem 2001; 276: 7630-7636.

97. Urner F, Leppens-Luisier G, Sakkas D. Protein tyrosine phosphorylation in sperm during gamete interaction in the mouse: the influence of glucose. Biol Reprod 2001; 64: 1350-1357.

98. Agustin JT, Wilkerson CG, Witman GB. The unique catalytic subunit of sperm cAMP-dependent protein kinase is the product of an alternative Calpha mRNA expressed specifically in spermatogenic cells. Mol Biol Cell 2000; 11: 3031-3044.

99. San Agustin JT, Witman GB. Differential expression of the C(s) and Calpha1 isoforms of the catalytic subunit of cyclic 3',5'-adenosine monophosphate-dependent protein kinase testicular cells. Biol Reprod 2001; 65: 151-164.

100. Aitken RJ, Paterson M, Fisher H, Buckingham DW, van Duin M. Redox regulation of tyrosine phosphorylation in human spermatozoa and its role in the control of human sperm function. J Cell Sci 1995; 108: 2017-2025.

101. Leclerc P, de Lamirande E, Gagnon C. Regulation of protein-tyrosine phosphorylation and human sperm capacitation by reactive oxygen derivatives. Free Radic Biol Med 1997; 22: 643-656.

102. Herrero MB, de Lamirande E, Gagnon C. Nitric oxide regulates human sperm

capacitation and protein-tyrosine phosphorylation in vitro. Biol Reprod 1999; 61: 575-581.

103. Zhang H, Zheng RL. Promotion of human sperm capacitation by superoxide anion. Free Radic Res 1996; 24: 261-268.

104. Herrero MB, Gagnon C. Nitric oxide: a novel mediator of sperm function. J Androl 2001; 22: 349-356.

105. Fisher HM, Aitken RJ. Comparative analysis of the ability of precursor germ cells and epididymal spermatozoa to generate reactive oxygen metabolites. J Exp Zool 1997; 277: 390-400.

106. de Lamirande E, Gagnon C. Human sperm hyperactivation in whole semen and its association with low superoxide scavenging capacity in seminal plasma. Fertil Steril 1993; 59: 1291-1295.

107. Eddy EM, O'Brien DA. The Spermatozoon. In: Knobil E, Neill JD (eds.), The Physiology of Reproduction, vol. 1. New York: Raven Press; 1994: 29-77.

108. Visconti PE, Johnson LR, Oyaski M, Fornes M, Moss SB, Gerton GL, Kopf GS. Regulation, localization, and anchoring of protein kinase A subunits during mouse sperm capacitation. Dev Biol 1997; 192: 351-363.

109. De Jonge CJ, Han HL, Lawrie H, Mack SR, Zaneveld LJ. Modulation of the human sperm acrosome reaction by effectors of the adenylate cyclase/cyclic AMP second-messenger pathway. J Exp Zool 1991; 258: 113-125.

110. Garde J, Roldan ER. Stimulation of Ca(2+)-dependent exocytosis of the sperm acrosome by cAMP acting downstream of phospholipase A2. J Reprod Fertil 2000; 118: 57-68.

111. Florman HM, Corron ME, Kim TD, Babcock DF. Activation of voltage-dependent calcium channels of mammalian sperm is required for zona pellucida-induced acrosomal exocytosis. Dev Biol 1992; 152: 304-314.

112. Beltran C, Zapata O, Darszon A. Membrane potential regulates sea urchin sperm adenylylcyclase. Biochemistry 1996; 35: 7591-7598.

113. Kawasaki H, Springett GM, Mochizuki N, Toki S, Nakaya M, Matsuda M, Housman DE, Graybiel AM. A family of cAMP-binding proteins that directly activate Rap1. Science 1998; 282: 2275-2279.

114. Shabb JB, Corbin JD. Cyclic nucleotide-binding domains in proteins having diverse functions. J Biol Chem 1992; 267: 5723-5726.

115. Gauss R, Seifert R, Kaupp UB. Molecular identification of a hyperpolarization-activated channel in sea urchin sperm. Nature 1998; 393: 583-587.

116. Weyand I, Godde M, Frings S, Weiner J, Muller F, Altenhofen W, Hatt H, Kaupp UB. Cloning and functional expression of a cyclic-nucleotide-gated channel from mammalian sperm. Nature 1994; 368: 859-863.

117. Suarez SS. Hyperactivated motility in sperm. J Androl 1996; 17: 331-335.

118. Jahn R, Sudhof TC. Membrane fusion and exocytosis. Annu Rev Biochem 1999; 68: 863-911.

119. Schulz JR, Wessel GM, Vacquier VD. The exocytosis regulatory proteins syntaxin and VAMP are shed from sea urchin sperm during the acrosome reaction. Dev Biol 1997; 191: 80-87.

120. Schulz JR, Sasaki JD, Vacquier VD. Increased association of synaptosome-associated protein of 25 kDa with syntaxin and vesicle-associated membrane protein following acrosomal exocytosis of sea urchin sperm. J Biol Chem 1998; 273: 24355-24359.

121. Ramalho-Santos J, Moreno RD, Sutovsky P, Chan AW, Hewitson L, Wessel GM, Simerly CR, Schatten G. SNAREs in mammalian sperm: possible implications for fertilization. Dev Biol 2000; 223: 54-69.

122. Michaut M, Tomes CN, De Blas G, Yunes R, Mayorga LS. Calcium-triggered acrosomal exocytosis in human spermatozoa requires the coordinated activation of Rab3A and N-ethylmaleimide-sensitive factor. Proc Natl Acad Sci U S A 2000; 97: 9996-10001.

123. Yunes R, Michaut M, Tomes C, Mayorga LS. Rab3A triggers the acrosome reaction in permeabilized human spermatozoa. Biol Reprod 2000; 62: 1084-1089.

124. Mandal A, Naaby-Hansen S, Wolkowicz MJ, Klotz K, Shetty J, Retief JD, Coonrod SA, Kinter M, Sherman N, Cesar F, Flickinger CJ, Herr JC. FSP95, a testis-specific 95-kilodalton fibrous sheath antigen that undergoes tyrosine phosphorylation in capacitated human spermatozoa. Biol Reprod 1999; 61: 1184-1197.

125. Vijayaraghavan S, Liberty GA, Mohan J, Winfrey VP, Olson GE, Carr DW. Isolation and molecular characterization of AKAP110, a novel, sperm-specific protein kinase A-anchoring protein. Mol Endocrinol 1999; 13: 705-717.

126. Pawson T, Scott JD. Signaling through scaffold, anchoring, and adaptor proteins. Science 1997; 278: 2075-2080.

127. Nakanishi T, Ikawa M, Yamada S, Toshimori K, Okabe M. Alkalinization of acrosome measured by GFP as a pH indicator and its relation to sperm capacitation. Dev Biol 2001; 237: 222-231.

128. Eisenbach M. Sperm chemotaxis. Rev Reprod 1999; 4: 56-66.

Chapter 15

SPERM-EGG INTERACTION AND EXOCYTOSIS OF ACROSOMAL CONTENTS

Daulat R.P. Tulsiani and Aida Abou-Haila
Vanderbilt School of Medicine, Nashville, Tennessee, USA and
Université René Descartes, Paris, France

INTRODUCTION

It is generally accepted that carbohydrate-binding proteins (glycosyltransferases/glycosidases/lectin-like molecules) on sperm plasma membrane (receptors) recognize and bind to their complementary glycan molecules (ligands) on the egg's extracellular coat, the zona pellucida (1-3). Thus, sperm-ovum interaction is a carbohydrate-mediated receptor ligand-binding event. This type of binding is analogous to cellular adhesion events of bacteria, viruses, and many pathogens to their respective host cells. Binding of opposite gametes initiates a calcium (Ca^{2+})-dependent signal transduction cascade resulting in the exocytosis of sperm acrosomal contents (i.e., induction of the acrosome reaction). This step is thought to be a prerequisite event that enables the acrosome-reacted spermatozoa to penetrate the zona pellucida and fertilize the egg.

There has been a longstanding interest in the basic biology of the fertilization process. The success of *in vitro* fertilization in humans and domestic animals is a result of the knowledge gained in small animal models. These events are best understood in the mouse, although there is considerable information in other species. Successful union of the opposite gametes in the mouse following mating involves: 1) Sperm capacitation in the female genital tract; 2) binding of the capacitated spermatozoa to the zona pellucida and induction of the acrosome reaction; 3) secondary binding of the acrosome-reacted sperm to the zona pellucida and its penetration; and 4) fusion of the sperm with the egg plasma membrane (Figure 1). This chapter focuses on the recent advances made toward elucidating the molecular basis of sperm-egg interaction and initiation of signal transduction cascade leading to acrosomal exocytosis.

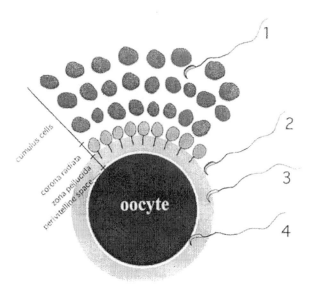

Figure 1. Progressive steps of sperm-egg interaction and induction of the acrosome reaction in mouse spermatozoa. 1, the acrosome-intact spermatozoon passes through cumulus cells and the innermost layer, the corona radiata; 2, receptor(s) on capacitated (acrosome-intact) spermatozoa binds to the complementary glycan chain(s) of mZP (mZP3) and starts a signal transduction cascade resulting in release of acrosomal contents at the site of sperm-zona binding; 3, the acrosome-reacted spermatozoon penetrates the zona pellucida; 4, the spermatozoon passes through the perivitelline space and fuses with the egg plasma membrane.

SPERM-ZONA (EGG) BINDING

The capacitated spermatozoa bind to the zona pellucida (ZP) in a highly precise manner. Extensive studies in the mouse and several other species suggest that the binding occurs in two steps: loose and reversible adhesion of spermatozoa to the zona-intact egg by means of plasma membrane overlying the acrosome, followed by a tight and irreversible binding. Most researchers agree that complementary molecules present on the surface of opposite gametes are involved in both types of bindings. Although the chemical nature of these molecules is far from clear, there is growing evidence that several steps in the fertilization process are mediated by carbohydrates in a receptor-ligand manner (2-5).

Zona Pellucida Glycoconjugates

The ZP in all species studied is a relatively simple structure composed of three glycoproteins designated ZP1, ZP2, and ZP3; and the pig has a fourth form as well. It is now clear that the ZP glycoproteins are highly conserved among mammalian species suggesting that their organization may also be conserved. The three glycoproteins, at least in the mouse, are synthesized and secreted by the growing oocytes during oogenesis (6). Two of the glycoproteins, mouse ZP2 and ZP3, interact noncovalently to form filaments of structural repeats that are interconnected by mZP1. Thus, the extracellular coat (matrix) surrounding the egg is made up of a three-dimensional network of cross-linked filaments. Such a structure may explain the elasticity of ZP and the relative ease of its penetration by the hyperactivated acrosome-reacted spermatozoon.

In recent years, considerable progress has been made in understanding structure-function of various zona components. In particular, work on mouse zona pellucida (mZP) has resulted in identification of primary (mZP3) and secondary binding sites (mZP2) for homologous spermatozoa (6). This conclusion was based on three lines of evidence. First, only mZP3 is able to inhibit sperm-egg binding in an *in vitro* assay in a concentration-dependent manner. The reported inhibition is apparently caused by competition of the added mZP3 for the recognition and binding to the complementary receptor site(s) on sperm head. Second, inclusion of mZP3 glycoprotein, purified from fertilized eggs, failed to inhibit sperm-egg binding in an *in vitro* assay system. This study is consistent with the suggestion that the loss of primary binding site(s) from the fertilized eggs is due to modification of mZP3. Finally, when the radioiodinated mZP2 or mZP3 was incubated with capacitated (acrosome-intact) or acrosome-reacted sperm cells, the former glycoprotein showed higher binding to the acrosome-reacted cells, whereas the latter glycoprotein bound to capacitated spermatozoa (6). Taken together, these results are consistent with the proposed role of mZP3 and mZP2 in primary and secondary binding, respectively.

Carbohydrates and Fertilization

Several lines of evidence given below strongly suggest that the glycan moiety (ligand) of mouse ZP is responsible for its binding activity (6).

1) Treatment of the purified mZP3 with denaturants, such as sodium dodecyl sulfate or high temperatures that alter the protein backbone, does not abolish its ability to inhibit sperm-egg binding in competiton assay system.

2) Inclusion of various lectins, a class of proteins/glycoproteins that bind to carbohydrate moieties with high affinity and specificity, in an *in vitro* assay system inhibit binding of spermatozoa to zona-intact egg. The added lectin presumably binds to the terminal sugar residue(s) on the bioactive glycan portion of zona-intact egg, and prevent sperm receptor(s) from recognizing the complementary ligand(s).

3) Addition of specific monosaccharides, disaccharides, oligosaccharides, or glycoproteins in an *in vitro* sperm-egg binding mixture inhibits binding of the opposite gametes in a concentration-dependent manner in several species. The added reagents likely bind to the sperm receptor(s) and prevent them from recognizing the ligand on zona-intact egg. Alternatively, these sugars may compete with the putative ligands and decrease the ability of spermatozoa to bind to the zona pellucida.

4) Treatment of zona-intact eggs with exoglycosidases inhibits or prevents sperm-egg binding. The enzyme treatment presumably cleaves putative sugar residue(s) from the surface of the ZP and prevents spermatozoa from recognizing and binding to the enzyme-treated egg.

5) Digestion of mZP3 with pronase causes proteolysis of the protein backbone and production of small glycopeptides ranging in size from 1.5 to 6.0 kDa. These glycopeptides with intact glycan units retain sperm binding activity as evident from their ability to inhibit sperm-egg binding *in vitro*.

6) Sperm binding activity of mZP3 is sensitive to trifluoromethane sulfonic acid treatment, an acid known to break glycosidic bonds between monosaccharide residues present in *N*-linked and *O*-linked glycan chains without altering the polypeptide backbone.

Taken together, these data provide strong evidence that the glycan portion of mZP3 glycoprotein is the ligand for capacitated spermatozoa.

What are *N*- and *O*-linked Glycans?

We will briefly discuss the chemistry and functional significance of glycoproteins to provide a better understanding of the glycan chains on ZP. Recognition of the biological importance of glycoproteins and glycolipids (complex carbohydrates) has been increasing since 1960 when it was first reported that sialic acid was the receptor site for the influenza virus on red blood cells (7). Glycan moieties of complex carbohydrates have now been documented to participate in many diverse biological functions, including antigenic determinants of blood group substances, recognition of receptor sites for hormones, targeting of glycohydrolases to lysosomes (8), and cell-cell interactions (9). Glycoproteins are formed *in vivo* following the covalent attachment of glycans to certain amino acid residues within the polypeptide chains. Depending on the linkage between the peptide (amino acid) and sugar moiety, the glycoproteins are classified into several groups. One group is asparagine (*N*)-linked, in which the linkage between asparagine (Asn) and *N*-acetylglucosamine is via *N*-glycosidic bond. The Asn-*N*-acetylglucosamine is linked to another *N*-acetylglucosamine through a β1,4 linkage forming a chitobiose unit (10). The fully processed mature *N*-linked glycan unit may be either high mannose, hybrid-type or complex-type (bi-, tri-, tetra-antennary) structure. In addition, *N*-linked complex-type (tri- and tetra-antennary) structures with (poly) lactosaminyl glycans are also present on several glycoproteins including mZP2 and mZP3. These glycans contain repeat units of disaccharide (3 Gal β1,4-GlcNAcβ1). Experimental evidence indicates that *N*-linked polylactosaminyl glycans may contain several structurally variable chains with different sugars including α-galactose-, β-galactose-, and α-sialyl residues at the non-reducing terminus (11).

All *N*-linked glycan units contain a basic structure composed of a branched trimannose unit linked to *N,N'*-diacetylchitobiose while the innermost *N*-acetylglucosamine is attached to the amide nitrogen of an Asn-residue in the tripeptide sequence Asn-X-Serine/Threonine where X cannot be proline. However, it is important to mention that the actual presence of the consensus sequence (tripeptide) within a protein does not mean that the site will be *N*-glycosylated, but that it is a potential site for accepting an oligosaccharide. Thus, there must be other factors, including the conformation of the protein, which determine whether a particular site can be glycosylated.

The biosynthesis of the *N*-linked class of glycoproteins proceeds via a common pathway referred to as the dolichol pathway (or cycle) because dolichyl phosphate serves as a carrier of the preassembled glycan chain with the structure $Glc_3Man_9GlcNAc_2$-pyrophosphate (PP)-dolichol (10). The preassembled lipid-linked oligosaccharide is transferred *en bloc* to Asn-residues of the nacent polypeptide chains in the endoplasmic reticulum where it undergoes a number of trimming reactions that result in the removal of three glucose residues, and some α1,2-linked mannose residues. The glycan chains are further modified in the Golgi apparatus that requires the removal of several mannose residues and the addition of sugars of complex chains (i.e., *N*-acetylglucosamine, galactose, fucose, and sialic acid). These sugar residues are added sequentially from the nucleotide sugar (donor) to the acceptor glycan moiety by membrane-bound glycosyltransferases in the Golgi apparatus. It is not known why a particular glycan chain retains a high mannose structure while others become complex-type or why some complex-type chains remain bi-antennary while others become tri- and tetra-antennary. The number of *N*-linked glycans that have been identified on various glycoproteins is large and may run in the hundreds. Although individual cells can synthesize many *N*-linked glycan chains, the process is highly specific and is controlled in such a way that the glycan units at a particular glycosylation site have one or a small number of closely related structures.

The second major type of glycosylation is O-linked. In this type of linkage, the sugar chains do not contain mannose. The O-glycosidic linkage present on β-hydroxyamino acids, serine or threonine, are found in mucous glycoproteins as well as non-mucous glycoproteins such as fetuin, thyroglobulin, immunoglobulins, ZP glycoconjugates (mZP2 and mZP3) and proteoglycans. The first sugar residue in O-linked glycoproteins is N-acetylgalactosamine which is linked to serine/threonine. As with the N-linked glycosylation, the protein structure may be important in determining as to which serine/threonine residues are glycosylated. In some mucous glycoproteins as many as 800 glycan units may be attached at intervals along the protein chain.

The biosynthesis of O-linked glycoproteins involves the sequential transfer of an individual sugar residue from a donor (nucleotide sugar) to the growing sugar chain. Usually, an N-acetylgalactosamine residue is added first, followed by a variable number of additional sugar residues, catalyzed by a group of glycosyltransferases that are generally integral components of cellular membranes. Each glycosyltransferase requires the product of the preceding transfer and, in turn, generates the substrate for the next enzyme. The net result is the addition of a variable number of sugar residues from a few to 10 or more.

Structural Diversity of Zona Pellucida Glycans

The fact that a large number of N- and O-linked structures are possible on ZP makes it difficult to identify and chemically characterize the bioactive (functional) glycan residue(s). The efforts are further hampered by the small amounts of mouse zona glycoproteins (approximately one nanogram of ZP2 or ZP3/egg) that can be purified and subjected to classical structure-function studies. Despite these limitations, considerable progress has been made to understand the qualitative characteristics of ZP glycoconjugates. Data from various laboratories indicate that both ZP2 and ZP3 in the mouse, as well as in other species, are highly glycosylated and, like other glycoproteins, show extensive microheterogeneity. The presence of sulfate and/or phosphate and a variety of terminal sugar residues on nonreducing terminus contributes to the extensive microheterogeneity and acidic nature of the glycoprotein (also see Chapter 12).

Structural analysis of N-linked glycan units of mZP2 and mZP3 glycoproteins has been attempted by several investigators using multiple approaches. These include radioiodination, fluorescent labeling, labeling with carbohydrate-specific lectins, release of N- and O-linked glycans followed by tritium labeling by reduction with NaB^3H_4, and high-sensitivity mass spectrometric techniques. Data have revealed the presence of a variety of high mannose and complex-type N-glycan chains in murine zona pellucida. The predominant high mannose-type glycan had the following composition $Man_5GlcNAc_2$; however, larger and smaller mannose-rich glycans have also been reported (11,12). In addition, poly-N-acetyllactosaminyl, bi-, tri-, and tetra-antennary complex-type N-glycans with terminal N-acetylneuraminic acid, galactose, N-acetylgalactosamine, and N-acetylglucosamine residues have also been reported. Combined, data from various studies provide evidence indicating the complex nature of glycan chains present on mZP2 and mZP3 glycoproteins. However, most studies have not examined the functional significance of various glycan chains.

Role of ZP3 in the Fertilization Process

The ability of mZP3, an 83 kDa glycoprotein, to serve as the primary ligand and an acrosome reaction inducer depends on the glycan moiety(ies) as well as the polypeptide portion of the molecule. First, the glycan portion serves as the ligand for the sperm surface molecules (receptors). Second, the polypeptide backbone of ZP3 is essential for inducing the acrosome reaction in the bound spermatozoa. Consistent with this possibility is the finding that, although monosaccharides, glycans and small peptides retain bioactivity as evident by their ability to inhibit sperm-zona binding in an *in vitro* assay system, the small molecules do not induce the acrosome reaction unless they are cross-linked on the sperm surface by anti-mZP3 IgG (13). Studies published from other laboratories are in agreement with the proposed aggregation of the sperm receptors located on or within the plasma membrane overlying the acrosome. First, a heat and acid stable protease inhibitor of seminal vesicle origin binds to the acrosomal cap region of the capacitated spermatozoa. The *in vitro* binding of the inhibitor to the sperm surface did not induce the acrosome reaction unless the bound protein is immunoaggregated with anti-inhibitor Fab-fragment (14). Second, studies published by another laboratory also suggest that binding of the sperm surface galactosyltransferase to multiple sugar residues on mZP3 causes its aggregation that triggers the acrosome reaction (15). Finally, our studies demonstrate that specific sugar residues induce the acrosome reaction, but only when covalently conjugated to a protein backbone (16). Taken together, these studies support the conclusion that the protein backbone of the mZP3 or synthetic glycoproteins (neoglycoproteins) plays a role in aggregation of the sperm surface receptors prior to the induction of the acrosome reaction.

It should be noted that the addition of structurally characterized glycans in the *in vitro* sperm-egg binding assay mixture does not completely prevent the binding; 10-20% binding persists even when higher concentrations of the oligosaccharides are included in the assay medium. This has led to the suggestion that protein-protein interactions may also be important in the binding of opposite gametes. However, there is no direct evidence to support this possibility.

In humans and pigs, the protein portion of the ZP3 appears to have a role in protein-protein interaction during sperm-zona binding. For instance, antibodies raised against the polypeptide sequence of the human or porcine ZP glycoprotein significantly inhibit sperm binding to the zona-intact egg in an *in vitro* assay system. In contrast, the antibodies against mZP3 polypeptide had no effect on the number of spermatozoa bound to the zona-intact eggs (17). These studies point out the possibility that the polypeptide portion of the human and pig ZP3 may possess biological activity; however, more direct evidence is needed to support this possibility.

Glycan Units that Regulate Sperm Binding in the Mouse

Despite numerous advances, considerable controversy remains regarding the precise identity of the glycan unit(s) responsible for the ligand activity of mZP3. Over 15 years ago, Wassarman and colleagues reported that sperm binding activity resides within *O*-linked glycan unit of an apparent molecular mass of 3.9 kDa and more precisely with an α-linked galactosyl residue at the non-reducing terminus of the oligosaccharide (2). Later the same group reported that when five serine residues located in the C-terminus region of mZP3 were mutated to either glycine, valine, or alanine, a biologically inactive mouse ZP3 was produced and secreted from transfected embryonal carcinoma cells (18). This result is consistent with the suggestion that serine-associated (*O*-linked) glycan units located in the C-terminus region of the polypeptide backbone are essential for the sperm binding activity.

Do the results of molecular biological approaches used by Wassarman's laboratory (18) clearly point to the identification of the bioactive glycan moiety(ies)? Two points have to be addressed. First, has the site-directed mutation altered the overall structure of an important domain of the ZP3 glycoprotein which could be essential for its biological function? Second, has the replacement of serine residues with other amino acids in mZP3 affected the processing of *N*-linked glycan chains? The last point is especially important since experimental evidence from other laboratories suggests that *N*-linked oligosaccharide chains also have a role in sperm-egg interaction: 1) treatment of zona-intact mouse eggs with almond glycopeptidase, an endoenzyme that hydrolyzes β-aspartyl-glucosamine of all classes of N-linked glycans, greatly reduces the number of sperm bound per egg (9); and 2) inclusion of N-linked glycans in an *in vitro* sperm-egg binding assay in the mouse (9), rat (9), and pig (19) greatly reduced the number of sperm bound per egg. These studies suggest that N-glycans may also be a part of the recognition/binding site(s) in several species (for review see 3, 9).

Wassarman and colleagues have also used glycobiological approaches to define the sugar and linkage specificity in murine sperm-egg binding. A total of 15 oligosaccharides, differing in chain length and configuration of the glycosidic linkage at the anomeric carbon of the terminal sugar residue, were synthesized and examined for their inhibitory effects in an *in vitro* sperm-egg binding assay. Bi- and tetra-antennary glycan chains containing six or more sugar residues with α- or β-linked galactose at the non-reducing terminus were found to inhibit sperm-zona interaction (20). Based on the result of this study, Wassarman's group suggests that glycan units containing α- or β-linked galactose residues participate in murine sperm-egg binding. More recently, carbohydrate specificity of the murine gamete interaction was also addressed using four synthetic oligosaccharides with the following structure: Galβ1,4GlcNAc-β1,4GlcNAc; Galα1,3Galβ1,4GlcNAc; Galβ1,4[Fucα1,3]-GalNAcβ1,4GalNAc; and Galα1,3Galβ1,4[Fucα1,3]GalNAc. Higher con-centrations of the first two non-fucosylated oligosaccharides (10-70 μM) were needed to achieve 40-60% inhibition of the number of murine sperm bound per egg in an *in vitro* assay. However, significantly lower concentrations of the last two oligosaccharides (1 μM) were needed when fucose was present. These results are consistent with the suggestion that, although a terminal fucose residue is not needed in initial binding of the opposite gametes, its presence is obligatory for an oligosaccharide to bind to spermatozoa with high affinity (21).

Experimental evidence from other groups suggests that a galactose α1,3 galactose(s) epitope is not necessary for sperm adhesion to the zona pellucida. First, Thall et al. (22) used a gene disruption approach to address the role of Galα1,Gal containing oligosaccharide. The authors generated knock-out mice lacking the gene (α1,3Galtase) encoding UDP-galactose: β-D-Galactose α1,3-galactose-galactosyltransferase, the enzyme responsible for the synthesis and expression of Galα1,3Gal epitopes. Female mice homozygous (α1,3GT,-/-) developed normally and produced oocytes devoid of Galα1,3Gal sequence. However, these mice were fully fertile, a result consistent with the investigators' conclusion that Galα1,3Gal sequences on O-linked oligosaccharides are not required for initial murine sperm-egg binding. Thus, if sperm-zona (egg) interaction is mediated by carbohydrates, other sugar epitopes must be recognized by the sperm receptors. Second, Shur's group has presented evidence suggesting that N-

acetylglucosaminyl residues on mZP3 are recognized by a galactosyltransferase (GalTase) present on mouse sperm plasma membrane overlying the outer acrosome membrane (4). In the absence of nucleotide sugar donor (UDP-galactose), the sperm enzyme acts as an ecto-enzyme and remains bound to the acceptor site (*N*-acetylglucosamine) until the next event in fertilization is triggered. Interestingly, terminal glucosaminyl residues have been reported on *O*- and *N*-linked glycans from mZP3 (11,23). mZP3 polypeptide cleavage experiments indicate that a small region of 16 amino acids (i.e., 328-343) in the C-terminal domain contains five serine residues that are presumably *O*-glycosylated (24). The peptide domain also has been identified in porcine and human homologues of the mZP3. Our own studies suggest that mZP3 contains 4-5 *O*-linked trisaccharides with the structure GlcNAc-Galβ1,3GalNAc (23). This result suggests that all these trisaccharides could be present in the C-terminus region of the mZP3.

Shur and associates have used a gene disruption approach to further address the *in vitro* and *in vivo* role of the sperm surface GalTase. The authors generated GalTase null mice through targeted mutation. Spermatozoa from these mice failed to undergo the ZP-induced acrosome reaction *in vitro* (25), a result suggesting the importance of the sperm surface GalTase in binding to the complementary glucosaminyl residues on the ZP before undergoing the acrosome reaction (see below). In spite of the impaired response of the spermatozoa *in vitro*, the GalTase null mice were fully fertile (26). This result, in conjunction with other evidence, strongly suggests the occurrence of multiple ligands on mZP3. The sugar residues suggested to have a role in initial binding of the opposite gametes include: α-D-galactose, β-D-galactose, *N*-acetylglucosamine, mannose and sialic acid (3,9).

This brief summary strongly suggests that a carbohydrate recognition process is involved in sperm-egg interaction. The mechanism underlying the binding of the opposite gametes remains one of the principal unresolved issues in reproductive biology. The evidence for the presence of a large diversity in the structure of glycans on mZP3 suggests that several ligands (glycans)-receptor interactions may be involved before successful interaction. This suggestion is consistent with the experimental evidence from several laboratories supporting the concept that sperm-egg interaction leading to the acrosomal exocytosis is a complex event that reflects interaction between multiple sperm surface receptors and multiple sugar residues on the ZP (27).

Bioactive Glycans in Other Species

It is important to keep in mind that the identification of a bioactive glycan(s) that regulates sperm-egg adhesion in humans has not been investigated as extensively as in mice or pigs because of the extremely limited number of eggs that can be obtained for *in vitro* studies as well as for ethical concerns. The situation is further complicated by the occurrence of structural variations in the glycan moieties of various human eggs. This may explain why eggs from the same donor demonstrate differential sperm binding activity. Although biologically active human ZP3 has been produced in Chinese hamster ovary cells through molecular biological approaches, the purification of material that gives reproducible results has been difficult (17). Thus, the progress in determining the chemical nature of the glycan unit(s) that mediates sperm-egg binding in humans has been painfully slow. Despite these drawbacks, some progress has been made largely due to the efficiency of binding assay where human eggs are microdissected to produce two equal halves (hemisphere); the hemizona assay uses one of the two halves as control (without addition) and the other half as experimental (with addition) to test the effects of added reagents on the number of spermatozoa bound per hemizona. Data from these studies suggest that several sugar residues may be involved in human sperm-zona interaction. These include fucoidan, a sulfated polymer of fucose, *N*-glycosylated glycodelin A, a glycoprotein of uterine origin, and sugars such as mannose and sialic acid (28).

On the question concerning the role of *N*- and/or *O*-linked glycans in other mammalian species, the pig zona pellucida has been chemically characterized by several investigators (for review see 3). The biological activity of ZP in this species is associated with ZP3α glycoprotein of an apparent molecular mass 55,000 (also see Chapter 12). The protein backbone is highly glycosylated containing *N*- and *O*-linked glycans. The sperm binding activity has been reported to be associated with the neutral N-linked carbohydrate chain fraction of bi-, tri- and tetra-antennary complex-type chains (19). Combined, data from various studies seem to suggest that multiple sugar residues present on *N*- and *O*-linked glycan units are involved in sperm-egg adhesion.

Species-Specificity during Sperm-Egg Interaction

With the exception of a very few closely related species, there is a high degree of species-specificity during fertilization. It has been known for several decades that ZP is responsible for this specificity. For instance, capacitated spermatozoa can fertilize the egg *in vitro* from other species only when ZP is first removed. This experimental evidence supports the conclusion that the egg's extracellular coat determines the species-specificity. Since the glycan portion of the ZP3 determines its primary binding site, it is reasonable to assume that bioactive glycan chains present on ZP3 from different species are different and contribute to species-specificity. This assumption is supported by experimental evidence in which the mZP3 gene was replaced with the human ZP3 gene (29). Although eggs from these animals interacted with the mouse (rather than human) spermatozoa *in vitro*, the rescued mZP3 had a human ZP3 core peptide; however, its glycan units, though not analyzed, are expected to be similar to the host animal. These studies further emphasize the importance of carbohydrate portion of the ZP3 in determining the species-specific sperm-egg interaction.

Sperm Molecules with Affinity for the ZP Glycan Unit(s)

It is important to emphasize that the sperm plasma membrane overlying the acrosome is a complex structure consisting of dozens of proteins/glycoproteins and lipids. Because of difficulties in preparing a region-specific sperm plasma membrane rich fraction, identification of receptors that recognize complementary ligand molecule(s) has been pains-taking process. To identify and chemically characterize putative receptor molecules on spermatozoa of mouse and other species with zona binding activity, researchers have used multiple approaches including: 1) fractionation of detergent-solubilized sperm membranes by column chromatography followed by testing the ability of various fractions to inhibit sperm-egg binding *in vitro;* 2) affinity purification of detergent-solubilized spermatozoa on a column of immobilized ZP; 3) assay for sperm surface enzyme activity in the presence of inhibitors that block sperm-egg binding in a concentration-dependent manner; 4) immobilizing electrophoretically-separated sperm proteins on a nitrocellulose paper followed by probing these molecules with iodinated ZP glycoproteins; 5) raising antibodies against sperm surface antigens that bind to the sperm head and prevent sperm-egg interaction in an *in vitro* assay system. These approaches have resulted in the identification of several putative receptors (Table 1). Some of the proposed receptors and their complementary ligands are discussed below.

β1,4-Galactosyltransferase (GalTase)

GalTase was the first enzyme suggested to have a role in sperm-egg interaction in the mouse (4). The enzyme is an integral plasma membrane component and binds to *N*-acetylglucosamine (GlcNAc) residue(s) on *O*-linked oligosaccharide units of mZP3 forming sperm (GalTase)-zona (GlcNAc) complex which remains stable until the next event of fertilization is triggered. Several lines of evidence support the proposed role for the

Table 1. Sperm surface proteins (receptors) and their complementary sugar residues (ligands) suggested to have a role in sperm-egg interaction (adopted from 9)

Proposed receptors	Potential Ligands	Species
β-1, 4-galactosyl transferase	N-Acetylglucosaminyl residue	Mouse[a]
Fucosyltransferase	Not known	Mouse
56-kDa sperm protein	α-Galactosyl residues	Mouse[b]
95-kDa sperm protein	Not known	Mouse[c]/Human[d]
Protease-sensitive site	Not known	Mouse
Sulfoglycolipid immobilizing protein	Sulfated glycolipid	Mouse
α-D-mannosidase	High mannose/hybrid type glycans	MouseRat
Mannose-binding protein	α-mannosyl residues	Human/Rat
Fucose-binding protein	α-fucosyl residues	Porcine/Hamster
Galactose-binding protein	β-galactosyl residues	Mouse
Fertilizing antigens (FA-1, NZ-1)	Not known	Human
Sperm agglutination antigen-1	Not known	Human
Posterior Head protein (PH-20) pig/Macque	Not known	Guinea
Zonadhesin (150 kDa protein)	Not known	Porcine/Mouse
Adhesion protein (AP$_z$)	Not known	Porcine
Spermadhesin	Not known	Porcine
Selectin-like molecules	Not known	Human
Rabbit sperm autoantigen (RSA)	Not known	Rabbit
Sperm protein-17 (SP17)	Not known	Rabbit

[a]Galactosyltransferase null mice are fully fertile despite the impaired response of their spermatozoa to the zona-induced AR *in vitro (26)*.

[b]Reported to be ortholog of AM67, an acrosomal matrix protein (9).

[c]Reported to be a unique hexokinase (9).

[d]Identical to proto-oncogene (9).

enzyme. First, inclusion of purified GalTase, its substrates or donor nucleotide sugar (UDP-Gal) in an *in vitro* sperm-egg-binding assay decreased the number of sperm bound per egg. Second, compounds such as UDP-dialdehyde or α-lactalbumin that inhibit sperm GalTase, also decreased the number of sperm bound per egg. Third, the removal of GlcNAc residues from the zona-intact eggs by digestion with *N*-acetylglucosaminidase inhibits sperm binding to the egg. Finally, GlcNAc residues covalently conjugated to a polypeptide backbone is a substrate for the sperm surface GalTase. Indeed, the mZP3 (and mZP2) has been demonstrated to contain a terminal-*N*-acetylglucosamine on an *O*-linked trisaccharide (23) and N-linked glycans (11,12). Combined, these data support a role for the sperm surface GalTase in initial sperm-zona binding. Gene disruption approaches demonstrate that spermatozoa from knock-out mice lacking GalTase activity failed to undergo the ZP-induced acrosome reaction *in vitro*. This result suggests the importance of this sperm enzyme in binding to its complementary GlcNAc residues on ZP before undergoing the acrosome reaction. Despite the impaired response of the spermatozoa to the ZP-induced acrosome reaction *in vitro*, the GalTase knock-out mice were fully fertile (26), a result supporting the presence of other receptors on spermatozoa.

α-D-Mannosidase

Sperm plasma membrane from several mammalian species contains an α-D-mannosidase activity. Unlike the sperm acrosomal "acid" mannosidase that has a pH optimum of 4.4, the sperm surface mannosidase is optimally active at pH 6.2 and 6.5 when assayed in the rat sperm plasma membrane and intact spermatozoa, respectively. In addition, the two enzymes have different substrate specificity; the acrosomal α-D-mannosidase is active mainly towards the synthetic substrate, p-nitrophenyl α-D-mannoside, whereas the sperm plasma membrane mannosidase is active primarily towards mannose-containing oligosaccharides (30).

The sperm surface mannosidase has all the characteristics required for an enzyme to have a role in sperm-egg interactions. First, the enzyme is active at or near physiological pH in the intact spermatozoa, a result suggesting that its catalytic domain is oriented towards the outer surface and that the enzyme would recognize its complementary high mannose/hybrid type glycans on ZP in physiological setting. Second, the sperm surface mannosidase is an integral component of the plasma membrane and is localized on the periacrosomal region of the epididymal cauda spermatozoa. This localization is of particular significance since the periacrosomal plasma membrane is thought to be involved in sperm-egg binding. Third, the mannosidase activity increased in spermatozoa as they transit from the caput to the cauda epididymidis, while several other surface enzyme activities either remained unchanged or decreased during epididymal transit. The increase in enzyme activity was demonstrated to be due to the conversion of an enzymatically inactive, less active precursor form of 135 kDa to an enzymatically active mature form of 115 kDa during sperm development in the testis and maturation in the epididymis. The maturation-associated increase in the sperm surface mannosidase activity correlated with an increased zona-binding ability of the rat spermatozoa. Finally, evidence available suggests the presence of *N*-linked high mannose glycans on mZP2 and mZP3 (11,12). These glycans are presumably the recognition/binding sites for the sperm surface mannosidase.

The functional significance of the sperm surface enzyme as a putative receptor was suggested by numerous studies in several species, including man:

1) Treatment of mammalian eggs with concanavalin A, a lectin known to bind high mannose/hybrid-type glycans, prevented sperm-egg binding presumably by blocking the potential ligand sites for the sperm surface mannosidase.

2) Treatment of zona-intact rat egg with jack bean α-mannosidase, an enzyme known to cleave α-linked mannose residues from high mannose/hybrid-type glycans, caused nearly complete inhibition of sperm-egg binding *in vitro*.

3) A significantly fewer number of sperm bound per egg in the presence of sugars that inhibited sperm surface mannosidase in mouse and rat than in their absence (32). Among the sugars examined, D-mannose and α-methyl mannoside were potent inhibitors of the sperm mannosidase activity.

4) Inclusion of a high mannose oligosaccharide (man$_5$GlcNAc) in an *in vitro* binding assay caused a dramatic reduction in the number of spermatozoa bound per egg and a similar inhibition of the sperm surface mannosidase activity. The oligosaccharide decreased sperm mannosidase activity by competing with the mannose-labeled substrate used for the enzyme assay, and decreased the ability of spermatozoa to bind to the zona-intact egg by competing with the putative ligand sites on the ZP.

5) Anti-mannosidase IgG (but not preimmune IgG) caused a concentration-dependent inhibition of the mannosidase activity and a similar concentration-dependent reduction in the number of sperm bound per egg.

Combined, these studies suggest that sperm surface mannosidase has a role in sperm-egg interaction.

Mannose-Binding Protein(s)

In addition to the novel α-D-mannosidase, there is evidence for the presence of a mannose-binding receptor(s) on human and rat spermatozoa. The receptor first appears uniformly distributed over the entire head region of the capacitated (acrosome-intact) spermatozoa. Benoff and associates have reported that an increase in the number of human spermatozoa capable of binding to mannose-BSA and undergoing the acrosome reaction are predictive of their fertilization rate (31). This result supports their argument that sperm surface mannose-binding protein(s) contributes to the initial sperm-zona (egg) binding in human. The protein(s) requires calcium for tight binding to mannose-containing oligosaccharides presumably present on human ZP suggesting that the molecule functions as a lectin. Similar mannose-binding protein appears to have a role in sperm-egg interaction in the rat where antibodies to the mannose- binding protein (but not preimmune IgG) caused a concentration-dependent reduction in the number of spermatozoa bound per egg (32).

β-Galactose-Binding Protein(s)

Galactose-binding protein(s), present on the rat testis and spermatozoa, has been suggested to be involved in sperm-zona interaction in multiple species (9). The molecule was immunolocalized to the dorsal surface of the rat sperm head overlying the acrosome that was lost in the acrosome-reacted spermatozoa. The role of galactose-binding molecule in sperm-egg interaction, at least in the rat, is still unclear since the complementary galactose residues have been reported mainly on the inner layer (rather than the externally exposed outer layer) of the ZP (33).

α-Galactose-Binding Protein(s)

A 56 kDa mouse sperm protein (Sp56) was preferably radiolabeled by a cross-linker covalently bound to purified mouse ZP3, a result suggesting that the protein may be a receptor for ZP. The expression of this protein is restricted to the mouse testicular germ cells (spermatids), and it is localized on the surface of acrosome-intact mouse sperm head (34). In addition, Sp56 binds tightly to immobilized mZP3 and α-galactose-affinity column. The location and its binding affinities are consistent with its proposed role in recognizing α-galactose residues on O-linked oligosaccharide. However, the fact that the mutant mouse eggs devoid of Galα1,3Gal epitopes are fully fertile suggests that these epitopes are not required for fertilization (see above). This study, in conjunction with the report that Sp56 is an orthologue of AM67 (35), an acrosomal matrix protein, raises serious concerns on the proposed role of Sp56 in sperm-egg interaction.

L-Fucose-Binding Protein(s)

L-fucose and its sulfated polymer, fucoidin, are strong inhibitors of sperm-egg binding in multiple species. Several fucose-binding proteins have been proposed including acrosin, a serine-like acrosomal protease (36), a fucose-binding protein (37,38) and a selectin-like molecule on human spermatozoa (39). Biochemical and immunohistochemical approaches using monoclonal antibody to leukocyte selectin have identified a 90 kDa molecule on human sperm head. Spermatozoa treated with the antibody to the molecule prior to inclusion in the hemizona assay blocked tight binding of the opposite gametes. Taken together, these results are consistent with the proposed role for fucose in sperm-egg interaction. However, a direct evidence for the involvement of sperm surface selectin-like molecule and/or acrosin in binding to the fucose residues on the ZP has not been presented. Furthermore, current evidence suggests that there are no exposed fucose residues on the surface of mouse ZP (40).

Fucosyltransferase

The synthetic enzyme, normally localized in the Golgi complex of mammalian cells, catalyzes transfer of fucose from donor sugar nucleotide (GDP-fucose) to terminal galactose residues on *N*- and *O*-linked glycans of complex carbohydrates. The enzyme is also present on the surface of mouse and rat spermatozoa. Inclusion of GDP-fucose, a sugar donor or asialofetuin, a fucose acceptor substrate for the enzyme in an *in vitro* sperm-egg-binding assay greatly reduced the number of mouse sperm bound per egg (41). These data suggest that, like galactosyltransferase, sperm surface fucosyltransferase may also be a zona-binding receptor.

Sulfoglycolipid-Immobilizing Protein (SLIP 1)

SLIP 1 is a 68 kDa protein localized on the plasma membrane of the mouse sperm head (42). The protein is of testicular origin and binds to the major sulfoglycolipid (see below) found in testicular germ cells and spermatozoa. Ejaculated mouse sperm have two forms of SLIP 1, one that is readily released and the other that is an integral membrane component. Several lines of evidence support the conclusion that SLIP 1 has a role in sperm-egg binding. First, inclusion of purified SLIP 1 in an *in vitro* binding assay inhibits sperm-egg binding in a dose-dependent manner. Second, pre-treatment of mouse eggs with purified SLIP 1 or pretreatment of mouse sperm with anti-SLIP 1 reduces the number of sperm bound per egg. Finally, heat treatment inactivates the protein, a result suggesting that the native conformation is important for its biological activity. Taken together, these results are consistent with the proposed role for the protein.

Sulfogalactosylglycerolipid (SGG)

SGG is demonstrated to be the sulphoglycerolipid which binds to the 68 kDa SLIP 1 (see above) *in vitro* (42,43). Synthesized in the zygotene spermatocytes, the SGG remains associated with spermatozoa during its development in the testis and maturation in the epididymis. Like other sulfoglycerolipids, SGG is thought to be involved in cell-cell adhesion. An affinity purified polyclonal antibody to SGG (anti-SGG) was used to demonstrate that the sulphoglycerolipid is localized on the convex and concave ridges of the mouse sperm head. Pretreatment of spermatozoa with anti-SGG immunoglobulins (IgG) or Fab fraction inhibited sperm-zona binding at primary level in a concentration-dependent manner. In addition, liposomes formed with the sulfated (but not unsulfated) glycolipid adhere to ZP of unfertilized eggs. This last result suggests that sulfated moiety of SGG likely interact with the ZP. Although the molecular mechanism of the interaction is not known, it has been suggested that the sulfated moiety of SGG may interact electrostatically with positively-charged amino acids, such as arginine of ZP glycoproteins in a manner similar to that suggested for the interaction between SGG and L-selectin (43). Alternatively, the investigators suggest that the galactosyl sulfate moiety of SGG may interact with ligand of ZP glycoproteins. This type of carbohydrate-carbohydrate interaction has been documented for other glycolipids as well as a ganglioside-mediated cell-substratum interaction (44).

It should be noted that both SLIP 1 and SGG are localized to the post-acrosome and the convex ridge of the mouse sperm head. Since SGG and SLIP 1 show affinity for each other *in vitro*, these two sperm components may act as a complex during the gamete interaction as suggested (43).

Other Receptors

In addition to carbohydrate-specific receptors, several other sperm surface molecules have been proposed to function as receptors (Table 1). Among the antigens suggested to be putative receptors are: a trypsin (protease)-sensitive site, a 95 kDa plasma membrane protein, a fertilizing antigen-1 (FA-1), zonadhesin, sperm protein 38, spermadhesin, an adhesion protein (APz), a rabbit sperm autoantigen (RSA), and a 17 kDa sperm surface protein. An integral membrane protein, PH-20, with a molecular mass of 64 kDa has been localized on both the plasma membrane and inner acrosomal membrane of spermatozoa. The antigen, localized on the posterior head region of sperm, has been proposed to be essential in sperm adhesion to ZP in multiple species. However, a potential complementary ligand for PH-20 and other putative receptors have not yet been identified. It is not yet known whether these putative receptors use carbohydrate recognition mechanism during sperm-egg adhesion.

Why are there so many putative receptors on the sperm plasma membrane? It is difficult to give a definitive answer that would satisfactorily explain the presence of several molecules. The following factors may have contributed to the long list of putative receptors and their complementary sugar residues (ligands) presented in Table 1. First, the use of multiple experimental approaches in various laboratories to identify receptors and ligands and the interpretation of *in vitro* results by investigators may in part be responsible for many putative molecules. Second, several receptors and ligands may participate either individually (Figure 2A) or as a multimeric complex (Figure 2B). The experimental evidence thus far presented is consistent with the possibility that the molecular mechanism of sperm-egg binding is a complex event that reflects interaction between multiple sperm proteins with multiple sugar residues on ZP3 as shown in Figure 2B. Finally, since the binding of opposite gametes is a species-specific event, it is reasonable to expect the presence and involvement of several receptor and ligand molecules in different mammalian species. The multiple receptors may interact with ligands in a highly orchestrated manner; the precise order of these interactions may vary among species and may contribute to the species-specificity in gamete interaction. It is not presently known whether several receptors act individually or form a multimeric complex, as has been suggested (4).

EXOCYTOSIS OF ACROSOMAL CONTENTS

The sperm acrosome plays an important role following species-specific sperm-zona (egg) binding. Clinical studies have identified a group of men whose infertility is associated with an abnormal acrosome reaction (31). Although the sequence of events leading to successful fertilization varies among species, the pre-fertilization events of sperm-egg adhesion and induction of the acrosome reaction share many similarities. In the mouse, after irreversibly binding to the ZP, the bound spermatozoon undergoes a calcium (Ca^{2+})-dependent signal transduction pathway resulting in the exocytosis of acrosomal contents. Morphologically, the exocytosis of acrosomal contents or the acrosome reaction occurs in several steps: 1) Fusion of the plasma membrane overlying the acrosome with the outer acrosomal membrane; 2) formation of hybrid vesicles and time-dependent release of hydrolytic enzymes and other components from the acrosome; and 3) disappearance of acrosomal contents and vesicles that are held together by acrosomal matrix (Figure 3). The hydrolytic action of glycohydrolases and proteases released at the site of sperm-zona binding makes it possible for

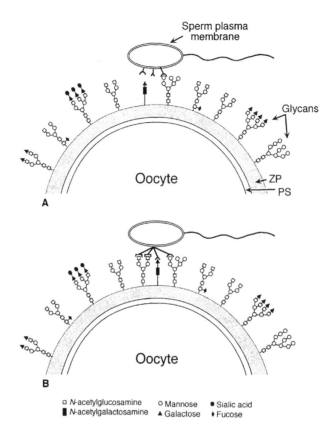

Figure 2. Models illustrating sperm-zona (oocyte) interaction. A, a single receptor on the sperm plasma membrane recognizes its complementary glycan (ligand) unit on zona-intact oocyte; B, multiple receptors on the sperm plasma membrane recognize multiple glycan (ligands) units for tight irreversible sperm-oocyte binding. ZP, zona pellucida; PS, perivitelline space.

the hyperactivated spermatozoon to penetrate the ZP and fertilize the egg (45,46).

As mentioned above, the signal that triggers the acrosome reaction in the mouse and other species is thought to be the recognition of multiple sugar residues (ligands) on bioactive glycan unit(s) of ZP3 by multiple sperm surface receptors. The net result is the opening of ion channels and the influx of Ca^{2+}. The transient rise in Ca^{2+} and other second messengers, such as cyclic AMP (cAMP) and inositol triphosphate (IP3), initiates a cascade of signaling events that elevate internal sperm pH and triggers the membrane fusion prior to the induction of the acrosome reaction (45,46).

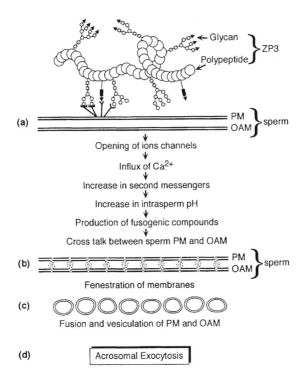

Figure 3. Model illustrating a possible sequence of events leading to acrosomal exocytosis and release of acrosomal contents. (a) multiple sugar residues (ligands) on ZP3 recognize and bind to complementary receptors on the sperm PM and starts a signal transduction cascade; (b) in response to increased Ca^{2+} and pH, the F-actin which provides a physical barrier between the PM and OAM depolymerizes to form soluble monomeric actin (G-actin which disperses bringing the PM closer to OAM); (c) the rise in Ca^{2+} also activates phospholipase A_2, an enzyme that cleaves fatty acids from phospholipids to form lysophospholipids promoting fusion and vesiculation of the sperm membranes; (d) the formation of PM and OAM vesicles (hybrid vesicles) allow acrosomal contents to be released at the site of sperm-oocyte binding. PM, plasma membrane; OAM, outer acrosomal membrane; IP3, inositol triphosphate.

Inducers of the Acrosome Reaction

Although terminal sugar residues on the bioactive glycan unit(s) of ZP provide the primary ligand sites for the surface receptors on capacitated acrosome-intact spermatozoa, a number of physiological and non-physiological compounds are known to induce the acrosome reaction in epididymal and ejaculated spermatozoa. The physiological inducers are the substances that spermatozoa will encounter during *in vivo* fertilization. Progesterone, a hormone produced during ovulation, has been reported to induce the acrosome reaction *in vitro* by interaction with the sperm plasma membrane in a receptor-mediated manner. Prostaglandins, sterol sulphate, and glycosaminoglycans present in the follicular fluid and cumulus cell secretions have also been reported to induce the acrosome reaction. Other agonists include epidermal growth factor, atrial natriuretic peptide, platelet activating factor and ATP (47). However, the putative receptors on the sperm plasma membrane that are recognized by these agonists *in vitro* or *in vivo* have not been identified. Thus, their functional significance in induction of the acrosome reaction is still unclear.

In addition to the physiological inducers, there is a long list of non-physiological substances that induce the acrosome reaction. These include calcium ionophore, synthetic glycoproteins (neoglycoproteins) and lectins. The Ca^{2+}-ionophore appears to induce the acrosome reaction by opening Ca^{2+} channels and allowing a Ca^{2+} influx. The increase in intracellular Ca^{2+} activates signaling pathways in the sperm by modulating the activities of enzymes, ion pumps, and proteins that regulate Ca^{2+}-dependent exocytosis. In addition, several neoglycoproteins have been demonstrated to mimic ZP (ZP3) and induce the acrosome reaction in the mouse (16) and human (31) spermatozoa.

In the mouse spermatozoa, mannose (Man)-BSA, N-acetylglucosamine (GlcNAc)-BSA and *N*-acetylgalactosamine (GalNAc)-BSA are capable of inducing the acrosome reaction, whereas glucose-BSA and galactose-BSA have no effect. The unconjugated monosaccharides (Man, GlcNAc, and GalNAc), even in millimolar concentrations, neither induce nor block the neoglycoprotein-induced acrosome reaction (16). These data emphasize the importance of protein backbone of the neoglycoproteins or mZP3 in induction of the acrosome reaction. Inclusion of L-type Ca^{2+} channel blockers verapamil or diltiazen (48) or calmodulin antagonists (49) in the assay mixture block the Man-/GlcNAc-/GalNAc-BSA and ZP-induced acrosome reaction, supporting the argument that the neoglycoproteins affect the same signal transduction pathway as mZP, the natural agonist. Since the protein conjugated Man, GlcNAc and GalNAc have similar effects in inducing the acrosome reaction as mZP3, these sugar residues are likely the functional components. Indeed, structural analysis of mZP3-derived glycans indicates the presence of Man, GlcNAc and GalNAc as terminal sugar residues (11,12).

Spontaneous Acrosome Reaction

Mammalian spermatozoa incubated in an appropriate capacitation medium undergo the acrosome reaction in a time-dependent manner. This type of acrosome reaction in motile spermatozoa without external stimulus is commonly known as the spontaneous acrosome reaction (1). The incidence of the non-physiological acrosome reaction depends on several factors including species and composition of the medium. In the mouse and several other species, spermatozoa that have undergone the spontaneous acrosome reaction are unable to bind to the zona-intact egg and fertilize it; however, the spontaneously acrosome-reacted spermatozoa are perfectly capable of fertilizing zona-free egg and producing normal offspring (1). The last result implies that spermatozoa that have undergone a non-physiological spontaneous acrosome reaction remain fertile.

Molecular Mechanisms of the Acrosome Reaction

The molecular mechanism by which solubilized ZP or purified ZP3 triggers the acrosome reaction has been the subject of numerous studies in the spermatozoa from mouse and other species. In the mouse, ion transport across the sperm membranes, especially Ca^{2+} influx, is essential prior to the acrosome reaction. The movement of Ca^{2+} ions from extracellular to intracellular space is a critical step that modulates the acrosome reaction. In several rodent species, spermatozoa incubated in calcium-free medium are unable to undergo the exocytotic event unless the medium is supplemented with calcium (50). Also, capacitated human spermatozoa fail to respond to Ca^{2+}-transporting agent (i.e., calcium ionophore) unless the agonist is added in the presence of external calcium. It is interesting that cAMP can induce the calcium ionophore-dependent acrosome reaction even when the Ca^{2+} is virtually absent; however, the cyclic nucleotide is not a substitute for Ca^{2+} during zona-induced acrosome reaction. These data suggest that cAMP can bypass Ca^{2+}-transporting requirement only for chemically induced acrosome reaction (50).

It is generally accepted that the first step following sperm capacitation is sperm-egg binding and an influx of exogenous Ca^{2+} ions into spermatozoa via ion channels that are regulated by guanine nucleotide (GTP-) binding proteins or G-proteins. Solubilized ZP or purified mZP3 activates G-proteins (G_0 and G_1-proteins) localized in the plasma membrane of the apical sperm head overlying the acrosome (51). Pertussis toxin, a bacterial toxin that inactivates G_1-proteins by ADP-ribosylation of α-subunit, blocks ZP-induced Ca^{2+} influx. This has led to the suggestion that G_1-like proteins influence one or more targets, including ion channels, activation of G-protein-linked receptors or other membrane associated enzymes. The net result is an influx of exogenous Ca^{2+} ions across the sperm membrane raising the intracellular concentration of free Ca^{2+} (52,53). Concurrent with or following Ca^{2+} influx, there is an activation of adenylyl cyclase, hydrolysis of adenosine triphosphate (ATP) and production of cAMP, the second messenger molecule that activates cAMP-dependent protein kinase and causes tyrosine phosphorylation of a subset of sperm protein(s) (50). There is also good evidence that activation of another membrane enzyme, such as phospholipase C, is important prior to the acrosome reaction. How the multiple molecular events that initiate a signal transduction pathway fit into a unified model is not yet known.

It should be noted that Ca^{2+} levels in spermatozoa before their binding to a ligand(s) are low, whereas it is relatively high in the extracellular fluid. The signal that induces the acrosome reaction opens Ca^{2+} channels and allows influx of Ca^{2+} between the sperm plasma membrane and outer acrosomal membrane, and triggers Ca^{2+} response proteins. Thus, the initial factor that triggers the acrosome reaction appears to be a rapid influx of Ca^{2+} followed by the activation of adenylyl cyclase and an increase in cAMP. The net result is the activation of protein kinases such as cAMP-dependent protein kinase and Ca^{2+} and phospholipid-dependent kinases (53).

Evidence accumulated over the years suggests that Ca^{2+} levels in spermatozoa are regulated by several ion channels on the plasma membrane and outer acrosome membrane (Table 2). Sperm membranes are known

Table 2. Channel proteins proposed to be involved in transport of ions across sperm membranes (adopted from 45)

Channel protein	Ions(s) transported
Transport ATPases	Ca^{2+}
	Mg^{2+}
	Na^+/K^+
Na^+/Ca^{2+} exchanger	Na^+/Ca^{2+}
Na^+/H^+ exchanger	Na^+/H^+
Cl^-/HCO^-_3 exchanger	Cl^-/HCO^-_3
K^+ channel	K^+
Ca^{2+} channels	
Voltage-dependent	Ca^{2+}
T-Type	Ca^{2+}
L-Type	Ca^{2+}
F-Type	Ca^{2+}, Co^{2+}, Mn^{2+}, Ni^{2+}
IP_3-gated channel	Ca^{2+}

to contain several ATPases that regulate the flow of ions (Ca^{2+}, Mg^{2+} and Na^+/K^+) across the sperm membranes. A Ca^{2+}-dependent ATPase associated with outer acrosomal membrane has been suggested to pump Ca^{2+}, thereby initiating the acrosome reaction. This ATPase may function as a calcium pump that keeps the Ca^{2+} levels in the acrosome relatively low until the activity of the pump is inhibited. Interestingly, inhibitors of sperm Mg^{2+}-ATPase (N,N'-dicyclohexylcarbodiimide) can stimulate the acrosome reaction by blocking exit of Ca^{2+} (Figure 4). There is also evidence that spermatozoa from several species contain Na^+-, K^+-ATPase (50). For instance, rat spermatozoa preincubated in medium containing a high K^+/Na^+

ratio enhances their fertilization rate. Also, hamster spermatozoa incubated in a solution that favors capacitation demonstrate high levels of Na^+-,K^+-ATPase. Moreover, an inhibitor of this enzyme (Ouabain) inhibits the acrosome reaction. Taken together, these data suggest a role for this ATPase in sperm exocytosis. The precise mechanism for the involvement of this enzyme in sperm function is still far from clear. It has been suggested that influx or efflux of Na^+ or K^+ ions alter the membrane potentials, thereby initiating their fusion (50).

The evidence thus far available also indicates the involvement of several other channels that regulate the entry of Ca^{2+}. These include exchanger channels, voltage-dependent and IP3-gated channels (Table 2). These channels control the flow of Ca^{2+} and other ions and form an early component of the signal transduction pathway (54). Regardless of the type of channel involved or the mode of its operation, it is important to note that ion specificity and the direction of their movement across the membranes are important factors that elevate levels of intrasperm Ca^{2+} ions and pH preceding the acrosome reaction. However, there is no unified view that can explain possible elevation in the intracellular Ca^{2+} in the spermatozoa from different species. All possible mechanisms are presented in Figure 4.

It is obvious from the above discussion that Ca^{2+} plays a central role in fusion of the sperm plasma membrane and the outer acrosomal membrane. Watson and associates (55) used pyroantimonate osmium fixation procedure to monitor the temporal and spatial localization of intrasperm Ca^{2+} during the acrosome reaction in ram spermatozoa. Calcium was initially associated with the outer acrosomal membrane. As the exocytotic event progressed, Ca^{2+} redistributes between the outer acrosomal membrane and the plasma membrane anterior to the equatorial segment where the sperm membranes fuse (Figure 4). This distribution suggests that Ca^{2+} may have a direct role in the membrane fusion during the acrosome reaction.

Phospholipases and Acrosome Reaction

As stated in Chapter 2, mammalian sperm acrosome contains phospholipases A_2 and C. It has been proposed that increased Ca^{2+} levels activate phospholipase C, which in turn causes hydrolysis of phosphatidyl-inositol biphosphate (PIP2) and an increase in inositol triphosphate (IP3) and diacylglycerol (DAG). Increased DAG activates DAG-dependent protein kinase C and opens up a voltage-dependent Ca^{2+} channel L in the

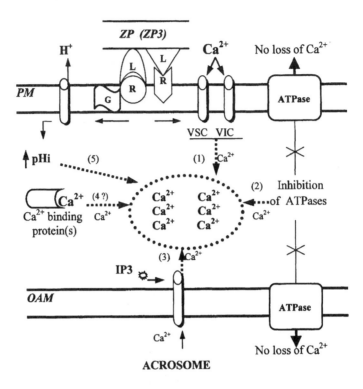

Figure 4. Model illustrating possible mechanisms that may explain an increase in sperm cytoplasmic Ca^{2+} preceding the acrosome reaction. During gamete interaction, ligands (L) on ZP3 bind to sperm surface receptors (R) and activate two major pathways (see arrows) under the PM. Pathway one (1) imports extracellular Ca^{2+} by voltage-sensitive channel (VSC) and voltage-insensitive channel (VIC). The other pathway (5) results from a transient rise in internal pH through a pertussis toxin-sensitive G-proteins (G) that promotes Ca^{2+} elevation. In addition, Ca^{2+} can be retained by the inhibition of Ca^{2+} transport ATPases (2), IP3-induced opening of Ca^{2+} channel(s) (3), and release of Ca^{2+} bound to a Ca^{2+}-binding protein(s) (4). There is no experimental evidence to suggest that Ca^{2+} is released from the Ca^{2+}-binding protein(s). Thus, the last possibility remains questionable (?). PM, plasma membrane; OAM, outer acrosomal membrane; IP3, inositol triphosphate.

plasma membrane (Table 2) leading to the increase in Ca^{2+}. This pathway has been demonstrated using pharmacological approach where the addition of synthetic DAG in the assay system activates protein kinase C and induces the acrosome reaction (56,57). It is interesting that protein kinase C also activates phospholipase A_2, an enzyme that acts on membrane phospholipids (especially phosphatidyl choline). The action of this enzyme generates lysophosphatidyl choline and fatty acid (arachidonic acid) known to be fusogenic (1).

Calcium-Binding Proteins and Acrosome Reaction

It is also known that Ca^{2+} exerts its effect through Ca^{2+}-binding protein(s) that undergo conformational changes upon interaction with free Ca^{2+}. Calmodulin, a 17 kDa acidic molecule, is one such Ca^{2+}-binding protein known to regulate many signaling pathways and membrane fusion events in the cell by modulating the activity of enzymes and ion pumps. Sperm cells contain high levels of calmodulin in the sperm head and flagellum. The head localization suggests that the protein may have a role in the induction of the acrosome reaction. Indeed, our group recently used a pharmacological approach to examine the role of this protein in ZP/neoglycoprotein(agonist)-induced acrosome reaction in mouse spermatozoa (49). Inclusion of calmodulin antagonists (calmodulin binding domain, calmidazolium, compound 48/80, ophiobolin A, W5, W7, and W13) either in *in vitro* capacitation medium or after sperm capacitation blocked the agonist-induced acrosomal exocytosis. Purified calmodulin reversed the acrosome reaction blocking effects of antagonists during capacitation. These results demonstrate that calmodulin has a role in induction of the acrosome reaction. Although the mechanisms underlying the action of calmodulin is not yet known, it has been suggested that the protein regulates sperm function by modulating sperm membrane component(s).

In addition to calmodulin, mammalian spermatozoa contain a calmodulin acceptor protein, the synaptical vesicle protein, rab3 A/a, a small GTPase, SNARE proteins, and angiotensin II receptor. Current evidence suggests that these sperm molecules also have a regulatory role in triggering membrane fusion and release of acrosomal contents. Thus, it is reasonable to suggest that multiple proteins are involved in modulating the contact and fusion of the membranes (Table 3). Whether various sperm components regulate the membrane fusion events individually or as a multimeric complex is not yet known.

CONCLUSIONS

This chapter summarizes the current state of research on sperm-egg interaction and induction of the acrosome reaction. This field is among the most active areas of research in reproductive biology. The major conclusion that emerges from the discussion above is that interaction of opposite gametes is a carbohydrate-mediated event. It is apparent that carbohydrate-binding molecules (receptors) on spermatozoa initiate sperm-egg interaction by a complex binding event reflecting interaction of multiple receptors with multiple ligands on ZP. The irreversible binding of opposite gametes starts a signal transduction pathway resulting in the exocytosis of acrosomal contents. In spite of numerous advances made in the identification and roles of multiple molecules, many intriguing discoveries remain to be made. One particular disadvantage of working with gametes is the availability of ample material for classical structure-function studies. Thus, the progress has been painfully slow. Future work in this area will undoubtedly be directed towards precise identity of sperm surface receptor(s) and its complementary ligand molecule(s) on ZP. Once identified, the functional significance of these molecules can be studied using molecular biological approaches. Any new information, we hope, will lead to a better understanding of the fertilization process.

Table 3. Proposed sperm molecules that are reported to contribute prior or on the on-set of acrosomal exocytosis

---

Non-Proteinous Components
    Anionic phospholipids, lysophospholipids
    cAMP, cGMP, IP3 and DAG
    Cholesterol
    Fatty acids (including archidonic acid)
    Unidentified small molecules
Plasma Membrane Proteins/Glycoproteins
    Adenylyl, and guanylyl cyclases
    Angiotensin II receptor
    ATPases responsible for the transport of ions (Ca^{2+}, Mg^{2+} and Na^+/K^+)
    Epidermal growth factor receptor
    Ion exchange channels such as Na^+/H^+-, Na^+/K^+-, Cl/HCO_3-
    IP3-gated Ca^{2+} channel
    G-proteins
    Phospholipases A_2 and C
    Progesterone receptor(s) ($GABA_A$?)
    Protein kinase C
    Rab3a/A (a small GTPase)
    SNARE proteins[a]
    Synaptic vesicle protein (synaptotagmin I) or very similar homolog
    Voltage-dependent Ca^{2+} channels
Outer Acrosomal Membrane and Intraacrosomal Proteins/Glycoproteins
    Actin
    Acrosin and acrosin inhibitor
    Cathepsin D
    Calmodulin
    Calmodulin acceptor proteins
    Calpain II
    Unknown Ca^{2+}-binding protein(s)

---

[a]Soluble N-ethylmaleimide-sensitive attachment protein receptors

ACKNOWLEDGEMENTS

The authors are indebted to Dr. Malika Bendahmane and Ms. Lynne Black for critically reading this chapter. The editorial assistance of Mrs. Loreita Little and Ms. Lynne Black is gratefully acknowledged. This research was supported by the National Institute of Child Health and Human Development (grants HD25869 and HD34041).

REFERENCES

1. Yanagimachi, R. "Mammalian Fertilization." In: The Physiology of Reproduction, E. Knobil, J.D. Neill, eds. New York: Raven Press, 1994, 189-317.
2. Wassarman P.M. Mammalian fertilization: molecular aspects of gamete adhesion, exocytosis, and fusion. Cell 1999; 96:175-183.
3. Tulsiani D.R.P., Yoshida-Komiya H., Araki Y. Mammalian fertilization: a carbohydrate mediated event. Biol Reprod 1997; 57:487-494.
4. Shur B.D. Is sperm galactosyltransferase a signaling subunit of a multimeric gamete receptor? Biochem Biophys Res Commun 1998; 250:537-543.
5. Topfer-Petersen E. carbohydrate-based interactions on the route of a spermatozoa to fertilization. Hum Repro Update 1999; 5:314-329.
6. Wassarman P.M. Zona pellucida glycoproteins. Annu Rev Biochem 1988; 57:415-442.
7. Gottschalk, A. "The Chemistry of Sialic Acid." In: The Chemistry and Biology of Sialic Acids and Related Substances, A. Gottschalk, ed. Cambridge: Cambridge University Press, 1960, 12-39.
8. Kornfeld S., Mellman I. The biogenesis of lysosomes. Annu Rev Cell Biol 1989; 5:483-525.
9. Tulsiani D.R.P., Abou-Haila A. Mammalian sperm molecules that are potentially important in interaction with female genital tract and egg vestments. Zygote 2001; 9:51-69.
10. Kornfeld R., Kornfeld S. Assembly of asparagine-linked oligosaccharides. Annu Rev Biochem 1985; 54:631-664.
11. Tulsiani D.R.P. Structural analysis of the asparagine-linked glycan units of the ZP2 and ZP3 from mouse zona pellucida. Arch Biochem Biophys 2000; 382:275-283.
12. Easton R.L., Patankar M.S., Lattanzio F.A., Leaven T.H., Morris H.R., Clark G.F., Dell A. Structural analysis of murine zona pellucida glycans: evidence for the expression of core 2-type O-glycans and Sd(a) antigen. J Biol Chem 2000; 63:355-360.
13. Leyton L., Saling P. Evidence that aggregation of mouse sperm receptors by ZP3 triggers the acrosome reaction. J Cell Biol 1989; 108:2163-2168.
14. Boettger-Tong H.L., Aarons D.J., Biegler B.E., George B., Poirier G.R. Binding of a murine proteinase inhibitor to the acrosome region of the human sperm head. Mol Rep Dev 1993; 36:346-353.
15. Gong X., Dubois D.H., Miller D.J., Shur B.D. Activation of a G protein complex by aggregation of beta-1,4-galactosyltransferase on the surface of sperm. Science 1995; 269:1718-1721.
16. Loeser C.R., Tulsiani D.R.P. The role of carbohydrates in the induction of the acrosome reaction in mouse spermatozoa. Biol Reprod 1999; 60:94-101.
17. Chapman N.R., Barratt C.L. The role of carbohydrate in sperm-ZP3 adhesion. Mol Human Reprod 1996; 2:767-774.

18. Kinlock R.A., Sakai Y., Wassarman P.M. Mapping the mouse ZP3 combining site for sperm by exon swapping and site-directed mutagenesis. Proc Natl Acad Sci USA 1995; 92:263-267.

19. Kudo K., Yonezawa N., Katsumata T., Aoki H., Nakano M. Localization of carbohydrate chains of pig sperm ligand in the glycoprotein ZPB of egg zona pellucida. Eur J Biochem 1998; 252:492-499.

20. Litscher E.S., Juntunen K., Seppo A., Pentlila L., Niemela R., Renkonen O., Wassarman, P.M. Oligosaccharide constructs with defined structures that inhibit binding of mouse sperm to unfertilized eggs in vitro. Biochemistry 1995; 34:4662-4669.

21. Johnson D.S., Wright W.W., Shaper J.H., Hokke C.H., Van den Eijnden D.H., Joziasse D.H. Murine sperm-egg binding, a fucosyl residue is required for a high affinity sperm-binding ligand. A second site on sperm binds a non-fucosylated, beta-galactosyl-capped oligosaccharide. J Biol Chem 1998; 273:1888-1895.

22. Thall A.D., Petr M., Lowe J.B., Oocytes galα1,3 gal epitopes implicated in sperm adhesion to the zona pellucida glycoprotein ZP3 are not required for fertilization in the mouse. J Biol Chem 1995; 270:31437-31440.

23. Nagdas S.K., Araki Y., Chayko C.A., Orgebin-Crist, M.-C., Tulsiani D.R.P. O-linked trisaccharide and N-linked poly-N-acetyllactosaminyl glycans are present on mouse ZP2 and ZP3. Biol Reprod 1994; 51:262-272.

24. Chen J., Litscher E.S., Wassarman P.M. Inactivation of the mouse sperm receptor, mZP3, by site-directed mutagenesis of individual serine residues located at the combining site for sperm. Proc Natl Acad Sci USA 1998; 95:6193-6197.

25. Lu Q., Shur B.D. Sperm from β1,4-galactosyl-null mice are refractory to ZP3-induced acrosome reaction and penetrate the zona pellucida poorly. Development 1997; 124:4121-4131.

26. Lu Q., Hasty P., Shur B.D. Targeted mutation in β1,4-galactosyltransferase leads to pituitary insufficiency and neonatal lethality. Dev Biol 1997; 181:257-267.

27. Thaler C.D., Cardullo R.A. The initial molecular interaction between mouse sperm and the zona pellucida is a complex binding event. J Biol Chem 1996; 271:23289-23297.

28. Dell A., Morris H.R., Easton R.L., Patankar M., Clark G.F. The glycobiology of gametes and fertilization. Biochem Biophys Acta 1999; 1473:196-205.

29. Rankin T.L., Tong Z.-B., Castle P.E., Lee E., Gore-Langton R., Nelson L.M., Dean J. Human ZP3 restores fertility in ZP3 null mice without affecting order-specific sperm binding. Development 1998; 125:2415-2424.

30. Tulsiani D.R.P. "Functional Significance of Sperm Surface Mannosidase in Mammalian Fertilization." In: Reproductive Immunology, S.K. Gupta, ed. New Delhi, India: Narosa Publishing House, 1999, 1-10.

31. Benoff S. Carbohydrates and fertilization: an overview. Mol Hum Reprod 1997; 3:599-637.

32. Yoshida-Komiya H., Tulsiani D.R.P., Hirayama T., Araki Y. Mannose-binding molecules of rat spermatozoa and sperm-egg interaction. Zygote 1999; 7:335-346.

33. Shalgi R., Maymon R., Bar-Shina B., Amihai D., Skutelsky E. Distribution of lectin receptor sites on the zona pellucida of follicular and ovulated rat oocytes. Mol Reprod Dev 1991; 29:365-375.

34. Cheng A., Le T., Palacios M., Bookbinder L.H., Wassarman P.M., Suzuki F., Bleil J.D. Sperm-egg recognition in the mouse: characterization of Sp56, a sperm protein having affinity for ZP3. J Cell Biol 1994; 125:867-878.

35. Foster J.A., Friday B.B., Maulit M.T., Blobel C., Winfrey V.P., Olson G.E., Kim K.S., Gerton G.L. AM67, a secretory component of the guinea pig sperm acrosomal matrix, is related to mouse sperm protein SP56 and the complement component 4-binding proteins. J Biol Chem 1997; 272:12714-12722.

36. Jones R., Brown C.R., Lancaster R.T. Carbohydrate-binding properties of boar sperm proacrosin and assessment of its role in sperm-egg recognition and adhesion during fertilization. Development 1988; 102:781-792.

37. Huang T.T.F., Ohzu E., Yanagimachi R. Evidence suggesting that L-fucose is part of a recognition signal for sperm-zona pellucida attachment in mammals. Gamete Res 1982; 5:356-361.

38. Topfer-Petersen E., Friess A.E., Nguyen H., Schill W.-B. Evidence for a fucose-binding protein in boar spermatozoa. Histochemistry 1985; 83:139-145.

39. Dell A., Morris H.R., Easton R.L., Panico M., Patankar M., Ochniger S., Koistinen R., Koistinen H., Seppala M., Clark G.F. Structural analysis of the oligosaccharides derived from glycodelin, a human glycoprotein with potent immuno-suppressive and contraceptive activities. J Biol Chem 1995; 270:24116-24126.

40. Aviles M., Okinaga T., Shur B.D., Ballesta J. Differential expression of glycoside residues in the mammalian zona pellucida. Mol Reprod Dev 2000; 57:296-308.

41. Ram P.A., Cardullo R.A., Millette C.F. Expression and topographical localization of cell surface fucosyltransferase activity during epididymal sperm maturation. Gamete Res 1989; 22:321-332.

42. Nongnug T., Smith J., Mongkolsirikieart S., Gradil C., Lingwood C.A. Role of a gamete-specific sulfoglycolipid immobilizing protein on mouse sperm-egg binding. Dev Biol 1993; 156:164-175.

43. White D., Weerachatyanukul W., Gadella B., Kamolvarin N., Attar M., Nongnug T. Role of sperm sulfogalactosylglycerolipid in mouse sperm-zona pellucida binding. Biol Reprod 2000; 63:147-155.

44. Roberts D.D., Ginsburg V. Sulfated glycolipid and cell adhesion. Arch Biochem Biophys 1988; 267:405-415.

45. Tulsiani D.R.P., Abou-Haila A., Loeser C.R., Pereira B.M.J. The biological and functional significance of the sperm acrosome and acrosomal enzymes in mammalian fertilization. Exp Cell Res 1998; 240:151-164.

46. Abou-Haila A., Tulsiani D.R.P. Mammalian sperm acrosome: formation, contents and function. Arch Biochem Biophys 2000; 379:173-182.

47. Baldi E., Luconi M., Bonaccorsi L., Muratori M., Forti G. Intracellular events and signaling pathways involved in sperm acquisition of fertilizing capacity and acrosome reaction. Front Biosci 2000; 5:E110-123.

48. Loeser C., Lynch C. II, Tulsiani D.R.P. Characterization of the pharmacological-sensitivity profile of neoglycoprotein-induced acrosome reaction. Biol Reprod 1999; 61:629-634.

49. Bendahmane M., Lynch C. II, Tulsiani D.R.P. Calmodulin signals capacitation and triggers the agonist-induced acrosome reaction in mouse spermatozoa. Arch Biochem Biophys 2001; 390:1-8.

50. Guraya S.S. Cellular and molecular biology of capacitation and acrosome reaction in spermatozoa. Int Rev Cytol 2000; 199:1-64.

51. Ward C.R., Storey B.T., Kopf G.S. Selective activation of Gi1 and Gi2 in mouse sperm by the zona pellucida, the ovum's extracellular matrix. J Biol Chem 1994; 269:13254-13258.

52. Florman H.M., Arnoult C., Kazam I.G., Li C., O'Toole C.M. A perspective on the control of mammalian fertilization by egg activated ion channels in sperm: a tale of two channels. Biol Reprod 1998; 59:12-16.

53. Ward C.R., Kopf G.S. Molecular events mediating sperm activation. Dev Biol 1993; 158:9-34.

54. Garcia M.A., Meizel S. Regulation of intracellular pH in capacitated human spermatozoa by a Na^+/H^+ exchanger. Mol Reprod Dev 1999; 52:189-195.

55. Watson P.F., Plummer J.M., Jones P.S., Bredl J.C.S. Localization of intracellular calcium during the acrosome reaction in ram spermatozoa. Mol Reprod Dev 1995; 41:513-520.

56. Roldan E.R.S. Role of phospholipases during sperm acrosomal exocytosis. Front Biosci 1998; 3:D1109-1119.
57. Breitbart H., Spungin B. The biochemistry of the acrosome reaction. Mol Hum Reprod 1997; 3:195-202.

Chapter 16

GAMETE MEMBRANE INTERACTIONS
The Cell-Cell Interactions between Sperm and Egg during Fertilization

Janice P. Evans
Johns Hopkins University, Baltimore, MD, U.S.A.

INTRODUCTION

Fertilization takes place in a series of discrete steps (Fig. 1). The sperm actually interacts with the egg on three separate levels: first with the cumulus cells and the hyaluronic acid extracellular matrix (ECM) in which they are embedded, secondly with the egg's own ECM, called the *zona pellucida* (ZP), and finally with the egg plasma membrane (Fig. 1). The binding of the sperm to specific ZP glycoproteins (ZP3 in the mouse) induces the sperm to undergo the "acrosome reaction" the exocytosis of the acrosome vesicle on the head of the sperm (1). The acrosome reaction allows the sperm to penetrate the ZP and exposes or modifies portions of the sperm membrane, including the inner acrosomal membrane and the equatorial segment (shown in Fig 1). These acrosome reaction-induced changes in the sperm are critical for fertilization success, as these modified regions of the sperm head participate in initial gamete membrane binding or subsequent sperm-egg membrane fusion (1-3). Thus, the acrosome reaction is a necessary prerequisite step for sperm-egg plasma membrane interactions to occur, as only acrosome-reacted sperm can bind and fuse with the egg plasma membrane (1). In addition, since capacitation is required for the sperm to be able to undergo the acrosome reaction, capacitation is therefore also a prerequisite for the sperm to be capable of interacting with the egg plasma membrane.

This chapter will focus on the molecules and processes that serve to bring sperm and egg together, as well as the biological and clinical applications of gamete membrane interactions. Gamete plasma membrane interactions occur in the perivitelline space, between the membrane and the ZP (Fig. 1). These interactions occur in a stepwise fashion, first with the sperm adhering to the egg (cell-cell adhesion), and then with the membranes of these two cells fusing to make the one-cell embryo (also known as a zygote). Following gamete membrane fusion, a change in the egg

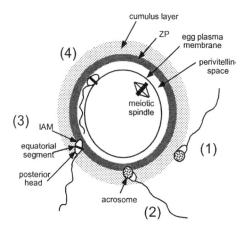

Figure 1. Schematic diagram of the steps of sperm-egg interactions during fertilization. The three levels with which the sperm interacts with the egg are labeled: the cumulus layer (including cumulus cells and the hyaluronic acid extracellular matrix in which they are embedded), the zona pellucida (ZP), and the egg plasma membrane. The perivitelline space is the space between the egg plasma membrane and the ZP. Step 1 shows the sperm interaction with the cumulus layer. In Step 2 the sperm proceeds to the ZP; this interaction induces the sperm to undergo acrosomal exocytosis (or acrosome reaction), and Step 3 shows the sperm surfaces that are exposed after acrosomal exocytosis: the inner acrosomal membrane (IAM), the equatorial segment, and the posterior head. In Step 4, the acrosome-reacted sperm gains access to the perivitelline space and interacts with the egg plasma membrane.

known as egg activation is initiated. Egg activation is a collective term, referring to the changes that occur within the egg that release it from its relatively quiescent state to initiate embryo development. One important change that occurs with egg activation is **cell cycle resumption**. As was discussed in Part II, female gametes are released from the ovarian follicles in response to the hormonal signal that triggers ovulation. This also triggers progression through meiosis, from prophase of meiosis I to metaphase of meiosis II in most mammalian species. At this metaphase II stage, the female gamete is often referred to as an **egg** rather than an oocyte, to indicate that it has progressed from prophase I into meiosis; the term "egg" will be used in this chapter. An egg will remain arrested at metaphase II unless it is fertilized. If the egg is fertilized, then it will continue through the cell cycle, completing meiosis and then entering into the first embryonic mitotic cycle. Another important change that occurs upon egg activation is the **prevention of polyspermy** (prevention of fertilization by additional sperm).

CELL BIOLOGY OF GAMETE MEMBRANE INTERACTIONS

Cell Adhesion

Cell-cell interactions occur throughout the body. Some cell-cell interactions hold tissues together (e.g., epithelia), and others occur under special conditions [e.g., platelet aggregation in response to wounding]. Gamete membrane interactions are frequently cited as being analogous to the interactions between leukocytes in the blood stream and endothelial cells that make up blood vessels (4,5). First, the interactions of the gamete plasma membranes appear to utilize multiple ligands and receptors (detailed below), as do leukocyte-endothelial interactions. Second, both sperm and leukocytes are under flow conditions – leukocytes from being in the blood stream and sperm from the motile force of the sperm tail [sperm tail beats are estimated to generate almost 3000 μdyns of shear force (6)]. Flow conditions put special pressures on cellular interactions. Initial molecular interactions must be sufficient to slow or "brake" the cell [the leukocyte or sperm] under flow, tethering it to the adhering surface [the endothelial cell or egg plasma membrane], so that it resist shear forces and ultimately stay attached. This initial attachment subsequently leads to moderate and then firm adhesion. Thus, the utilization of multiple ligands and receptors in these examples of cell-cell interactions might be necessary because different ligand-receptor pairs serve certain specific roles.

In the leukocyte-endothelium model, one example is L-selectin on leukocytes; L-selectin binds to specific carbohydrates on endothelial cells, allowing the leukocyte roll along the surface of the endothelium lining the blood vessel. The L-selectin-carbohydrate interaction slows the leukocyte along the endothelium but is not of sufficiently high affinity to allow leukocyte to stop and firmly adhere to the endothelium. This task is mediated by other ligands and receptors, including specific integrins on leukocytes and their ligands on endothelial cells. The molecules that mediate the specific steps of leukocyte-endothelial interactions are known (Fig. 2). While the specific molecules that mediate the specific steps of gamete membrane interactions are not known, molecules on both the sperm and egg that participate in these interactions are being identified [see below]. As we learn more about these molecules and the properties of their interactions, we will begin to be able to assign specific gamete molecules to specific roles.

292

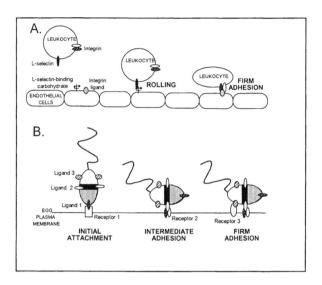

Figure 2. Schematic diagrams of leukocyte-endothelium and sperm-egg interactions and a hypothetical model for the mechanism of sperm-egg adhesion. **A,** the interaction of a leukocyte in the bloodstream with endothelial cells, forming the vessel wall, is shown. This interaction is initiated by L-selectin on the leukocyte binding to a carbohydrate on the endothelial cells; this molecular interaction is not tight but is sufficient to slow the leukocyte to roll along the endothelial cells. This leads to firm adhesion, mediated by different molecules [an integrin on the leukocyte and its ligand on the endothelial cells]. Firm adhesion allows the leukocyte to flatten on the endothelial surface. **B,** by analogy to the leukocyte-endothelium interaction, the gametes are shown with multiple ligands on the sperm and receptors on the egg. These ligands and receptor are hypothetical. Sperm Ligand 1 is on the anterior head surface exposed after the acrosome reaction [the inner acrosomal membrane, IAM shown in Figure 1]. Ligand 2 is on the equatorial segment, and Ligand 3 is on the posterior head. Based on electron microscopic studies, sperm in several species interact with the egg membrane "tip first," via the inner acrosomal membrane, and then rotate down to interact with egg membrane via the equatorial segment and the posterior head. These latter two regions of the sperm head also mediate sperm-egg fusion.

Membrane Fusion

Gamete membrane interactions culminate in membrane fusion between the plasma membranes of the sperm and the egg, making one cell [the zygote] from two. Some other cells in the body undergo cell-cell fusion, including myoblasts during muscle development and trophoblast cells to form the syncytiotrophoblast of the placenta in some species. While relatively little is known about how membrane fusion between cells occurs, other membrane fusion events are better understood. From studies of other systems such as the fusion of specific viruses to cells or of intracellular vesicles and intracellular or plasma membranes, we have the following model for the step-wise fashion by which membrane fusion occurs, shown in Figure 3 (7). First, the two membrane-bound entities that are destined to fuse come into close contact, through molecular interactions between ligands and receptors that allow tight adhesion and membrane apposition. Next, facilitated by interactions between protein and then lipid components of the membranes, the facing leaflets of two lipid bilayers intermingle, a state called **hemifusion**. This leads to the formation of an opening, called a **fusion pore**, that connects the two membranes. This fusion pore can then expand, ultimately incorporating one membrane into the other. It should be noted that the egg incorporates the entire sperm, including the tail.

There are different models for how these steps are modulated on a molecular level. One model comes from studies of the membrane fusion events between cells and virus particles, especially influenza. In virus-cell interactions, membrane fusion is mediated by a viral fusion protein in the membrane of the viral envelope. The virus's fusion protein contains a hydrophobic subdomain, called a **fusion peptide**. Fusion peptides are normally conformationally "concealed," folded within the fusion protein so that its hydrophobic amino acids are not exposed to the aqueous environment. Following the appropriate stimulus, such as a pH change or a molecular interaction or other signal, a conformational change in the fusion protein occurs, unmasking the fusion peptide. The fusion peptide can then insert into the lipid bilayer of the plasma membrane of the cell to which the virus particle is attached. The fusion peptide insertion can lead to a conformation change in the fusion protein, which bends the target membrane, bringing it into close apposition with the viral membrane envelope. The insertion of the fusion peptide into the target cell membrane may also disrupt the "recipient" lipid bilayer, which then intermingles with the phospholipid components of the bilayer with the fusion peptide, leading to hemifusion and then the formation of a fusion pore.

How this fusion peptide model might apply to sperm-egg interactions is shown in Figure 4, although it should be emphasized that this is only a hypothetical model for how sperm-egg fusion occurs. In the hypothetical model pictured, the fusion protein is in the sperm membrane, although other possibilities exist, such as the fusion protein being a component of the egg membrane that then inserts into the sperm membrane. In reality, relatively little is known about membrane fusion events between cells, but some insights into this cellular process could come from studies of sperm-egg fusion.

A second model for membrane fusion comes from studies of membrane fusion events that occur inside cells, i.e., the fusion of membrane-bound intracellular vesicles with other membranes [plasma membrane or other intracellular vesicles and organelles]. Membrane fusion events in this case are mediated by transmembrane proteins known as Soluble N-

294

ethylmaleimide-sensitive factor (NSF) Attachment protein Receptor (SNARE), along with cytoplasmic accessory proteins that regulate SNARE function. A SNARE protein on a vesicle (a v-SNARE) interacts with a SNARE on the target membrane (a t-SNARE). These two SNARE proteins are believed to intertwine and form bundles of α-helices, bringing the vesicle and target membranes into close apposition, analogous the way a viral fusion protein brings membranes close after insertion of the fusion peptide. This close proximity of the membrane bilayers leads to hemifusion (mixture of the outer leaflets of the lipid bilayers) and then formation of the fusion pore. How

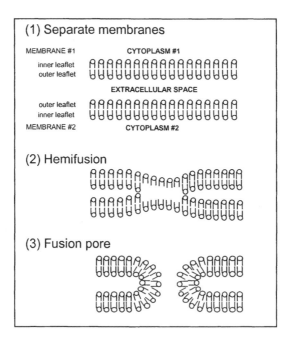

Figure 3. Membrane fusion. The schematic diagram shows two phospholipid bilayers [Membrane #1 and Membrane #2], made up of outer leaflets, facing the extracellular space, and inner leaflets, facing the cytoplasm. Membrane fusion begins with hemifusion [Step 2], or the merging of the outer leaflets, and then continues with the formation of the fusion pore [Step 3], which serves as an opening connecting the two cytoplasms.

How might this model be applied to sperm-egg fusion? It is possible that there are SNARE-like proteins on the extracellular surface of the sperm and egg that mediate gamete fusion; the sperm SNARE and egg SNARE could intertwine and lead to fusion of the gamete membranes. However, it should be emphasized that the SNARE model is based on membrane fusion events that occur on the interior of cells, with fusion being initiated by the membrane proteins and lipids that face the cytoplasm. Sperm-egg [and other

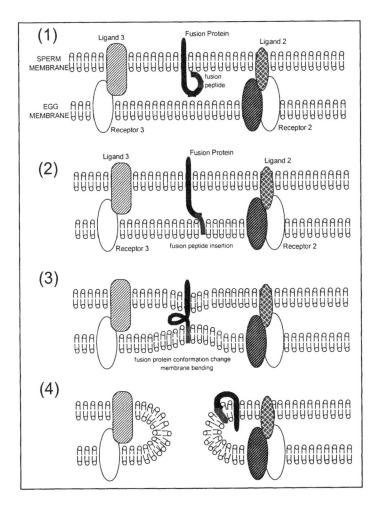

Figure 4. A hypothetical model for the mechanism of sperm-egg fusion. This diagram shows a magnified view of the sperm and egg in Figure 2, continuing from the stage where firm adhesion has occurred, mediated by Ligands 2 and 3 on the sperm and Receptors 2 and 3 on the egg. Following establishment of firm adhesion [Step (1)], some signal induces the fusion protein to undergo a conformational change, so that the fusion peptide is exposed and then can insert in the lipid bilayer of the egg membrane [Step (2)]. A conformational change in the fusion protein then occurs [Step (3)], bending the lipid bilayers. This ultimately then induces mixture of the lipid bilayer components, leading to hemifusion and then formation of the fusion pore [Step (4)]. In this hypothetical model, the fusion protein is in the sperm membrane, although other possibilities exist [see text].

cell-cell] fusion events are different in this regard, occurring on the extracellular surfaces of membranes and thus initiated by membrane proteins and lipids that face the extracellular space.

EXPERIMENTAL METHODS USED TO STUDY MOLECULES INVOLVED IN FERTILIZATION

Cell-cell interactions and other cell adhesion processes have been traditionally studied using *in vitro* cell adhesion assays, i.e., allowing cells to adhere to a substrate or to other cells, and manipulating the conditions or including reagents that are hypothesized to inhibit or enhance cell adhesion. In studies of gamete membrane interactions, this has been done using *in vitro* fertilization (IVF) assays, i.e., mixing sperm and eggs in experimentally manipulated conditions. One important example of such an experimental manipulation is shown in Figure 5, with the use of experimental reagents that might be hypothesized to inhibit sperm-egg interactions. These reagents include antibodies that cross-react with certain sperm or egg molecules, and also peptides or proteins that mimic parts of sperm or egg proteins and thus competitively inhibit gamete interactions.

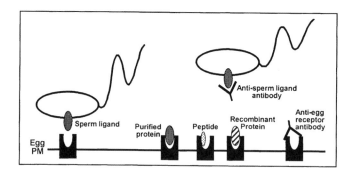

Figure 5. Experimental assays used to assess function of sperm or egg molecules. Shown at the left side of the diagram is a simplified example of one sperm ligand interacting with its receptor on the egg surface. Subsequent images show reagents that can be used experimentally to examine the roles of molecules in sperm-egg interactions. One experimental approach is doing a competitive inhibition assay using purified, isolated sperm ligand protein or mimetics of it. Examples of mimetics include synthetic peptides with sequences matching key portions of the sperm molecule or recombinant proteins, expressed from the cDNA encoding the sperm ligand in an expression system utilizing prokaryotic or eukaryotic cells. In a competitive inhibition assay, the egg is incubated with a molar excess of sperm ligand; these isolated sperm ligand molecules block many or all of the egg receptor sites. Thus, when sperm are added to the eggs, egg receptor sites are occupied and sperm-egg interactions mediated by that particular sperm ligand cannot occur. Alternatively, antibodies that disrupt the function of either the sperm ligand or the egg receptor could be used to inhibit sperm-egg interactions mediated by those molecules.

But how does a researcher decide what specific antibodies or proteins to test in an IVF assay? Key insights into gamete membranes interactions have come from studies of cell-cell interactions in other experimental systems or cells other than gametes [i.e., cultured cells], which have characterized cell adhesion molecules that function in somatic cells to mediate cell-cell or cell-environment interactions. One example of this is the integrins, a family of cell adhesion molecules that are widely expressed by many different cell types in the body and are known to mediate interactions of cells with extracellular matrix substrates and/or with other cells. Recent studies have focussed on integrins as candidates to mediate gamete membrane interactions. Results from these studies have demonstrated that integrin family members are expressed by eggs and that anti-integrin antibodies can perturb gamete membrane interactions [see below].

Another approach that has been used to study gamete surface proteins involved in sperm-egg membrane interactions has been to immunize mice with intact gametes or with gamete membrane proteins, to generate a battery of monoclonal antibodies to the gamete surface antigens. The antibodies are then screened for their abilities to bind to surface antigen(s) and to inhibit sperm-egg binding or fusion, with IVF assays similar to the type shown in Figure 5. One sperm antigen has been well characterized and has emerged as one of the major players in sperm-egg membrane interactions. This protein is known as fertilin, and will be discussed in more detail below.

The generation of knockout mice has also shed tremendous light on the field of sperm-egg membrane interactions. Some of these knockout mice have been generated for the purpose of testing the hypothesis that a particular molecule is involved in fertilization [i.e., fertilin β or cyritestin; discussed below]. Some other knockout mice have been found to have the unanticipated phenotype of male or female infertility, and when the gametes were analyzed in *in vitro* fertilization assays, it was found that there were defects in some part of the fertilization process.

The vast majority of what we know about fertilization, including the process of gamete membrane interactions, comes from studies of experimental animal models, including mouse, rat, bovine, and porcine. Although human sperm are often available for study, it is extremely difficult to obtain human eggs for research studies, and those that can be obtained are often donated after a failed *in vitro* fertilization attempt and thus may be of questionable quality. Moreover, laws in the United States and many other countries have set up strict restrictions on the use of human gametes in research studies; much of what is done experimentally with animal gametes cannot be done with human gametes. However, the cellular and molecular processes of fertilization have proved to be highly conserved between mammalian species, and thus what has been learned from non-human mammalian species translates well to insights into human fertilization. In addition, heterologous systems [i.e., sperm from one species and eggs from another species] are occasionally used. ZP-free hamster eggs are commonly used for these sorts of studies, because hamster eggs are capable of fusing with sperm from nearly every mammalian species tested (3). [The ZP needs to be removed, as sperm-ZP interactions occur in a species-specific manner.]

This is frequently referred to as the "zona-free hamster egg penetration test." It is not clear how biologically relevant the zona-free hamster egg penetration test is, since fusion with a hamster egg may occur via a mechanism that is unique to the hamster egg. Nevertheless, the zona-free hamster egg penetration test has proved to be useful, particularly in studies of sperm from species (such as human) for which homologous eggs are not readily obtainable.

MOLECULES OF GAMETE MEMBRANE INTERACTIONS

Sperm Molecules

ADAMS: Fertilin α (ADAM1), Fertilin β (ADAM2), and Cyritestin (ADAM3)

Three of the best characterized sperm proteins with roles in gamete membrane interactions are related molecules with significant amino acid sequence homology, and thus are members of the same molecular family. This family is known as ADAM proteins, with the name ADAM being an acronym representing the domain structure of these proteins [A Disintegrin and A Metalloprotease, see Figure 6]. This molecular family has also been called "MDCs" [for Metalloprotease / Disintegrin / Cysteine-rich domains] and "cellular disintegrins," and thus these names are seen in the literature. The sperm ADAMs that are known to participate in sperm-egg membrane interactions are fertilin α (ADAM1), fertilin β (ADAM2), and cyritestin (ADAM3).

To date 28 different vertebrate ADAMs have been identified, and there are ADAMs in invertebrates (*Drosophila* and *C. elegans*) as well. Understanding the significance of ADAM proteins requires a little lesson in some other cell biology topics. Some extracellular matrix proteins (e.g., fibronectin, vitronectin) utilize small amino acid motifs with the sequence Arg-Gly-Asp [RGD in single letter amino acid code] to interact with their receptors, which are known as integrins. Integrin-binding proteins have also been identified in snake venoms. These snake venom polypeptides also have RGD amino acid sequences, and can competitively inhibited integrin-mediated adhesion of cells to RGD sequences in extracellular matrix proteins. Because of this integrin-perturbing activity, these snake venom proteins are called disintegrins (8). When cDNA sequences for four ADAM family members were first isolated in the early 1990s (two on sperm, one in the epididymis, and one as the gene product of a putative breast cancer tumor suppressor), one of the most striking features was the significant sequence homology between one region of the ADAM proteins and snake venom disintegrins – and so this region of ADAM proteins came to be called a "disintegrin domain."

The discovery that fertilin β is involved in sperm-egg membrane interactions came long before its cDNA sequence and domain structure were known. Fertilin β was identified as the antigen of one antibody characterized from a battery of monoclonal antibodies made in the early 1980s by immunizing mice with guinea pig sperm membrane proteins; these

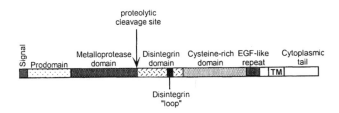

Figure 6. Domain structure of ADAM proteins. This schematic diagram shows the generalized domain structure of all ADAM proteins, based on the consensus amino acid sequences (primary protein structure). All ADAM proteins have the following functional domains: a signal sequence ("signal"); a prodomain; a metalloprotease domain; a disintegrin domain; a cysteine-rich domain; an EGF-like repeat; a transmembrane domain (TM), and a short cytoplasmic tail. The signal sequence allows the protein to be inserted into the endoplasmic reticulum during protein synthesis. The metalloprotease domain is an active protease in some ADAM family members that contain the required amino acid sequence that mediates the binding of zinc ions, which is required for proteolytic activity. [Not all ADAM proteins have this sequence, and thus not all ADAM proteins have protease activity.] The disintegrin domain has homology to the integrin-binding polypeptides in snake venoms known as disintegrins. The disintegrin loop is a short amino acid sequence within the disintegrin domain. In snake venom disintegrins, this disintegrin loop has the sequence RGD (or similar sequence KGD), but in ADAM proteins, this sequence is almost never an RGD. In the sperm ADAM proteins fertilin α, fertilin β, and cyritestin, the protein is proteolytically cleaved between the metalloprotease domain and the disintegrin domain. This releases the prodomain and the metalloprotease domain, and leaves only the disintegrin domain, the cysteine-rich domain, and the EGF-like repeat exposed on the sperm surface.

immunized mice made many different antibodies cross-reacting with many different molecules on the sperm membrane. One of these antibodies, PH-30, cross-reacts with the posterior head of the guinea pig sperm, a region of the sperm head that interacts with the egg membrane [see Fig. 2]. PH-30 was also found to inhibit fertilization of guinea pig eggs (9). Since that time, the antigen recognized by PH-30, fertilin β and its associated protein fertilin α, have been cloned in many mammalian species (mouse, rat, rabbit, bovine, human, and some non-human primates). How these molecules function in fertilization has been characterized, using the mouse as an experimental model. Biochemical studies show that fertilin α and fertilin β pair together, existing as a heterodimeric complex in the sperm membrane (9-12). Coincident with these research studies of fertilin β and fertilin α, another member of the family was discovered and characterized; cyritestin is so-named because it was a cysteine-rich protein identified in the testis (13) (see Fig.7).

There are many experimental data that indicate that fertilin α, fertilin β, and cyritestin participate in gamete membrane interactions. Sperm treated with antibodies that bind to fertilin β (9,14,15) and cyritestin (14) have a reduced ability to fertilize eggs in *in vitro* fertilization assays. In com-

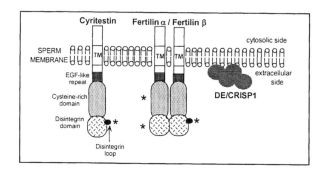

Figure 7. Major molecular players on the sperm involved in sperm-egg adhesion. Three different ADAMs are known to participate in the adhesion of the sperm to the egg membrane: cyritestin, fertilin α, and fertilin β. All three molecules are integral membrane proteins. Fertilin α and fertilin β are known to form heterodimers, and so are shown together. However, these three ADAMs use different domains to achieve this adhesion; asterisks indicate the functional domains (the disintegrin loop in cyritestin and fertilin β, and the disintegrin and cysteine-rich domains in fertilin α). In addition, the epididymis secretes the protein DE/CRISP-1, which then associates with the sperm surface. It is not known how DE/CRISP-1 binds to the sperm membrane. It is shown associated with the lipid bilayer, although it is possible that DE/CRISP-1 binds its receptor protein in the sperm membrane as well.

plimentary experiments, eggs are treated with peptides and recombinant proteins that mimic parts of the ADAM proteins on sperm; these peptides or proteins bind to the receptors on the egg surface, blocking the receptors and leading to reduced sperm-egg binding (14,16-21). These studies have been done primarily with mouse gametes, although there are similar data from rabbit (15), baboon (22), and human (23). Finally, male mice with the fertilin β or cyritestin genes knocked out have been demonstrated to have reduced fertility (24-26); this will be discussed in more detail below.

Fertilin α, fertilin β, and cyritestin have been additionally characterized by the determination of functional domains, or what parts of the molecules mediate interactions with the egg membrane. This was addressed by analyses of the sperm ADAM proteins by generating recombinant versions of the sperm proteins with either specific domains deleted or with specific amino acids mutated. These studies revealed that fertilin β and cyritestin seem to function by similar molecular mechanisms. Both of these proteins utilize similar sequence in a specific region of the disintegrin domain, called the disintegrin loop. In fertilin β, this sequence is glutamic acid-cysteine-aspartic acid [ECD] (27,28), and in cyritestin, this sequence is glutamine-cysteine-aspartic acid [QCD] (29). Fertilin α, however, utilizes different domain regions from fertilin β and cyritestin. The fertilin α disintegrin domain and also the cysteine-rich domain (with the EGF-like repeat) can mediate interactions with the egg surface (21,30). It has not been definitively identified what amino acids of the disintegrin domain are involved, although initial studies suggest that the disintegrin loop is not involved (14,30).

Another interesting finding regarding these proteins pertains to their domain structure and proteolytic processing. As shown in Figure 5, some ADAM proteins are cleaved by a proteolytic enzyme between the metalloprotease and disintegrin domain, removing the signal sequence, the prodomain, and the metalloprotease domain from the protein anchored in the membrane. This appears to happen to fertilin α while sperm are developing in the testis, and to fertilin β and cyritestin as the sperm move through the epididymis (10,14,31,32). As is discussed in Chapter 6, epididymal transit is a critical maturational process for sperm. The proteolytic processing of fertilin β and cyritestin during epididymal transit may be an important molecular change that renders sperm capable of fertilizing an egg. In guinea pig sperm, the proteolytic processing of fertilin β during epididymal transit correlates with the relocalization of fertilin β from the entire head to the posterior head of the sperm (10).

DE / CRISP-1

DE is not an actual sperm protein per se, but is synthesized by and secreted from the epididymal epithelium and associates with the surface of sperm during epididymal transit (33). DE gets its name from two bands [band D and band E] present in a lysate of rat epididymal proteins run on a denaturing polyacrylamide gel. Proteins D and E have very similar amino acid sequences, might be products of the same gene, and are routinely purified together, and thus are referred to collectively as "DE." When the cDNA encoding rat DE was cloned, it was found to be highly homologous to members of family of cysteine-rich secretory proteins (34), specifically CRISP-1 in the mouse and human. DE is also referred to as Acidic Epididymal Glycoprotein [AEG or AEG-1] in the literature (35). The human homolog has been called ARP, for AEG-related protein; this human

DE-like cDNA shares 40% homology with rat DE/CRISP-1. Thus, the terms DE, CRISP-1, AEG-1, and ARP all refer to the same protein.

As noted above, DE is synthesized by the epididymis. This contrasts with the sperm proteins previously discussed, fertilin α, fertilin β, and cyritestin, which are integral components of the sperm membrane and are synthesized by the developing male germ cells during spermatogenesis. DE, a small (37 kD) glycoprotein, is secreted from epididymal cells into the epididymal lumen, where it becomes associated with the membrane covering the sperm head as they pass through the epididymis from the testis. Acquisition of DE on the sperm surface may be an important component of the epididymal maturation process.

What is the evidence that DE is involved in gamete membrane interactions? Anti-DE antibodies inhibit fertilization of rat eggs, either when mixed with rat sperm before they are used for artificial insemination (36) or in *in vitro* fertilization (37). Additionally, DE protein purified from rat epididymal extracts binds to the plasma membrane of rat and mouse eggs. Rat and mouse eggs treated with purified rat DE show reduced levels of fertilization when mixed with sperm in *in vitro* fertilization (38,39). DE may also participate in human gamete membrane interactions based on some recent preliminary data. A recombinant form of human DE binds to the plasma membrane of human eggs and human sperm treated with anti-human DE antibodies show a reduced ability to fertilize hamster eggs (40).

Egg Molecules

Integrins

Integrins as well as other families of cell adhesion molecules have lingered in the background for decades as candidates to mediate cell-cell interactions during fertilization. As integrins, cadherins, and immunoglobulin superfamily cell adhesion molecules were discovered and found to mediate many different cell adhesion processes, gamete biologists speculated about the possibility that these molecules might also be functioning in sperm-egg interactions. However, integrins became strong candidates to mediate gamete membrane interactions with the identification of the disintegrin domain in fertilin β in 1992 (41) and then subsequently in other sperm ligands, cyritestin and fertilin α. The presence of integrin ligand-like domain in a sperm protein raised the possibility that this sperm protein could bind to an egg receptor. The identity of the integrin that mediates sperm-egg adhesion has not been conclusively determined, however. Integrins are heterodimeric molecules, made up of an α subunit and a β subunit (Fig. 8). To date, 18 different α subunits and 8 different β subunits have been identified in vertebrates, and these make 24 different combinations, each of which recognizes different ligands. Of these 24 different integrins, one candidate, $\alpha_6\beta_1$, has been studied extensively as a receptor for fertilin β, with complicated and conflicting experimental results. A function-blocking monoclonal antibody that cross-reacts with the α_6 integrin subunit reduces the binding of sperm and fertilin β to eggs in some (28,42) but not all experiments (19,43). A synthetic peptide corresponding to the fertilin β disintegrin loop can be cross-linked to an egg surface protein that is immunoprecipitated with an anti-α_6 antibody (44), suggesting that fertilin β and α_6 interact. Non-egg cells that express $\alpha_6\beta_1$ can support sperm binding (42). However, eggs that lack α_6 or its functional ligand binding motif can still support the binding of sperm and fertilin β (19,45). Taken together, these data indicate that $\alpha_6\beta_1$ can participate, directly or perhaps indirectly, in fertilin β-mediated sperm adhesion to the egg, but it is not required for these adhesions. More recent studies implicate other integrins

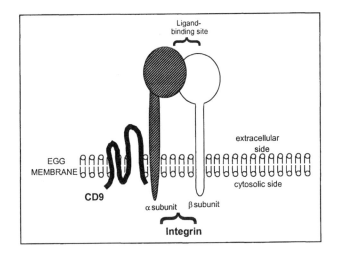

Figure 8. Major molecular players on the egg involved in sperm-egg adhesion. Because of the involvement of the integrin ligand-like disintegrins domains of cyritestin, fertilin β, and fertilin α in fertilization, it has been hypothesized that an integrin on the egg serves as the receptor for these sperm ADAMs. The exact identity of the integrin α and β subunits have not been definitively identified, although α_6, α_v, α_9, and β_1 are good candidates. The ligand binding domain of integrins is known to be located in the globular head domains of the α and β subunits, atop extended stalks. Shown associated with the integrin is the integrin-associated protein CD9, which is known to be critical for sperm-egg membrane interactions. CD9 is a member of the tetraspanin family, proteins that span the plasma membrane four times. Although CD9 is not known to be a receptor for a sperm ligand, it likely serves as an accessory protein to the integrins, facilitating their function as receptors.

as additional candidates to be involved in gamete membrane interactions (46) or specifically in fertilin β binding to the egg membrane (43). It should also be noted that an egg integrin(s) could be a receptor for other sperm ligands, including cyritestin and fertilin α. A function-blocking monoclonal antibody that cross-reacts with the α_6 integrin subunit reduces the binding of cyritestin to eggs (29), although this same antibody has no effect on the binding of fertilin α to eggs (30). The egg integrin(s) that binds to sperm ADAMs has yet to be definitively determined.

CD9

CD9 is a protein that associates with a subset of integrin family members within the plane of the plasma membrane (Fig. 8). CD9 is a member of another protein family known as tetraspanins, so named because they span the membrane four times. The discovery that CD9 is involved in gamete membrane interactions came from two different studies. One was the generation of CD9 knockout mice, which was done by three different research groups. CD9 -/- female mice were found to be infertile, and further analysis of this defect revealed that sperm could not interact with the plasma membranes of eggs from these CD9 -/- mice (47-49). Additional evidence for the role of CD9 on the egg came from studies using anti-CD9 antibodies in *in vitro* fertilization experiments, and observing that these antibodies have a significant inhibitory effect on gamete membrane interactions (50). Anti-CD9 antibodies have some inhibitory activity on the binding of fertilin β (43,50), fertilin α (30), and cyritestin (29) to eggs. It is not clear, however, exactly what role CD9 plays in fertilization. CD9 and other members of the tetraspanin family appear to play important roles in several events mediated by the cell surface, including cell adhesion, cell migration, and even some membrane fusion events (51,52). Because of this, tetraspanin proteins have been dubbed "molecular facilitators," not serving as direct receptors for ligands involved in adhesion or migration or fusion, but facilitating the function of molecules that do serve as receptors. The field continues to advance, however, as other tetraspanin knockout mice have been generated and analyses of their phenotypes have shed light on tetraspanin function (51,52). The distinctive role of CD9 in the egg should also provide important insights into the molecular functions of tetraspanin family members. Although we lack a precise picture of CD9's role in fertilization, all current data point to CD9 being a very important regulator of egg membrane function. For example, the eggs from knockout mice deficient in the α_6 integrin subunit can be fertilized, but fertilization of these α_6-lacking eggs can be inhibited by anti-CD9 antibodies (45). This demonstrates that CD9 is still expressed on the plasma membrane of α_6-lacking eggs and appears to have a critical role in regulating the ability of the egg membrane to be penetrated by sperm.

INTERPRETING ADDITIONAL DATA AND REFINING THE MODEL FOR GAMETE MEMBRANE INTERACTIONS

Phenotype Interpretation and Understanding some Causes of Infertility

We have learned a great deal about sperm-egg membrane interactions and the participating molecules not just from the cell biological and biochemical experiments discussed above, but also from knockout mice, engineered to lack a gene of interest. In addition, knockout mice can be excellent models of infertility. It is important to note, however, that interpretation of knockout mouse phenotypes and diagnosis of a pathological condition like infertility can be complicated. To illustrate this, we will focus on the fertilin β knockout mouse as a case study.

Fertilin β knockout mice were generated by traditional means, using homologous recombination to replace a portion of the fertilin β gene with a different piece of DNA, disrupting the gene so that no functional protein could be made. Male and female fertilin β -/- mice (with two copies of the disrupted gene) are viable and healthy, and females are fertile. However, male fertilin β -/- mice have very low fertility, based on mating trials demonstrating that very few wild type females mated with fertilin β -/- males got pregnant and produced pups (24). This would be analogous to a couple being unable to get pregnant [the clinical definition of infertility is the inability to conceive after one year of unprotected intercourse]. More detailed analyses are then undertaken looking for reasons for the infertility. In the case of the fertilin β -/- mice, the investigators knew to examine the males specifically. Since the animals looked phenotypically male and showed normal mating behavior, it appeared that hormone levels would be normal. Therefore, investigators proceeded to analyze sperm functions.

Sperm from fertilin β -/- mice are produced in normal numbers, and have normal motility and undergo the acrosome reaction. However, sperm from fertilin β -/- mice bind to the egg plasma membrane very poorly. In the experimental *in vitro* fertilization conditions used in these studies, eggs incubated with sperm from wild type males had ~9 sperm bound per egg whereas eggs incubated with sperm from fertilin β -/- males only have ~1 sperm bound per egg (24). Sperm-egg fusion, however, was not as adversely affected as sperm-egg binding was. Eggs incubated with sperm from wild type males had ~0.7-0.8 sperm fused per egg while eggs incubated with sperm from fertilin β -/- have ~0.3-0.4 sperm fused per egg (24), suggesting that the few sperm from fertilin β -/- males that are able to bind to eggs are indeed capable of fusing with the egg membrane.

It would appear that these data clearly and simply implicate fertilin β in sperm-egg adhesion, and not sperm-egg fusion. However, in truth, it turned out that there was more to this fertilin β knockout mouse. Some critical insights come from analysis of not only how the sperm function in fertilization, but also the protein expression profile. Obviously, since the fertilin β gene is disrupted, the sperm from these animals do not express fertilin β protein. However, knowing that fertilin β forms a heterodimer with fertilin α, researchers also looked to see if fertilin α was present in sperm from fertilin β -/- mice – and it is not (12). In addition, the levels of cyritestin are very low on sperm from fertilin β -/- mice (26). Therefore, one has to be careful about data interpretation or the diagnosis of the cause of infertility in these mice. We can say that there is a defect in the ability of sperm from fertilin β -/- mice to adhere to the egg plasma membrane, but these defects could be due to the absence of fertilin β, the absence of fertilin α, and/or the reduced levels of cyritestin.

Interestingly, cyritestin knockout mice have similar fertilization function phenotypes as the fertilin β -/- mice. Sperm from cyritestin -/- mice have a reduced ability to bind to the egg plasma membrane, although the few of these mutant sperm that do bind are able to fuse with and fertilize an egg (26). This membrane adhesion defect makes sense because, based on data from biochemical and cell biological studies, cyritestin is involved in sperm adhesion to the egg membrane. Furthermore, sperm from cyritestin -/- males have very little fertilin α and moderately reduced levels of fertilin β (26); these molecular deficiencies also might contribute to reduced membrane binding.

There is an added layer of complexity. It turns out that the sperm from fertilin β -/- mice have some additional, unanticipated defects. In addition to not adhering very well to the egg plasma membrane, they also do not adhere well to the zona pellucida (ZP). In the case of the fertilin β -/- animals, ZP-intact eggs incubated with sperm from wild type males had an average of ~15 sperm bound per ZP whereas eggs incubated with sperm from fertilin β -/- males have an average of less than 1 sperm bound per ZP (24). A similar defect was observed in sperm from cyritestin -/- mice (25,26).

In addition, the oviducts of females mated with fertilin β -/- males have far fewer sperm in them than do the oviducts of females mated with wild type males (24), suggesting a defect in sperm migration to the oviduct or interaction with the oviduct walls. Does this mean that fertilin β is involved with ZP binding or with migration to or interaction with the oviduct? This is one possible interpretation, but not the only one. Particularly with the knowledge that the protein expression profile is impaired in sperm from fertilin β -/- males, one can also speculate that the ZP binding defect and/or oviductal transit are disrupted due to protein expression abnormalities. In other words, sperm from fertilin β -/- males lack fertilin β, fertilin α, and cyritestin, and so theoretically could also lack yet another protein (or many proteins) critical for interaction with the ZP or the oviduct.

Finally, it should be noted that these concepts apply not only to sperm but to eggs as well. As noted above, CD9 is an important player in gamete interactions. However, CD9 also associated with other membrane proteins, and therefore eggs from CD9 -/- mice might have an abnormal protein expression profile. For example, these eggs might be lacking not only CD9 but also another important molecule (such as an ADAM-binding integrin), just as sperm from fertilin β -/- mice lack not only fertilin β but also fertilin α with reduced levels of cyritestin. These and other related questions remain to be investigated.

What are the take-home messages from all this? Careful and thorough analysis is necessary to gain full insight into the molecular cause(s) of infertility, particularly since multiple molecular pathways can lead to similar defects. In the cases illustrated here, the fertilization function defects can be traced back to both the sperm themselves [back to when the male germ cells were developing in the testis and synthesizing proteins]. Other defects in sperm function have been traced to abnormalities in the epididymis and its ability to support sperm maturation (53-56). Additionally, in nearly all cases of an abnormality, be it something generated in the laboratory or a patient presenting a pathological condition, there can be inter-related phenotypes: the "functional" phenotype [in this case, how the gametes function in fertilization], and the "molecular" phenotype [here, what gene is disrupted, and how that affects protein expression that affect function]. How a molecular phenotype translates to a cellular and/or physiological function phenotype may be elusive or complex, such as in the case illustrated here with the fertilin β knockout mouse, but is one of the keys to understanding causes and effects of disease states and other pathologies. It is important for researchers and clinicians to be cognizant of these complexities, as such awareness is critical for data analysis and for diagnosis and development of treatments for infertility.

Role Assignment – Which Molecule Plays which Role?

In the Introduction, the steps by which sperm-egg membrane interactions occur were described; these include initial attachment, intermediate adhesion, firm adhesion, and membrane fusion. We have also discussed candidate sperm and egg molecules that appear to mediate gamete interactions. Although we do not have sufficient experimental data to assign definitively specific roles to certain molecules, we can make some preliminary assignments.

Fertilin α, fertilin β, and cyritestin are likely to be involved in an adhesion step, perhaps multiple adhesion steps (initial, intermediate, and/or firm). In turn, this implicates integrins on eggs (exact ones as yet unidentified) in adhesion of the sperm to the egg, but not in gamete membrane fusion. Why can we say fertilin α, fertilin β, and cyritestin seem to be gamete adhesion molecules? These proteins, purified or recombinant forms have adhesive activity, having the ability to bind the egg plasma membrane. Furthermore, soluble forms of these proteins can block sperm-egg binding, but the few sperm that bind can go on and fuse with the egg plasma membrane. This is supported by the "sperm function" phenotype of sperm from fertilin β -/- mice (24). Although fertilin α was initially hypothesized to be involved in gamete membrane fusion (41), the more recent finding that sperm from fertilin β -/- mice lack fertilin α and yet are still able to fuse with the egg membrane (12) argues against a requirement for fertilin α for membrane fusion.

DE/CRISP-1 may be involved in firm adhesion that leads to membrane fusion or perhaps fusion itself. DE/CRISP-1 protein purified from rat epididymal extracts binds to the egg plasma membrane. In IVF assays, DE/CRISP-1 inhibits sperm-egg fusion in ZP-free eggs, but apparently has no effect on sperm-egg binding (38). This is of interest because DE/CRISP-1 is one of the few reagents that affects sperm-egg fusion without any apparent effect on sperm-egg adhesion. However, there is nothing in the amino acid sequence of DE/CRISP-1 that has homology to proteins known to be involved in membrane fusion events, and so it is possible that DE/CRISP-1 mediates firm adhesion that leads to membrane fusion, and that the true fusion-mediating molecule still has yet to be identified. Although much has been made about a region in fertilin α that has some similarity to viral fusion peptides, this region is only highly hydrophobic in guinea pig fertilin α and not in other species. Moreover, the finding that sperm from fertilin β and cyritestin -/- mice have little or no fertilin α and yet are still able to fuse with eggs suggests that fertilin α is not required for gamete membrane fusion. To date, the molecule(s) on sperm and/or egg that mediates membrane fusion is still not known. Based on pharmacological studies, metalloprotease activity has been implicated in sperm-egg fusion (57).

Similar concepts regarding firm adhesion versus fusion apply to CD9 on the egg. CD9 is clearly important based on the knockout phenotype (47-49), but it is not completely clear what role in gamete membrane interactions CD9 plays. Eggs from CD9 -/- mice support some sperm binding, but virtually none of these sperm fuse with the egg, which might be interpreted as suggestive of a specific role of CD9 in membrane fusion and not sperm adhesion. However, anti-CD9 antibodies inhibit sperm-egg binding (50), as well as the binding of fertilin α, fertilin β, and cyritestin in certain binding assays (29,30,43), suggestive of a role for CD9 in sperm adhesion mediated by these three sperm ligands. As noted above, there is no evidence that CD9 or any tetraspanin have activity as a receptor for extracellular ligands; instead, these proteins might help modulate or regulate cell adhesion, cell migration, and similar processes. One way by which they might do this is by regulating multimolecular complexes in the egg plasma membrane; these complexes are sometimes referred to as "tetraspanin webs." Recent evidence also suggests that CD9 may play a role in the strengthening of adhesions mediated by fertilin α and fertilin β, rather than initial molecular interactions (43).

Interesting Caveats in the Human: Fertilin α and Cyritestin are Non-Functional Genes

Fertilin α and cyritestin have been cloned in humans (58). Surprisingly, no open reading frame could be found in the cDNA sequence; instead, potential open reading frames were disrupted with numerous premature stop codons and frame shifts, so that any mRNA would not produce any functional protein. This makes the human fertilin α and cyritestin genes what are known as "pseudogenes." A survey of other primate species revealed that the gorilla fertilin α gene is also a pseudogene, whereas chimpanzee, orangutan, baboon, macaque, and tamarin genes have full-length reading frames that could encode functional fertilin α protein (59). The lack of functional fertilin α and cyritestin genes in the human thus suggests at least two possibilities. Some other protein, perhaps another ADAM protein, might play the same role and thus substitutes for fertilin α and cyritestin in human sperm. Many additional ADAMs have recently been identified to be expressed in the testis (and possibly by sperm), and so it is conceivable that some of these could participate in sperm-egg adhesion as well. An alternative possibility is that fertilin α and cyritestin are not required for human fertilization. With regard to fertilin α, which forms heterodimers with fertilin β in the guinea pig, bovine and mouse sperm (9-12), it could be that fertilin β functions by itself in human sperm, as a monomer (or homomultimer), without fertilin α.

Other Molecules

This chapter has emphasized on the gamete molecules which are the best-characterized. The most is known about ADAMs on sperm. Somewhat less is known about egg receptors for ADAMs, although evidence is accumulating which indicates that some members of the integrin family on the egg serve as receptors for sperm ADAMs, and the integrin-associated tetraspanin protein, CD9, facilitate these interactions. It is not known what the egg receptor for DE/CRISP-1/ARP is. Additionally, there are several other less well-characterized candidate molecules that have been hypothesized to be involved in gamete membrane interactions, although relatively little is known about these or the roles they may play in fertilization. These include several extracellular matrix proteins [fibronectin, laminin, and vitronectin]; sulfated glycolipids [specifically sulfogalactosyl-glycerolipid] and a protein that interacts with them; components of the complement pathway, C3b and C1q, and their receptors; and antigens of antibodies that have inhibitory activity against gamete membrane interactions [reviewed in (60)]. Despite the paucity of information about the roles that these and other molecules may play in fertilization, it is worth noting that the process of getting sperm and egg together is complex. In future years, it may be discovered that gamete membrane interactions involve much more than sperm ADAMs and egg integrins and integrin-associated proteins.

EGG ACTIVATION

Sperm-egg membrane interactions lead to a series of signal transduction events in the egg, known collectively as egg activation (addressed in more detail in the next chapter). The events associated with egg activation include the initiation of oscillations in intracellular calcium concentration, the exit from meiosis, the entry into the first embryonic mitosis, and the formation of blocks to polyspermy. There are two main hypotheses for how egg activation is initiated. One is based on models of signal transduction in other cells, called a receptor-effector mechanism. Many signal transduction events in other cell types are initiated by the binding of an extracellular ligand [such as growth factors, a cytokines, peptide hormones, etc.] to a membrane receptor that transduces the ligand signal to intracellular effector molecules. These effector molecules eventually lead to the activation of specific changes within the cell [such as increase in intracellular calcium, activation of certain enzyme activities, changes in gene expression, etc.]. It has been hypothesized that egg activation is initiated by an analogous mechanism, via the binding of a ligand on the sperm surface to an egg receptor. In fact,

ADAM proteins were attractive candidates for this, since it is know that integrins [the putative receptors for ADAMs] are involved in signal transduction in other cells. However, although there are data that indicate that some egg activation responses can be induced and mimicked by some experimental extracellular ligands, there is little evidence that sperm-induced egg activation operates by or requires this receptor-effector mechanism. Instead, there is the alternative hypothesis [dubbed the "fusion hypothesis" or "sperm factor hypothesis"] that egg activation is initiated by the diffusion of an as yet unidentified factor from the interior of the sperm into the cytoplasm of the egg upon membrane fusion. The hypothesis is rooted in the finding that extracts of soluble sperm proteins can be injected into eggs and this injection induces oscillatory calcium release very similar to the patterns of calcium oscillations induced by sperm itself. However, the identity of the egg-activating factor in sperm has remained elusive; several candidates have been proposed but none has been confirmed. At present, it is unclear if sperm-egg binding, sperm-egg fusion, or a combination of both processes initiate egg activation. The reader is referred to the following chapter and review articles for additional information (61-63).

Blocks to Polyspermy

In response to fertilization by one sperm, eggs establish blocks to prevent fertilization by additional sperm, which would result in aneuploidy and eventual death of the embryo. Mammalian eggs establish blocks to polyspermy at two different levels: on the extracellular egg coat (the zona pellucida) and on the plasma membrane. The zona block to polyspermy occurs as a result of exocytosis of cortical granules in the cortex of the egg. In response to the calcium signals initiated by fertilization, the cortical granules fuse with the egg plasma membrane, releasing their contents into the perivitelline space and causing the conversion of the ZP to a form that cannot support sperm binding (1).

Mammalian eggs also appear to establish some sort of block to polyspermy at the plasma membrane. This is based on a number of observations. If the extent of fertilization is observed over time, a plateau in the number of sperm fusing with ZP-free eggs is observed with increased time (64,65). Fertilized ZP-free mouse eggs are unable to be penetrated by additional sperm by 1-2 hr after the first insemination (64,66,67), and there is a decrease in sperm-egg binding with increased time post-insemination or to previously fertilized eggs (66,68). Sperm can be found in the perivitelline space of fertilized ZP-intact eggs; these supplemental sperm do not fuse with the egg plasma membrane, suggesting that a membrane block to polyspermy prohibits the interaction [adhesion or fusion] of sperm with the egg membrane [numerous studies, summarized in (1)]. On a molecular level, however, very little is known about the membrane block, although differences in membrane protein composition (69,70) and membrane fluidity (71,72) between unfertilized and fertilized eggs have been observed. The eggs of invertebrates and non-mammalian vertebrates [primarily echinoderms and amphibians] utilize a transient change in the membrane's electrical potential as a membrane block to polyspermy [often referred to as the "fast block" to polyspermy]. However, no change in membrane potential has been detected in mammalian eggs following fertilization (73). It should be noted that a membrane block could occur at multiple levels of gamete membrane interaction, i.e., reducing sperm-egg adhesion and/or sperm fusion with the egg plasma membrane. It should also be emphasized that the membrane block in mammalian eggs occurs over a period of ~1 hr, contrasting the sea urchin egg membrane block (via depolarization) that is established within seconds. Therefore, the mammalian egg blocks to polyspermy are distinguished spatially [membrane and zona] rather than temporally, as the echinoderm and amphibian blocks to polyspermy are.

CLINICAL IMPLICATIONS AND APPLICATIONS FROM UNDERSTANDING SPERM-EGG INTERACTIONS

It is important to understand the mechanisms underlying in gamete membrane interactions because these processes have significant implications for infertility treatment and for potential future contraceptive therapeutics. Clearly, if we understand the molecular basis for how sperm and egg get together [including cumulus interactions, ZP interactions, and membrane interactions], then we have the ability to understand a failure in these processes that could cause infertility and the potential to develop reagents that might inhibit these processes that could be used to contracept an individual.

One possibility for contraceptive development is the design of molecules that can block and inhibit sperm-egg interactions, thus preventing fertilization. In a similar fashion, anti-HIV-1 therapeutics have been designed based on reagents that block a receptor or fusion intermediates that HIV-1 virus particles use to infect these cells (74). In vitro, protein or peptide reagents that mimic sperm ligands bind to eggs and block sperm binding (Fig. 5), leading to inhibition of fertilization. The next step for gamete interaction-blocking contraceptive development would be to design non-peptide analogs that would be stable in the female or male reproductive tracts, as well as an appropriate drug delivery system. In addition, molecules that are involved in these gamete interactions are potential targets for immunocontraception. Immunocontraception in this case would involve immunization with a fertilization-involved antigen. The immunized animals mount an immune response to the antigen, making antibodies that will cross-react with it; if these antigens are secreted into the reproductive tract, they could prevent fertilization. Fertilin β and DE/CRISP-1 have both been investigated as candidate immunogens to generate infertility (37,75,76). Because of issues regarding delivery, efficacy, and reversibility, immuno-contraception may be more appropriate for use in over-populated animal species, than for individual human patients who want to be contracepted for a period of time but then recover fertility when parenthood is desired.

CONCLUSIONS

To summarize the key points of this chapter:

• Gamete membranes interact in a step-wise fashion, starting with cell adhesion and ending with membrane fusion. Cell adhesion probably occurs in multiple steps (initial attachment, intermediate adhesion, and firm adhesion), based on the model of leukocyte interactions with endothelial cells. We do not know exactly how the membranes of gametes fuse, although models are based on other membrane fusion systems; these include the fusion of virus particles with cell plasma membranes, and the fusion of intracellular membrane-bound vesicles.

• Although we cannot identify exactly what molecules on the sperm and egg are involved in each of these discrete steps, a number of molecules have been identified to participate, in some way, in gamete membrane interactions. Some of the best characterized include three sperm ADAM proteins (fertilin α, fertilin β, and cyritestin), the sperm-associated protein DE/CRISP1, and egg integrins and the integrin-associated tetraspanin CD9. Data that provide evidence for the roles of these proteins in gamete membrane interactions come from molecular, biochemical, and cell biological studies, as well as knockout mice.

• Great insights into fertilization have come from analysis of knockout mice, and these mice also serve as excellent models for some causes of infertility in humans. In this chapter, we examined the fertilin β knockout mouse as a case study, highlighting the connections between the "functional" phenotype (how the sperm behaved in *in vitro* fertilization assays and other assays of sperm function) and the protein expression phenotype. The fact that sperm from fertilin β knockout mice lack not only fertilin β but also fertilin α and cyritestin highlights the importance of protein complexes for proper membrane function.

• Insights into the molecular mechanisms of fertilization, including gamete membrane interactions, have implications and applications to the diagnosis and treatment of infertility and to the development of new contraceptive therapeutics.

314

ACKNOWLEDGEMENTS

I am grateful to my students and colleagues, Xiaoling Zhu, Genevieve Wortzman, and Eugene Oh, for their critical reading of this chapter. I also thank them for reminding me everyday of the joys of research and discovery in this field.

REFERENCES

1. Yanagimachi R. Mammalian fertilization. In The Physiology of Reproduction, E. Knobil and J.D. Neill., eds. New York: Raven Press, Ltd., 1994. pp. 189-317.
2. Huang Jr.,T.T.F., Yanagimachi R. Inner acrosomal membrane of mammalian spermatozoa: Its properties and possible functions in fertilization. Am. J. Anat. 1985; 174:249-268
3. Yanagimachi R. Sperm-egg fusion. Current Topics in Membranes and Transport 1988; 32:3-43.
4. Springer T. A. Traffic signals for lymphcyte recirculation and leukocyte emigration: The multistep paradigm. Cell 1994; 76:301-314.
5. Brown E. J. Adhesive interactions in the immune system. Trends Cell Biol. 1997; 7:289-295.
6. Drobnis E. Z., Yudin A. I., Cherr G. N., Katz D. F. Hamster sperm penetration of the zona pellucida: kinematic analysis and mechanical implications. Dev. Biol. 1988; 130:311-23..
7. Lentz B. R., Malinin V., Haque M. E., Evans K. Protein machines and lipid assemblies: current views of cell membrane fusion. Curr. Op. Struct. Biol. 2000; 10:607-615.
8. Gould R. J., Polokoff M. A., Friedman P. A., Huang T. -F., Holt J. C., Cook J. J., Niewiarowski S. Disintegrins: A family of integrin inhibitory proteins from viper venoms. Proc. Soc. Exp. Biol. Med. 1990; 195:168-171.
9. Primakoff P., Hyatt H., Tredick-Kline J. Identification and purification of a sperm surface protein with a potential role in sperm-egg membrane fusion. J. Cell Biol. 1987; 104:141-149.
10. Blobel C. P., Myles D. G., Primakoff P., White J. M. Proteolytic processing of a protein involved in sperm-egg fusion correlates with acquisition of fertilization competence. J. Cell Biol. 1990; 111:69-78.
11. Waters S. I., White J. M. Biochemical and molecular characterization of bovine fertilin α and β (ADAM 1 and ADAM 2): A candidate sperm-egg binding/fusion complex. Biol. Reprod. 1997; 56:1245-1254.
12. Cho C., Ge H., Branciforte D., Primakoff P., Myles D. G. Analysis of mouse fertilin in wild-type and fertilin β -/- sperm: Evidence for C-terminal modification, α/β dimerization, and lack of essential role of fertilin α in sperm-egg fusion. Dev. Biol. 2000; 222:289-295.
13. Heinlein U. A. O., Wallat S., Senftleben A., Lemaire L. Male germ cell-expressed mouse gene TAZ83 encodes a putative, cysteine-rich transmembrane protein (cyritestin) sharing homologies with snake toxins and sperm-egg fusion proteins. Dev. Growth Differ. 1996; 36:49-58.
14. Yuan R., Primakoff P., Myles D. G. A role for the disintegrin domain of cyritestin, a sperm surface protein belonging to the ADAM family, in mouse sperm-egg plasma membrane adhesion and fusion. J. Cell Biol. 1997; 137:105-112.

15. Hardy C. M., Clarke H. G., Nixon B., Grigg J. A., Hinds L. A., Holland M. K. Examination of the immunocontraceptive potential of recombinant rabbit fertilin subunits in rabbit. Biol. Reprod. 1997; 57:879-886.

16. Myles D. G., Kimmel L. H., Blobel C. P., White J. M., Primakoff P. Identification of a binding site in the disintegrin domain of fertilin required for sperm-egg fusion. Proc Natl Acad Sci U S A 1994; 91:4195-4198.

17. Evans J.P., Schultz R. M., Kopf G. S. Mouse sperm-egg membrane interactions: analysis of roles of egg integrins and the mouse sperm homologue of PH-30 (fertilin) β. J. Cell Sci. 1995; 108:3267-3278.

18. Pyluck A., Yuan R., Galligan Jr.,E., Primakoff P., Myles D. G., Sampson N. S. ECD peptides inhibit in vitro fertilization in mice. Bioorg. Med. Chem. Lett. 1997; 7:1053-1058.

19. Evans J. P., Kopf G. S., Schultz R. M. Characterization of the binding of recombinant mouse sperm fertilin β subunit to mouse eggs: Evidence for adhesive activity via an egg β_1 integrin-mediated interaction. Dev. Biol. 1997; 187:79-93.

20. Evans J. P., Schultz R. M., Kopf G. S. Characterization of the binding of recombinant mouse sperm fertilin α subunit to mouse eggs: Evidence for function as a cell adhesion molecule in sperm-egg binding. Dev. Biol. 1997; 187:94-106.

21. Evans J. P., Schultz R. M., Kopf G. S. Roles of the disintegrin domains of mouse fertilins α and β in fertilization. Biol. Reprod. 1998; 59:145-152.

22. Mwethera P. G., Makokha A., Chai D. Fertilin β peptides inhibit sperm binding to zona-free eggs in a homologous baboon in vitro fertilization system. Contraception 1999; 59:131-135.

23. Bronson R. A., Fusi F. M., Calzi F., doldi N., Ferrari A. Evidence that a functional fertilin-like ADAM plays a role in human sperm-oolemmal interactions. Mol. Human Reprod. 1999; 5:433-440.

24. Cho C., Bunch D. O., Faure J. -E., Goulding E. H., Eddy E. M., Primakoff P., Myles D. G. Fertilization defects in sperm from mice lacking fertilin β. Science 1998; 281:1857-1859.

25. Shamsadin R., Adham I. M., Nayernia K., Heinlein U. A. O., Oberwinkler H., Engel W. Male mice deficient for germ-cell cyritestin are infertile. Biol. Reprod. 1999; 61:1445-1451.

26. Nishimura H., Cho C., Branciforte D. R., Myles D. G., Primakoff P. Analysis of loss of adhesive function in sperm lacking cyritestin or fertilin β. Dev. Biol. 2001; 233:204-213.

27. Zhu X., Bansal N. P., Evans J. P. Identification of key functional amino acids of the mouse fertilin β (ADAM2) disintegrin loop for cell-cell adhesion during fertilization. J. Biol. Chem. 2000; 275:7677-7683.

28. Bigler D., Takahashi Y., Chen M. S., Almeida E. A. C., Osburne L., White J. M. Sequence-specific interaction between the disintegrin domain of mouse ADAM2 (fertilin β) and murine eggs: Role of the α_6 integrin subunit. J. Biol. Chem. 2000; 275:11576-11584.

29. Takahashi Y., Bigler D., Ito Y., White J. M. Sequence-specific interaction between the disintegrin domain of mouse ADAM3 and murine eggs: Role of the β_1 integrin-associated proteins CD9, CD81, and CD98. Mol. Biol. Cell 2001; 12:809-820.

30. Wong G. E., Zhu X., Prater C. E., Oh E., Evans J. P. Analysis of fertilin α (ADAM1)-mediated sperm-egg cell adhesion during fertilization and identification of an adhesion-mediating sequence in the disintegrin-like domain. J. Biol. Chem. 2001; 276:24937-24945.

31. Lum L., Blobel C. P. Evidence for distinct serine protease activities with a potential role in processing the sperm protein fertilin. Dev. Biol. 1997; 191:131-145.

316

32. Linder B., Bammer S., Heinlein U. A. Delayed translation and posttranslational processing of cyritestin, an integral transmembrane protein of the mouse acrosome. Exp Cell Res 1995; 221:66-72.

33. Kohane A. C., Gonsalez Echverria F. M. C., Piniero L., Blaquier J. Interaction of proteins of epididymal origin with spermatozoa. Biol. Reprod. 1980; 23:737-742.

34. Haendler B., Kratzschmar J., Theuring F., Schleuning W. D. Transcripts for cysteine-rich secretory protein-1 (CRISP-1; DE/AEG) and the novel related CRISP-3 are expressed under androgen control in the mouse salivary gland. Endocrinology 1993; 133:192-198.

35. Lea O. A., Petrusz P., French F. S. Purification and localization of acidic epididymal glycoprotein (AEG): A sperm coating protein secreted by the rat epididymis. Int. J. Androl. Suppl. 1978; 2:592-607.

36. Cuasnicu P. S., Gonzalez-Echeverria M. F., Piazza A. D., Cameo M. S., Blaquier J. A. Antibody against epididymal glycoprotein blocks fertilizing ability in rats. J. Reprod. Fert. 1984; 72:467-471.

37. Cuasnicu P.S., Conesa D., and Rochwerger L. Potential contraceptive use of an epididymal protein that participates in fertilization. In Gamete Interaction: Prospects for Immunocontraception, N.J. Alexander, D. Griffin, J.M. Speiler, and G.M.H. Waites., eds. New York: Wiley-Liss, 1990. pp. 143-153.

38. Rochwerger L., Cohen D. J., Cuasnicú P. S. Mammalian sperm-egg fusion: The rat egg has complementary sites for a sperm protein that mediates gamete fusion. Dev. Biol. 1992; 153:83-90.

39. Cohen D. J., Ellerman D. A., Cuasnicú P. S. Mammalian sperm-egg fusion: evidence that epididymal protein DE plays a role in mouse gamete fusion. Biol. Reprod. 2000; 63:462-468.

40. Cohen D. J., Ellerman D. A., Busso D., Morgenfeld M. M., Piazza A. D., Hayashi M., Young E. T., Kasahara M., Cuasnicu P. S. Evidence that human epididymal protein ARP plays a role in gamete fusion through complementary sites on the surface of the human egg. Biol. Reprod. 2001; 65:1000-1005.

41. Blobel C. P., Wolfsberg T. G., Turck C. W., Myles D. G., Primakoff P., White J. M. A potential fusion peptide and an integrin ligand domain in a protein active in sperm-egg fusion. Nature 1992; 356:248-252.

42. Almeida E. A. C., Huovila A. -P. J., Sutherland A. E., Stephens L. E., Calarco P. G., Shaw L. M., Mercurio A. M., Sonnenberg A., Primakoff P., Myles D. G., White J. M. Mouse egg integrin $\alpha_6\beta_1$ functions as a sperm receptor. Cell 1995; 81:1095-1104.

43. Zhu X., Evans J. P. Analysis of the roles of RGD-binding integrins, α_4/α_9 integrins, α_6 integrins, and CD9 in the interaction of the fertilin β (ADAM2) disintegrin domain with the mouse egg membrane. Biol. Reprod. 2002; 66:1193-1202.

44. Chen H., Sampson N. S. Mediation of sperm-egg fusion: evidence that mouse egg $\alpha_6\beta_1$ integrin is the receptor for sperm fertilin β. Chem. Biol. 1999; 6:1-10.

45. Miller B. J., Georges-Labouesse E., Primakoff P., Myles D. G. Normal fertilization occurs with eggs lacking the integrin $\alpha_6\beta_1$ and is CD9-dependent. J. Cell Biol. 2000; 149:1289-1295.

46. Linfor J., Berger T. Potential role of α_v and β_1 integrins as oocyte adhesion molecules during fertilization in pigs. J. Reprod. Fert. 2000; 120:65-72.

47. Le Naour F., Rubinstein E., Jasmin C., Prenant M., Boucheix C. Severely reduced female fertility in CD9-deficient mice. Science 2000; 287:319-321.

48. Miyado K., Yamada G., Yamada S., Hasuwa H., Nakamura Y., Ryu F., Suzuki K., Kosai K., Inoue K., Ogura A., Okabe M., Mekada E. Requirement of CD9 on the egg plasma membrane for fertilization. Science 2000; 287:321-324.

49. Kaji K., Oda S., Shikano T., Ohnuki T., Uematsu Y., Sakagami J., Tada N., Miyazaki S., Kudo A. The gamete fusion process is defective in eggs of CD9-deficient mice. Nat. Genet. 2000; 24:279-282.

50. Chen M. S., Tung K. S. K., Coonrod S. A., Takahashi Y., Bigler D., Chang A., Yamashita Y., Kincade P. W., Herr J. C., White J. M. Role of the integrin associated protein CD9 in binding between sperm ADAM 2 and the egg integrin $\alpha_6\beta_1$: Implication for murine fertilization. Proc. Natl. Acad. Sci. U. S. A. 1999; 96:11830-11835.

51. Hemler M. E. Specific tetraspanin functions. J. Cell Biol. 2001; 155:1103-1107.

52. Boucheix C., Rubinstein E. Tetraspanins. Cell. Mol. Life Sci. 2001; 58:1189-1205.

53. Sonnenberg-Riethmacher E., Walter B., Riethmacher D., Godecke S., Birchmeier C. The c-ros tyrosine kinase receptor controls regionalization and differentiation of epithelial cells in the epididymis. Genes. Dev. 1996; 10:1184-1193.

54. Ramaraj P., Kessler S. P., Colmenares C., Sen G. C. Selective restoration of male fertility in mice lacking angiotensin-converting enzymes by sperm-specific expression of the testicular isozyme. J. Clin. Invest. 1998; 102:371-378.

55. Hagaman J. R., Moyer J. S., Bachman E. S., Sibony M., Magyar P. L., Welch J. E., Smithies O., Krege J. H., O'Brien D. A. Angiotensin-converting enzyme and male fertility. Proc. Natl. Acad. Sci. U. S. A. 1998; 95:2552-2557.

56. Hellsten E., Evans J. P., Bernard D. J., Janne P. A., Nussbaum R. L. Disrupted sperm function and fertilin β processing in mice deficient in the inositol polyphosphate 5-phosphatase inpp5b. Dev. Biol. 2001; 240:641-653.

57. Correa L. M., Cho C., Myles D. G., Primakoff P. A role for a TIMP-3-sensitive Zn^{2+}-dependent metalloprotease in mammalian gamete membrane fusion. Dev. Biol. 2000; 225:124-134.

58. Jury J. A., Frayne J., Hall L. The human fertilin α gene is non-functional: Implications for its proposed role in fertilization. Biochem. J. 1997; 321:577-581.

59. Jury J. A., Frayne J., Hall L. Sequence analysis of a variety of primate fertilin α genes: Evidence for non-functional primate genes in the gorilla and man. Mol. Reprod. Dev. 1998; 51:92-97.

60. Evans J. P. Sperm disintegrins, egg integrins, and other cell adhesion molecules of mammalian gamete plasma membrane interactions. Front. Biosci. 1999; 4:D114-D131.

61. Schultz R. M., Kopf G. S. Molecular basis of mammalian egg activation. Curr. Topics in Dev. Biol. 1995; 30:21-62.

62. Fissore R. A., Gordo A. C., Wu H. Activation of development in mammals: is there a role for a sperm cytosolic factor? Theriogenology 1998; 49:43-52.

63. Swann K., Parrington J., Jones K. T. Potential role of a sperm-derived phospholipase C in triggering the egg-activating Ca^{2+} signal at fertilization. Reproduction 2001; 122:839-846.

64. Wolf D. P. The block to sperm penetration in zona-free mouse eggs. Dev. Biol. 1978; 64:1-10.

65. Sengoku K., Tamate K., Horikawa M., Takaoka Y., Ishikawa M., Dukelow W. R. Plasma membrane block to polyspermy in human oocytes and preimplantation embryos. J. Reprod. Fertil. 1995; 105:85-90.

66. Wolf D. P., Hamada M. Sperm binding to the mouse egg plasmalemma. Biol. Reprod. 1979; 21:205-211.

67. Horvath P. M., Kellom T., Caulfield J., Boldt J. Mechanistic studies of the plasma membrane block to polyspermy in mouse eggs. Mol. Reprod. Dev. 1993; 34:65-72.

68. Redkar A. A., Olds-Clarke P. J. An improved mouse sperm-oocyte plasmalemma binding assay: Studies on characteristics of sperm binding in medium with or without glucose. J. Androl. 1999; 20:500-508.

69. Johnson L. V., Calarco P. G. Electrophoretic analysis of cell surface proteins of preimplantation mouse embryos. Dev. Biol. 1980; 77:224-227.

70. Boldt J., Gunter L. E., Howe A. M. Characterization of cell surface polypeptides of unfertilized, fertilized, and protease-treated zona-free mouse eggs. Gamete Res. 1989; 23:91-101.

71. Wolf D. E., Edinin M., Handyside A. H. Changes in the organization of the mouse egg plasma membrane upon fertilization and first cleavage: Indications from the lateral diffusion rates of fluorescent lipid analogues. Dev. Biol. 1981; 85:195-198.

318

72. Wolf D. E., Ziomek C. A. Regionalization and lateral diffusion of membrane proteins in unfertilized and fertilized mouse eggs. J. Cell Biol. 1983; 96:1786-1790.
73. Jaffe L. A., Sharp A. P., Wolf D. P. Absence of an electrical polyspermy block in the mouse. Dev. Biol. 1983; 96:317-323.
74. Eckert D. M., Kim P. S. Mechanisms of viral membrane fusion and its inhibition. Annu. Rev. Biochem. 2001; 70:777-810.
75. Ramarao C. S., Myles D. G., White J. M., Primakoff P. Initial evaluation of fertilin as an immunocontraceptive antigen and molecular cloning of the cynomolgus monkey fertilin β subunit. Mol. Reprod. Dev. 1996; 43:70-75.
76. Ellerman D. A., Brantua V. S., Martinez S. P., Cohen D. J., Conesa D., Cuasnicu P. S. Potential contraceptive use of epididymal proteins: immunization of male rats with epididymal protein DE inhibits sperm fusion activity. Biol. Reprod. 1998; 59:1029-1036.

Chapter 17

ACTIVATION OF MAMMALIAN OOCYTES: PRINCIPLES AND PRACTICE

L Liu, M Deng, XC Tian, X Yang
University of Connecticut, Storrs, CT, USA

INTRODUCTION

Oocyte activation refers to the release of the meiotically arrested oocytes by the entry of sperm during fertilization. One of the initial events of activation is the triggering of Ca^{2+} oscillations in the oocytes. Understanding of physiological and biochemical events during fertilization allows artificial stimuli to be used to induce oocyte activation without the entry of sperm. Recently, artificial activation of oocyte has regained the interest of scientists because of its implications in somatic cloning by nuclear transfer, intracytoplasmic sperm injection (ICSI) and genomic imprinting. Artificial stimulus also permits precise control of the timing of activation providing synchronized embryos for studies of early development while avoiding polyspermy. It is known that for a portion of the sperm-injected human oocytes as well as all bovine, pig, or even mouse oocytes, ICSI alone is insufficient to activate oocytes unless artificial stimulations are applied. Since the cloning procedure process does not itself activate oocytes and full activation of the recipient cytoplast is required, understanding the underlying molecular components as well as the morphological changes during the initial stages of oocyte activation is critical for the effective use of cytoplasts in cloning by nuclear transfer. Herein we review some of the key issues that are associated with effective activation of oocytes.

PRINCIPLES OF FERTILIZATION-INDUCED OOCYTE ACTIVATION

In the arrested metaphase II (MII) oocytes, cortical granules (CG) are localized beneath the plasma membrane; chromosomes are aligned across the metaphase plate of meiotic spindles. Oocytes at MII are characterized by high levels and activity of c-Mos, mitogen-activated protein kinases (MAPK), maturation or metaphase promoting factor (MPF), whose up-regulation coincide with the assembly of the new spindles. Both c-Mos and MAPK are supposedly components of the cytostatic factors (CSF) that contribute to the arrest of oocytes at meiosis II. MAPKs are serine/threonine kinases that require phosphorylation to become fully activated. Furthermore, most of the MAPK activity in bovine oocytes can be attributed to p42ERK2 (1). MAPK activity is required for MII arrest and meiotic spindle organization or stabilization (2-4). The MPF, composed of cyclin B and p34^{cdc2} kinase, displays a cyclic activity that peaks at metaphase (5, 6). The activity of histone H1 kinase is known to reflect the activity of MPF (6). Sperm penetration of mammalian oocytes during fertilization initiates a series of signal transduction events that correspond to structural, morphological and biochemical changes.

Ca^{2+} Oscillations

In mammalian oocytes, repetitive increases in Ca^{2+} (Ca^{2+} oscillations) in the cytoplasm were first reported during fertilization of mouse oocytes (7) and later it was found to occur in all mammalian species studied so far, including hamsters, pigs, bovine, rats, rabbits, and humans. In mice, within a few minutes of the sperm-oocyte interaction, Ca^{2+} oscillations begin with the initial transient lasting longer (> 2 min) than the subsequent short transients (~1 min; Figure 1). Inside the oocytes, the initial Ca^{2+} increase spreads as a wave, originating exclusively from the point where a sperm fuses with the oocyte and subsequent waves initiate from the vegetal region of the oocyte (8). The frequency of the Ca^{2+} oscillations varies among species. Furthermore, a high frequency of Ca^{2+} oscillations is also associated with polyspermy in oocytes without a zona pellucida (9).

Ca^{2+} utilized for each transient in mammalian oocytes appears to be discharged from an intracellular store (10) and the endoplasmic reticulum (ER) is the recognized cellular organelle that stores and releases Ca^{2+} in oocytes (11). Inositol 1,4,5-triphosphate (IP3) receptors are the primary Ca^{2+}-releasing channels in the ER and play a central role in Ca^{2+} signaling. Ryanodine receptors present on the ER membrane may also participate in mediating Ca^{2+} release (25).

Extensive evidence demonstrates that sperm-induced intracellular Ca^{2+} release and oscillations are essential for complete oocyte activation and embryo development in mammals. Induced increases in intracellular Ca^{2+} can artificially activate oocytes and initiate embryo development (12, 13).

Extensive evidence demonstrates that sperm-induced intracellular Ca^{2+} release and oscillations are essential for complete oocyte activation and embryo development in mammals. Induced increases in intracellular Ca^{2+} can artificially activate oocytes and initiate embryo development (12, 13). In contrast, preventing the Ca^{2+} rises with the Ca^{2+} chelator BAPTA inhibits oocyte activation (10).

It should be noted that in most non-mammalian oocytes, the fertilizing sperm causes a single large transient increase in the cytosolic Ca^{2+} concentration, whereas mammalian oocytes exhibit sustained Ca^{2+} oscillations at fertilization (12). Additionally, the cell cycle progression of non-mammalian oocytes is much faster after fertilization than that of mammalian oocytes. Therefore, data collected from non-mammalian

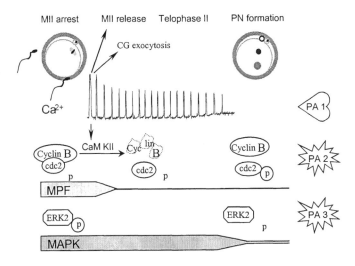

Figure 1. Activation of mammalian oocytes during fertilization and proposed methods for parthenogenetic activation (PA). Sperm penetration triggers a Ca^{2+} transient, followed by Ca^{2+} oscillations. The first Ca^{2+} wave is critical for both cortical granule (CG) exocytosis and activation of CaM KII, which decomposes cyclin B through "cyclin degradation machinery" and resulting in inactivation of MPF. Decline in MPF activity is associated with oocyte meiotic release. Low MPF activity can be maintained by the lack of cyclin B or phosphorylation of cdc2. The subsequent Ca^{2+} oscillations are indispensable for continuous degradation of cyclin B and maintenance of the low MPF activity, until dephosphorylation (p) of ERK and thus inactivation of MAPK, which is associated with pronuclear (PN) formation. Parthenogenetic activation can be achieved either by mimicking sperm-induced Ca^{2+} oscillations (PA1) or by inhibition of MPF (PA2) and/or MAPK activities (PA3). Cyclin B degradation is responsible for the instant inactivation of MPF activity. Low MPF activity can be maintained by either inhibition of continuous synthesis of cyclin B (cycloheximide) or by dephosphorylation of ERK2 (6-dimethylaminopurine, 6-DMAP). The balance of phosphate seems to play an important role in keeping MPF activity low and MAPK inactivated.

oocytes on Ca^{2+} signaling should be carefully interpreted when applying to Ca^{2+} signaling in mammalian oocytes.

Upstream of Ca^{2+} Modulators

Although the sperm-triggered Ca^{2+} oscillations have been widely recognized, it is not yet completely understood how penetration of sperm causes Ca^{2+} transients. It is generally accepted that the release of Ca^{2+} at fertilization from intracellular stores of mammalian oocytes is mediated by inositol 1,4,5-trisphosphate (IP3). It has been shown that injection of IP3 can trigger Ca^{2+} release and oscillations in mouse and hamster oocytes (12), but the pattern of Ca^{2+} oscillations produced by injection of IP3

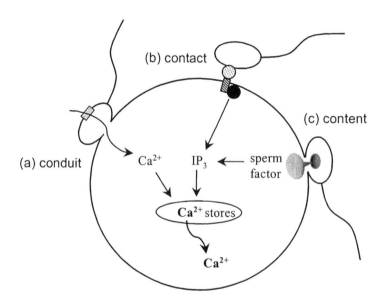

Figure 2. Models for Ca^{2+} signaling in mammalian oocytes at fertilization. (a) The sperm fuses with oocyte membrane and acts as conduit to transport Ca^{2+} into the oocyte. (b) The sperm binds to membrane receptors which stimulate the production of IP3. (c) The sperm fuses with the oocyte and introduces sperm factor (s) that initiate Ca^{2+} release.

do not match those induced by the sperm under the same conditions. In particular the Ca^{2+} oscillations produced by IP3 injection tend to be smaller and of higher frequency than those seen during natural fertilization (12, 14). These data do not preclude IP3 from playing the key role at fertilization but they suggest that a sustained IP3 increase is not the best explanation for sperm-induced Ca^{2+} oscillations. A recent study, however, showed that a continuous small amount of IP3 release over an extended period of time, rather than the large amount of IP3 infusion at one time by microinjection, mimics the distinctive pattern of Ca^{2+} release in mammalian oocytes at fertilization (15). The mechanism by which the sperm initiates IP3 production, however, is still not understood.

Three models or hypotheses have been proposed to explain the signal transduction by which sperm triggers Ca^{2+} oscillations and oocyte activation. The three models are abbreviated as "the 3Cs": conduit, contact and contents (Figure 2).

a). The "conduit" model: This model proposes that the fusion of sperm with the oocyte plasma membrane transfers Ca^{2+} into oocytes and triggers a cascade of Ca^{2+}-induced Ca^{2+} release (CICR, 16). During mouse fertilization, it was noted that the initial Ca^{2+} transient is generally more prolonged than the subsequent transients (Figure 1). As mentioned above, microinjection of IP3 or sperm extracts does not mimic sperm-induced Ca^{2+} oscillations in most studies, particularly in respect of the first Ca^{2+} transient, whose duration is distinct from that of subsequent ones. This unique first Ca^{2+} transient could result from sperm entry which carries Ca^{2+} into the oocyte. Further supporting evidence comes from experiments showing that MPF inhibitor, abolishes subsequent Ca^{2+} oscillations, probably through disruption of Ca^{2+} release and refilling of Ca^{2+} store, without suppressing the first one or two waves after fertilization (17). Another kinase inhibitor, 6-DMAP, also has similar effects (our unpublished data).

A prerequisite for the conduit hypothesis is that sperm-oocyte fusion should precede Ca^{2+} release and oocyte activation, which is shown to be the case in mice (18). The first Ca^{2+} influx mediated by sperm entry probably functions to sensitize the oocyte to Ca^{2+}-induced Ca^{2+} release (CICR) (12). Sensitization of CICR, on the other hand, is both necessary and sufficient to explain oscillations (12). This first extended Ca^{2+} transient could cause a gradual filling of an internal Ca^{2+} store, sustaining subsequent Ca^{2+} oscillations. However, Ca^{2+} itself cannot trigger Ca^{2+} oscillations. This has been shown in mice and hamsters that Ca^{2+} could be introduced into the oocyte in a number of different ways, but none of these led to sustained Ca^{2+} oscillations (12).

b). The "contact" model or the receptor hypothesis: This model states that a ligand on the surface of the sperm interacts with an oocyte surface receptor and the resultant sperm-oocyte binding activates receptors affiliated with GTP-binding proteins (G proteins). The activated G proteins in turn stimulate the phosphoinositide cascade and an increase in the activities of a phospholipase C (PLC), generating IP3 (13, 19). The contact hypothesis is consistent with the finding that sperm-oocyte fusion precedes or GDPβS, which inhibits G-proteins, blocks Ca^{2+} release at fertilization in hamster oocytes (20). Heterotrimeric G proteins may be involved in mouse oocyte activation since inhibition of G protein's beta-gamma subunits partially inhibits sperm-induced cell cycle resumption. In addition, specific events of activation can be initiated in the absence of sperm by acetylcholine stimulation of mouse oocytes overexpressing the human m1 muscarinic receptor, a G protein-coupled receptor (21). The initiation of Ca^{2+} release at fertilization of mammalian oocytes requires IP3 (19), which also induces an intracellular release of calcium in a variety of cell types, indicating that an enzyme of the PLC family is probably activated. Furthermore, release of Ca^{2+} at fertilization in echinoderm oocytes is initiated by Src-homology 2 (SH2) domain-mediated activation of PLCγ (22). Although activation of a G family protein-mediated signaling pathway can result in oocyte activation, the specific G protein subtype involved in sperm-induced oocyte activation is still not known. This is because at least in the mouse none of the G protein families examined, Gs, Gi or Gq, seem to be involved in oocyte activation during fertilization. Furthermore, this model does not explain the observation that injection of sperm or the soluble sperm extracts triggers oocyte activation, suggesting that the binding of sperm to the oocyte membrane is not essential for activation.

c). *The "content" model or the sperm factor hypothesis:* This model states that the fusion of sperm-oocyte membranes introduces a soluble cytosolic sperm factor into the oocyte which triggers Ca^{2+} release (23-26). This sperm factor hypothesis is consistent with the observation that successful fertilization and pregnancy can result from intra-cytoplasmic sperm injection (ICSI) which cause Ca^{2+} oscillations and oocyte activation similar to those seen in natural fertilization (27). Injecting soluble sperm extracts also triggers oocyte activation and development of the resulting embryo to at least the blastocyst stage. These data suggest that content, not the binding, of the sperm is important for activation. The sperm factor that causes this response has been shown to be protein based. It is also well conserved across species, phyla and perhaps even kingdom. Recently, a flowering plant sperm was shown to contain a cytosolic soluble protein factor which triggered calcium oscillations in mouse oocytes (28).

In mammals, sperm factors defined by their ability to induce oocyte activation are referred to as the sperm-borne oocyte-activating factors (SOAFs) (29). However, a number of artificial stimuli that can parthenogenetically stimulate oocytes do not cause Ca^{2+} oscillations (12). Therefore, sperm factor has also been referred to as 'oscillogen' for its ability to cause sustained Ca^{2+} oscillations. The precise nature of the sperm factor, however, remains unresolved despite its potential role in fertilization. Oscillin, a 33 kDa protein and an isoform of glucosamine-6-phosphate deaminase (25, 30), was believed to be oscillogen. Subsequent work, however, demonstrated that the recombinant 33 kDa protein does not trigger Ca^{2+} oscillations when injected into oocytes (31, 32), suggesting that it is not a component of the oscillogen or sperm factor.

Recent data suggest that the sperm factor may have phospholipase C (PLC) activity that leads to the generation of IP3 and Ca^{2+} oscillations during fertilization (15, 33). Another SOAF that has been proposed is tr-kit, a truncated c-kit tyrosine kinase, which causes parthenogenetic activation when microinjected into mouse oocytes (34). It is specifically expressed in the post-meiotic stages of spermatogenesis and is present in mature spermatozoa, suggesting that it might play a role in gamete fertilization. Furthermore, tr-kit activates PLCγ1 and the SH3 domain of PLCγ1 is essential for tr-kit-induced oocyte activation. However, tr-kit has not been demonstrated to cause Ca^{2+} oscillations upon injection into oocytes, suggesting that tr-kit may not be essential, if indeed involved in oocyte activation at natural fertilization.

Despite evidence from the above experiments supporting the possibility that sperm factor has PLC activity or PLCγ is involved in tr-kit-induced oocyte activation, Ca^{2+} release at fertilization in mouse oocytes does not require SH2-domain-mediated activation of PLCγ. The sperm's PLC is not responsible for initiating Ca^{2+} release at fertilization in mouse oocytes (35). Whether sperm factor is itself a PLC or whether it acts upstream of the oocyte's PLCs remains to be elucidated.

To further add to the confusion, nuclei from fertilized embryos or fertilizing sperm introduces Ca^{2+}-releasing activity and causes Ca^{2+} transients and oocyte activation (36). This observation challenges the cytosolic sperm factor theory for triggering Ca^{2+} oscillations and suggests the presence of an insoluble form of sperm factor. Such a SOAF has been found to be associated with the sperm perinuclear material (29). Whether or not the perinuclear factor is different from the soluble cytosolic one is unclear.

Collectively, all three of the above models have been challenged by experimental observations and no definitive conclusion can yet be made. It is possible that sperm utilize multiple pathways to ensure that sufficient Ca^{2+} oscillations will be induced. These multiple pathways may be combined in such a synergistic fashion that the sperm can induce the specific Ca^{2+} profiles during natural fertilization.

The three models presented here have been proposed to explain how Ca^{2+} is released, even though they remain to be validated. By contrast, very few experiments have been carried out to address how the released Ca^{2+} is restored to a basal level and how the Ca^{2+} oscillations are sustained. These questions are equally important to understanding Ca^{2+} release. The declining phase of the Ca^{2+} transients is probably attributable to the release of extracellular milieu, re-uptake by ER or sequestering by other cellular components. The sustained Ca^{2+} oscillations have been proposed to be a consequence of activation of the capacitative Ca^{2+} influx pathway (37). In this hypothesis, depletion of ER's Ca^{2+} stores is proposed to activate capacitative Ca^{2+} entry to refill the stores. However, what signals link store depletion and capacitative Ca^{2+} entry in oocytes is not known.

The persistent influx of Ca^{2+} is responsible for maintaining the repetitive Ca^{2+} transients in the oocyte (11). Conflicting observations have been reported on the sources of Ca^{2+} for oscillations. Both IP3 and strontium have been demonstrated to induce Ca^{2+} oscillations in the absence of extracellular Ca^{2+} (10, 14, 38). ER also has been proposed to sequester released Ca^{2+}. These data suggest that Ca^{2+} can be recycled intracellularly without influx from extracellular sources. Sperm-induced Ca^{2+} transients in hamster oocytes, however, disappear following perfusion of Ca^{2+} free medium. The frequency of Ca^{2+} transients is shortened or prolonged by increasing or reducing extracellular Ca^{2+} concentrations, respectively. Further study is necessary to identify the Ca^{2+} stores in oocytes for oscillations.

Oscillations of Ca^{2+} in the mammal oocytes usually persist until just before pronuclear formation, when MAPK is inactivated. When a microtubule inhibitor was used to maintain oocytes arrested at metaphase, Ca^{2+} oscillations persist without cessation. Pronuclear formation, however, does not cause and is not required for cessation of Ca^{2+} oscillations (39). Enucleated oocytes exhibit Ca^{2+} oscillations that cease around the time of pronucleus formation. Reduced Ca^{2+} influx is unlikely to be the sole mechanism for cessation of Ca^{2+} oscillation (39). The cessation of Ca^{2+} oscillations may be associated with inactivation of MAPK, microtubule network formation, and/or the down regulation of type I IP3 receptors, or possibly transient disruptions of ER (11).

Function of Ca^{2+} Oscillations - Downstream Targets

What is the consequence of Ca^{2+} oscillations? Why do mammalian oocytes acquire Ca^{2+} oscillations at fertilization? Oscillations of Ca^{2+} are sufficient to trigger activation of oocytes and initiate development of mammalian embryos. Regardless of the mechanisms underlying Ca^{2+} release and uptake, it is established that Ca^{2+} waves and oscillations are crucial for inactivating MPF and cytostatic factor (CSF) as well as maintaining low MPF activity, followed by inactivation of MAPK. Low MPF activity is associated with meiotic release while low MAPK activity is important for further cell cycle progression and pronuclear formation, transforming meiosis to interphase of mitosis.

Calmodulin-dependent protein kinase II (CaM KII) acts as a switch in the transduction of Ca^{2+} signals to mammalian oocyte activation. Activation of CaM KII seems to be responsible for the inactivation of MPF (40). The rise in intracellular Ca^{2+} initiates a Ca^{2+}-calmodulin-CaM KII process that is required as an early event following fertilization to destroy cyclin B by activating ubiquitin-dependent cyclin degradation machinery , thus inactivating MPF. In contrast, p39mos degradation is not required for this process. Similar to MPF inactivation, CaM KII activation is associated with meiotic release, but it also may be implicated in second polar body extrusion (41).

A relatively high MPF and CSF activity in mammalian oocytes prior to sperm-oocyte fusion is perhaps required to prevent premature activation and to develop blockage to polyspermy. A single Ca^{2+} transient can inactivate MPF and release meiotic arrest but MPF activity will recover and the meiotic release will be aborted, resulting in arrest of oocytes at MIII or MIII-Anaphase III stage (2, 42), where a reduced number of chromosomes spread over a slightly elongated spindle after extrusion of a second polar body. Additionally, the first Ca^{2+} transient is likely to be responsible for the exocytosis of cortical granules (CG). A prolonged period of Ca^{2+} signaling is perhaps required to maintain the low MPF activity and to ensure a transition from metaphase to interphase (Figure 1). However, sustained elevation of Ca^{2+} can damage cells, causing cell death. To avoid this possible damage, mammalian oocytes develop a mechanism of Ca^{2+} oscillations for completion of oocyte activation and initiation of development, with Ca^{2+} returning to a basal level between spikes.

Metaphase II (MII) arrest of oocytes is probably caused by at least two factors: MPF and CSF. Active MPF is composed of a certain amount of cyclin B and a dephosphorylation state of p34cdc2 on Tyr-15 and 14 and phosphorylation state on Thr161. The phosphorylation of p34cdc2 on Thr-14 and Tyr-15 prevents the pre-mature activation of p34cdc2/cyclin B complex. During MII arrest, cyclin degradation is initiated in oocytes, but new synthesis of cyclin replenishes the degraded cyclin and maintains the activity of MPF. Inhibition of protein synthesis induces oocyte activation, demonstrating the role of new protein synthesis in maintaining MII arrest. Activity of MPF is maintained during MII arrest by the active CSF, which prevents degradation of the cyclin B in MPF (43). Activity of CSF is in turn dependent on the activities of the c-Mos protein, its substrate MAPK kinase and MAPK. The catalytic component of CSF is known to be c-Mos, the product of a c-mos proto-oncogene (43). MAPK mediates the MII-arresting CSF activity of Mos and disruption of c-Mos activity abolishes MAPK activity (44). A large amount of evidence indicates that a positive feedback loop exists between MAPK and Mos accumulation. The Mos/MAPK pathway plays an important role for meiotic spindle assembly and thus maintains oocytes at MII stage.

Following fertilization-induced Ca^{2+} transients, MPF activity is rapidly inactivated in MII arrested mammalian oocytes (45, 46), which is associated with meiotic release. MAPK activity remains high until chromosomes de-condense and a pronucleus begins to form. These occur several hours following the inactivation of MPF. Recently, it has been proposed that CaM KII activation observed immediately following parthenogenetic activation serves to potentiate MAPK activity, thus preventing pronuclei formation (47). However, MAPK activity remained high before and after activation of CaM KII, perplexing a direct causal link between CaM KII and MAPK. Direct inhibition of MAPK by 6-DMAP, a phosphorylation inhibitor, coincides with rapid formation of the interphase microtubule network and pronuclei (2), suggesting the significance of MAPK inactivation in inducing the transition toward an interphase morphological organization and in promoting pronuclear formation.

It has been reported that protein kinase C (PKC) activity rose at the same time (40 min) as the second polar body formation and then subsided over the next 5 hours of post-activation (48). Although several different isotypes of PKC have been identified in both mouse oocytes and early embryos, their role in oocyte activation remains to be defined. On the one hand, activation of PKC by diacylglycerol or phorbol diester can result in CG exocytosis and zona pellucida (ZP) modifications of mouse oocytes. Nonetheless, second polar body extrusion and pronuclear formation are not observed upon PKC activation. These results suggest that PKC does not seem to regulate meiotic release (13). Consistently, PKC activation does not induce the resumption of meiosis and pronuclear formation in porcine oocytes but can induce CG exocytosis, which is independent of a Ca^{2+} rise. Moreover, treatment with PKC activators induces CG exocytosis and ZP modifications in both GV-intact and MII arrested oocytes, further demonstrating that PKC activation is not specifically associated with MII oocyte activation. It is possible that PKC may not be a key signaling component for oocyte activation but rather a by-product generated during oocyte activation.

MORPHOLOGICAL AND MOLECULAR MARKERS OF OOCYTE ACTIVATION

Ca^{2+} Increase

The first observed event at oocyte activation is the Ca^{2+} transients that occur at certain intervals, i.e. Ca^{2+} oscillations, the frequency of which varies from 5 to 30 min among species and individuals. Sperm penetration-triggered Ca^{2+} oscillations have been found in all mammalian species studied so far, including human.

a) *CG exocytosis, second polar body extrusion and pronuclear formation:* The rise in Ca^{2+} concentration is followed by critical morphological changes in the oocytes. The first Ca^{2+} transient is the primary trigger that induces CG exocytosis. Exocytosis of CG has been shown to cause structural changes in the zona pellucida (49). It is Ca^{2+} dependent and MPF and MAPK independent. Typically, membrane enclosing CGs fuses with the plasma membrane of the oocytes and releases enzymes. The released enzymes modify zona pellucida structures and the fusion of membranes changes membrane potential and possibly membrane structures. Both of these events prevent subsequent penetration by additional sperm (polyspermy). Consequently, the oocytes undergo a series of pre-programmed processes: spindle elongation and spindle birefringence increase, chromosome separation, progression into anaphase and telophase II, and extrusion of a second polar body. If inadequately activated, oocytes will arrest at MIII-anaphase III. If oocytes are fully activated, the chromosomes de-condense and pronuclei form, followed by DNA synthesis and the first mitosis of development. The formation of a pronucleus, or pronuclei, but not a second polar body extrusion, is the better end-point for early signs of development (12). In mammals, this can take 4-8 hours depending upon the species and the conditions of stimulation. Pronuclear formation has thus been commonly used to assess the completeness of oocyte activation (50-53).

b) *Inactivation of MPF and MAPK:* At the molecular level, the rise in Ca^{2+} is believed to be a trigger for MPF inactivation and probably indirect inactivation of MAPK as well. Following parthenogenetic activation or fertilization in mammalian MII oocytes, MPF is inactivated through degradation of cyclin B and the low level of MPF is sustained by continuous degradation of newly synthesized cyclin B, and subsequently by phosphorylation of $p34^{cdc2}$. These molecular events are followed by MAPK inactivation through dephosphorylation of ERK2. Interestingly, evident differences in the kinetics of MPF and MAPK activities have been observed in bovine, ovine, and porcine oocytes following parthenogenetic activation

(2, 54, 55). Major changes in the kinetics are as follows. Firstly, MPF inactivation precedes MAPK inactivation. Secondly, MPF activity drops and increases again during the MII/MIII-AIII transition whereas MAPK remains active, which may play a role in microtubule assembly into spindles as well as chromatin condensation. Therefore, changes in microtubule organization correlated with high MAPK activity during MII/MIII-AIII transition (2, 45). In oocytes undergoing "full" activation, defined by pronuclear formation, MAPK is inactivated. A decrease in MPF activity correlated with MII exit and a decrease in MAPK activity correlated with pronuclear formation. These correlations have been shown to be true regardless of the speed of nuclear progression (our unpublished observation). Inactivation of MAPK was independent of MPF inactivation and its low activity persisted throughout the pronuclear formation stage. In general, low MPF but high MAPK activity is indicative of partial oocyte activation, resulting in arrest at MIII. However, the only reliable assessment for genuine full activation of oocytes is probably full term development, an impossibility at present upon parthenogenetic activation of oocytes.

In addition, oocytes of different species may exhibit distinctive protein changes after sufficient activation. For instance, activated porcine oocytes at pronuclear stage exhibit typical changes in the shift of a 25 kDa protein to 22 kDa (56, 57), possibly attributable to dephosphorylation (see Figure 4). In bovine, two protein molecules of 138 and 133 kDa present in oocytes at the metaphase stage when a spindle is present (MII and MIII) disappear after activation. They might be microtubule-associated proteins that are crucial for spindle organization or stabilization (53). Their disappearance might be a prerequisite for the full activation. Further studies are needed to fully characterize these specific proteins and their functions during oocyte activation.

PRACTICAL CONSIDERATIONS IN ARTIFICIAL ACTIVATION

Oocytes are naturally activated by sperm. Oocyte activation and parthenogenetic development can also be induced by artificial means. Two definitive and readily measurable signals associated with fertilization and full activation of oocytes, Ca^{2+} oscillations and inactivation of both MPF and MAPK, are frequently used to evaluate practical applications of artificial activation.

Many different agents or methods that elicit a rise in intracellular Ca^{2+} can induce some degree of parthenogenetic activation and development. The stimuli used for parthenogenetic activation seem to increase intracellular free Ca^{2+} levels either by promoting Ca^{2+} influx, by mobilizing intracellular Ca^{2+} or by a combination of both. Agents or methods that induce multiple Ca^{2+} transients that mostly mimic sperm-induced Ca^{2+} oscillations have been found suitable for sufficient parthenogenetic activation of newly matured mammalian oocytes. The major finding in recent years is that the process mimicking molecular changes coupled with Ca^{2+} oscillations can also be simulated, for instance, by inhibition of MPF or MAPK.

A Single Agent or Method that Induces a Single Rise in Cytosolic Ca^{2+}

Ethanol: Ethanol was first used for parthenogenetic activation in mouse oocytes (58). Subsequently, treatments with 6-10 % ethanol for 5-10 minutes have been extensively used for activating oocytes of large domestic animals as well. Effective activation by ethanol treatment is oocyte age dependent. Newly arrested oocytes do not respond well to ethanol. The treatment induces the extrusion of the second polar body, but the oocyte does not enter interphase. Instead, it arrests again at metaphase (MIII-arrest) and this is termed aborted activation (42). Ethanol disrupts the organization of cytoskeletal elements and the resulting aborted activation may result in significant increases in the incidence of aneuploidy (42, 59). The lack of a good activation response by ethanol treatment might be because it causes a single large rise in intracellular Ca^{2+}, which appears to derive from both intracellular Ca^{2+} release and extracellular Ca^{2+} influx (58).

Ca^{2+} ionophore: Similarly, Ca^{2+} ionophore A23187 or ionomycin also induces a single Ca^{2+} rise in mammalian oocytes. The single Ca^{2+} rise by ionophore A23187 is mostly derived from that released by internal stores (60). The duration of the rise in Ca^{2+} by ethanol or Ca^{2+} ionophore is significantly longer than that by sperm at fertilization, although the peaks are smaller.

Electrical stimulation: An appropriate electrical direct current (DC) pulse in the presence of Ca^{2+}-containing medium brings a transient Ca^{2+} increase in the cytosol and initiates oocyte activation in all species examined so far. Electrical pulse causes the formation of transient pores in the oocyte plasma membrane, and Ca^{2+} in the external medium enter the oocytes through these transient pores. A single electrical pulse causes a single large Ca^{2+} transient (56, 61).

However, a single Ca^{2+} rise by ethanol, Ca^{2+} ionophore or electrical pulse is inadequate to fully activate oocytes, as evidenced by incomplete

CG exocytosis (52), MIII arrest without pronuclear formation of newly matured oocytes (2, 42, 53), and poor preimplantation development (50, 53, 62). The initial response of oocytes to the single Ca^{2+} rise is the inactivation of MPF. However, this inactivation is short-lived and MPF activity increases again after the extrusion of the second polar body approximately 4 hours post-treatment, and at 15 hours when the new spindle forms at MIII-AIII stage or at MIII in mouse oocytes (42, 45). A single Ca^{2+} rise also fails to inactivate MAPK, whose activity stays at a high level throughout this progression.

Aged oocytes are more responsive to electro-stimulation than young oocytes that have just completed the maturation process. Both MPF and MAPK behave differently in young and aged bovine oocytes following a single activation treatment. Inactivation of MPF occurs in both young and aged oocytes, whereas MAPK is inactivated only in aged oocytes not young oocytes (2, 46). Furthermore, mature oocytes have a limited temporal window for normal fertilization, within just a few hours after ovulation or maturation. Aging of oocytes can lead to an increase in the incidence of chromosomal abnormality. Aged oocytes also have altered structures and physiology, which are incompatible with normal fertilization, activation, and development.

Approaches that Mostly Resemble the Physiological Processes of Fertilization- Induced Oocyte Activation

Methods That Induce Ca^{2+} Oscillations

Multiple electrical pulses: Multiple electrical stimulation induces multiple Ca^{2+} rises (56, 61, 62). Repetitive electrical stimulations in Ca^{2+}-containing medium triggers Ca^{2+} oscillations and pronuclear formation in nearly 100% of freshly ovulated mouse oocytes. Electrical stimulation itself is not sufficient to induce oocyte activation, because electrically stimulated oocytes in electroporation medium devoid of Ca^{2+} do not exhibit Ca^{2+} rises and fail to be activated. Multiple electrical stimulations induce full suppression of MPF activity similar to that recorded after normal fertilization. Strength, duration, and numbers of electrical pulses as well as electroporation medium all influence the effectiveness of oocyte activation. Multiple pulses are effective only when the interval between pulses simulates the frequency of Ca^{2+} oscillations triggered by sperm. For example, three electric pulse separated by a one second interval are no better than a single pulse. Multiple electrical stimulations at ~22 minutes intervals have been successfully used to activate cytoplasts during nuclear transfer in rabbits and sheep oocytes.

Strontium: Strontium (Sr^{2+}) effectively induces Ca^{2+} rises at a frequency similar to that of sperm-induced Ca^{2+} oscillations (10, 38, 63, 64) (also see Figure 3). Sr^{2+} has been shown to induce Ca^{2+} release from intracellular stores in other cell types as well. Chromosome abnormalities do not result from Sr^{2+} activated oocytes (65). Instead, preimplantation development of diploid parthenogenetic embryos derived from oocytes activated by Sr^{2+} plus cytochalasin D exhibit a rate similar to that of IVF embryos (66). Moreover, cloned mice have been produced by using Sr^{2+} activated cytoplasts in nuclear transfer, demonstrating its suitability to be used in studies when full-term development is required (67). Whether Sr^{2+} can effectively activate oocytes of other species remains to be determined.

Thimerosal: Thimerosal, a sulfhydryl reagent, has been shown to increase the sensitivity of CICR by oxidation of critical cysteine residues in hamster and mouse oocytes (68), and induce the repetitive transient rises in intracellular Ca^{2+} in mammalian oocytes (69, 70). However, thimerosal alone could not stimulate mouse oocyte activation (71). The peak and the duration of the rises in Ca^{2+} induced by thimerosal are smaller and shorter than those induced by sperm. Moreover, thimerosal appears to have many undesirable side effects, such as spindle disruption (71), making it of little practical use in activation (12). Dithiothreitol (DTT) might enhance the transduction event that normally occurs at fertilization to initiate both the influx of external Ca^{2+} and the onset of CICR (71). Also, DTT was found to reverse the side effects of thimerosal and regenerate typical metaphase spindles (71). The combinations of thimerosal with DTT have been found effective in the activation of pig oocytes (70, 72).

Combination Of Chemical Activation - Targeting On MPF And MAPK

A single Ca^{2+} increase can induce early activation events, such as resumption of meiosis, CG exocytosis and the modifications of zona pellucida glycoproteins, but it cannot induce late events of mRNA recruitment, pronuclear formation, DNA synthesis and cleavage (13, 51, 73). Although multiple electrical pulses can improve parthenogenetic development, the procedures are complicated and labor consuming. Moreover, repeated electrical stimulations induce Ca^{2+} oscillations which differ from those induced by sperm (56), particularly the much prolonged decline of Ca^{2+}. Multiple Ca^{2+} oscillations are believed to maintain low activity of MPF by degradation of cyclin B and/or phosphorylation of $p34^{cdc2}$ and inactivation of MAPK by ERK2 dephosphorylation (Figure 1). Thus, an alternative approach for activation of oocytes that mostly mimics fertilization would be combinations of a Ca^{2+} rise and an inhibition of either protein synthesis with cycloheximide or protein phosphorylation with 6-

dimethylaminopurine (6-DMAP). The chemical 6-DMAP was shown to enhance the activation of young mouse and bovine oocytes (73-75). The first Ca^{2+} increase is indispensable for full activation of oocytes, because inhibiting either proteins synthesis with cycloheximide or kinase activation with 6-DMAP alone fails to promote subsequent parthenogenetic development (53). The key is whether or not MAPK is inactivated. Prolonged inactivation by $p34^{cdc2}$ phosphorylation is probably contributed by MAPK dephosphorylation and inactivation of MAPK.

Inhibition of protein synthesis is an effective way to enhance oocyte activation since it induces temporal changes in both MPF and MAPK similar to those that occur following fertilization (76). Cycloheximide treatment does not change the intracellular Ca^{2+} profile (38), but can cause instant decline in cyclin B levels which maintain low activity of MPF. Moreover, dephosphorylation of MAPK occurs several hours later when most oocytes reach pronuclear stage after cycloheximide treatment (46). Different from cycloheximide or IVF induced activation, 6-DMAP induces dephosphorylation of MAPK, and therefore earlier pronuclear development. Agents that increase intracellular Ca^{2+} or prevent protein synthesis/phosphorylation alone result in low cleavage and development rates. Combined treatments of these agents were more effective in inducing activation and development of young bovine oocytes than any single treatment alone (50, 53), demonstrating that both the first Ca^{2+} increase and the subsequent inhibition of certain kinases or proteins are indispensable for normal cell cycle progression. Parthenogenetic activation can lead to either haploid or diploid development. Treatment of Ca^{2+} ionophore A23187 or ionomycin sequentially combined with 6-DMAP leads to one pronuclear formation in bovine oocytes without a second polar body extrusion (53, 73), resulting in diploid development. In the mouse, 6-DMAP has been shown to impair the contact between spindle-pole and the cortex as well as to inhibit contractile activity in the cortex after activation (74). Treatment with Ca^{2+} ionophore A23187 sequentially combined with cycloheximide, however, leads to the formation of one pronucleus and extrusion of a polar body, resulting in haploid development. The latter activation regime can be used in combination with cytochalasin D, an inhibitor of actin filament, to induce formation of two pronuclei by preventing the extrusion of the second polar body (53). A high percentage of oocytes activated in this manner develop to the blastocyst stage.

Through the use of combined artificial stimulators (ethanol, ionophore, ionomycin, or an electrical pulse) and cycloheximide, puromycin or kinase inhibitors, parthenogenetic activation and development have been significantly improved to levels comparable to sperm induced activation in a variety of species including rabbit, bovine, pig, goat, equine and human.

Reagents that specifically target proteins in the activation pathways would be beneficial to activate oocytes parthenogenetically (77).

FACTORS AFFECTING EFFECTIVENESS OF ARTIFICIAL ACTIVATION

Species-Specific Differences

Even for the same activation regime, modifications in the treatment are required when applied to different species. For example, a 5 min exposure to 5 μM Ca^{2+} ionophore followed by 10 μg/ml cycloheximide treatment for 6 h is effective for bovine oocyte activation (53), but ineffective for equine oocyte activation, for which 50 μM Ca^{2+} ionophore and 24 h exposure to cycloheximide is required (78). Interestingly, the requirement of relatively intense stimulation and/or longer exposure to cycloheximide for activation of equine oocytes may be related to very low rates of spontaneous activation in this species (78). It is possible that MPF activity is maintained higher in equine oocytes than bovine oocytes. Pig oocytes are also quite resistant to activation stimulation, and high concentrations of Ca^{2+} ionophore are required for their meiotic release (79). The inefficient activation of pig oocytes contribute to the extremely low efficiency of cloning in this species, compared to cloning in bovine and sheep. Therefore, a strategy of double nuclear transfer, the second round being the transfer of pronuclei of reconstituted activated oocytes into enucleated in vivo-fertilized zygotes, was used to compensate for the ineffective artificial activation (80). Many different methods, chemical, electrical, and combinations, have been used for pig oocyte activation during cloning procedures. So far, successful production of cloned piglets, in spite of low efficiency, has been achieved by using electrical activation alone or in combination with chemical activation (81, 82). In addition, conditions used for maturing porcine oocytes *in vitro* are important to their subsequent response to artificial activation.

Interestingly, a single Ca^{2+} rise is sufficient to activate oocytes from many non-mammalian species such as frog, fish, and sea urchin. In addition, the short cell cycle length may also play a role. For instance, the frog oocyte forms the female pronucleus within 20 min of fertilization which induces a single Ca^{2+} increase to release meiotic arrest and progress into interphase. Mouse oocytes, however, form pronuclei around 4 h after sperm penetration, while 7-8 h are required for bovine. Mouse oocytes are more readily activated artificially and also frequently undergo spontaneous sperm penetration, while 7-8 h are required for bovine. Mouse oocytes are more readily activated artificially and also frequently undergo spontaneous activation, while human oocytes are relatively resistant to parthenogenetic activation by Ca^{2+} ionophore treatment (83).

It seems that the less time required for maturation of oocytes generally correlates with the ease of artificial activation. *In vitro* maturation requires 12-14 h in the mouse, 22-24 h in bovine and sheep, 40-48 h in pigs and horses, and 44-48 h in humans. Correspondingly, mouse oocytes are easily activated and frequently undergo spontaneous activation; bovine and sheep oocytes are also readily activated, while pig, horse and human oocytes are relatively resistant to activation stimulation. Whether the longer period of maturation leads to or is required for building up higher levels of MPF activity is not known. A direct comparison of kinase activity among species would be necessary to correlate kinase activities with effectiveness of oocyte activation.

Oocyte Competence

The development of normal "activation competence' likely involves changes in the oocyte's ability to respond to signals for intracellular Ca^{2+} release as well as the ability to respond to this Ca^{2+} increase (49). It has been shown that between the germinal vesicle (GV) and metaphase II stages, there may be a 3- to 4-fold increase in the capability of ionomycin-inducible Ca^{2+} release (84). Oocytes acquire the ability to undergo full Ca^{2+} release at the end of maturation. Many factors such as maturation medium and the sources of the oocytes can affect the "activation and developmental competence". For example, equine oocytes matured in TCM-199 are activated more easily than oocytes matured in follicular fluid. Consistently, the rate of fertilization and early embryo development is lower when their oocytes are matured in follicular fluid (78). Cytoplasmic immaturity could compromise competence of oocyte activation and development. Oocytes from small follicles of young prepubertal mice are deficient for further development (85). Similarly, oocytes derived from calves also manifest insufficiency in "activation and developmental competence", as evidenced by low MPF and MAPK activities, defects in Ca^{2+} oscillations, and poor activation and developmental rates (86, 87). On the other hand, oocytes collected from older individuals may also have defects in both nuclear status and cytoplasmic components due to aging, leading to abnormalities in activation and development.

Oocyte Age

There is an age dependency on the ability of oocytes to be artificially activated, particularly by a single Ca^{2+} rise. Young oocytes, whether freshly ovulated or newly matured *in vitro*, are readily fertilized, but are resistant to activation by treatments that produce a single Ca^{2+} transient. On the contrary, aged oocytes are more easily activated by a single Ca^{2+} transient, which may be due to partial decrease in MPF activity in older oocytes (2). Low MPF activity in aged oocytes is possibly due to the degradation but slow synthesis of cyclin B during oocyte aging *in vitro* or *in vivo*. Old oocytes, however, are not suitable for use as nuclear recipients in nuclear transfer, not only because developmental potential of aged oocytes is possibly compromised but also because lower MPF activity is insufficient to remodel a donor nucleus.

A single Ca^{2+} rise induced by Ca^{2+} ionophore can induce a decline in MPF activity by inactivating the existing MPF component, but is insufficient to maintain low activity of MPF in young oocytes. A single intracellular Ca^{2+} rise allows young bovine oocytes to resume meiosis and emit the second polar body. However, active protein synthesis and phosphorylation events in the young oocytes would quickly restore the MPF activity (61) and thus force the nuclear materials to enter a new metaphase arrest (MIII) (42, 60, 73). Further treatment of oocytes with protein synthesis inhibitors, such as cycloheximide, can prevent MPF reactivation after initial calcium transient (50, 76).

NUCLEAR INDEPENDENCE OF OOCYTE ACTIVATION

Enucleated oocyte cytoplasts have been used in the cloning procedures. Many key proteins and kinases are reportedly associated with meiotic spindles (47, 88). It has been a concern whether removal of metaphase chromosomes and/or spindles affects oocyte activation. Moreover, whether or not cytoplasm alone could be induced to sufficiently activate has not been tested. We removed chromosomes and spindles by micromanipulation and confirmed complete enucleation by either Pol-Scope spindle imaging or by the absence of DNA staining by Hoechst 33342 (89). Since chromosomes are removed, the nuclear dynamics cannot be followed. However, molecular changes in the cytoplasm can be compared between intact and enucleated oocytes. Here we present data from several animal species with selected markers, showing that Ca^{2+} profiles and levels of key proteins and kinases in the enucleated cytoplasts are comparable to those of intact oocytes and that cytoplasm activation of oocytes is independent of nuclear chromosomes.

Ca^{2+} Oscillations Do Not Change in Enucleated Oocytes (Mouse)

Ca^{2+} oscillations are associated with normal activation of oocytes triggered by sperm. As stated earlier, Sr^{2+} also elicits Ca^{2+} oscillations that mimic sperm-induced Ca^{2+} oscillations, although the first transient and the duration of subsequent transients by Sr^{2+} generally last longer. It is interesting to determine whether removal of nuclear chromosomes and spindle affects Ca^{2+} profiles. As shown in Figure 3, Ca^{2+} oscillations are induced in enucleated mouse oocyte cytoplasts. The frequency of Ca^{2+} oscillations is comparable to that of intact oocytes with chromosomes and spindles (13.4 ± 3.2 and 13.3 ± 3.0 min/interval, respectively). The first Ca^{2+} transient lasts much longer than subsequent ones and there are no obvious differences in peak duration (average of 3.3 min) between enucleated and intact mouse oocytes after Sr^{2+} activation. The frequency of Ca^{2+} oscillations does not differ between these two types of oocytes.

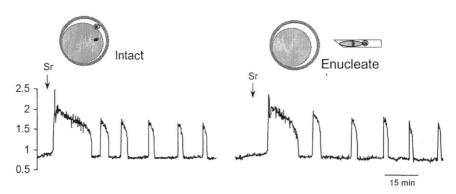

Figure 3. Mouse oocyte cytoplasts without chromosomes and spindles exhibit Ca^{2+} oscillations, detected by Fura-2/AM labeling and ratio-metric imaging, similar to those of intact oocytes following Sr^{2+} treatment.

DEPHOSPHORYLATION OF SPECIFIC PROTEINS IN PORCINE OOCYTE ACTIVATION

As stated earlier, electrical activation gives more consistent activation of porcine oocytes. In one study, we used repeated electrical stimulations to induce artificial activation of intact and enucleated porcine oocytes, and the protein changes were compared to those of oocytes after IVF.

Both electrical stimulations and sperm induced the shift of a 25 kDa protein in intact MII oocytes to 22 kDa in activated oocytes (Figure 4). In addition, spontaneous shifting of the 25 kDa protein to 22 kDa occurs in oocytes during aging *in vitro*. The shift of this protein from 25 kDa to 22 kDa also is found in enucleated cytoplasts after activation, suggesting that the protein is not attached to the spindle and the change in its molecular mass is nuclear independent. The nature of this shift remains to be determined and my be indicative of dephosphorylation of the 25 kDa to 22 kDa protein (56, 57).

ENUCLEATED OOCYTES EXHIBIT SIMILAR CHANGES IN MPF AND MAPK TO THOSE IN INTACT OOCYTES (BOVINE)

High levels of MPF and MAPK activities exist in enucleated, intact, and micromanipulated bovine oocytes before activation, indicating that active MPF and MAPK are mostly located in the cytoplasm and that removal of chromosomes does not remove key kinases. It is now known that independent inactivation of MPF and MAPK occurs in bovine oocytes following either artificial or sperm-induced activation, with the inactivation of MPF preceding the inactivation of MAPK. Not unexpectedly, MPF and MAPK in enucleated cytoplasts exhibit dynamics similar to that in intact oocytes following parthenogenetic activation.

These results demonstrate that the absence of chromosomes and/or meiotic spindles do not affect the normal dynamics of both MPF and MAPK after oocyte activation. The demonstration of nuclear independency of MPF and MAPK activation also helps explain the success of nuclear transfer.

CONCLUSIONS

During fertilization in mammals, the sperm triggers a series of Ca^{2+} oscillations in the mature oocyte. These Ca^{2+} oscillations activate the

Figure 4. Protein synthesis in activated porcine oocytes. At 44 hours of maturation, denuded oocytes or cytoplasts were activated by three electrical pulses of 1.0 kV/cm, 50 μsec at 5 min intervals in 0.28 M inositol supplemented with 0.1 mM Ca^{2+} and Mg^{2+}. Twenty hours after treatment, oocytes were labeled with ^{35}Smethionine for 3 h, then run on 8-15% SDS-PAGE gradient gels, and processed for autoradiography.

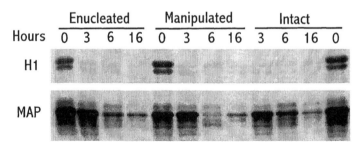

Figure 5. MPF (Histone H1 kinase, H1) and MAPK activities in enucleated, manipulated, and intact bovine oocytes stimulated with one electrical pulse followed by incubation in cycloheximide. Experiments were repeated three times. In manipulated oocytes, a small cell cloning by nuclear amount of cytoplasm similar to that of a karyoplast was removed as micromanipulation control.

development of the oocyte into an embryo. The first Ca^{2+} transient is required for CG exocytosis and meiotic release, but multiple Ca^{2+} transients are required for cell cycle progression. Molecular mechanisms under-lying sperm-triggered oocyte activation include the inactivation of both

MPF and MAPK, coupled with Ca^{2+} oscillations. These are important considerations when applying artificial activation. Indeed, the combination approach of inducing Ca^{2+} transient and inhibiting both MPF and MAPK for oocyte activation has been employed widely in somatic transfer and may be important in improving the cloning efficiency. Continued search for sperm factors and molecular mechanisms underlying oocyte activation should provide further insights to enhance our understating of initiation of development as well as facilitate the application of oocyte activation in the related biotechnology and medicine.

ACKNOWLEDGEMENTS

The authors are indebted to Marina Julian for critical editing of this report, Katherine Hammar and Peter Smith for calcium imaging experiments, David Keefe for mouse oocyte activation experiments, and Yanfeng Dai and Robert Moor for pig oocyte maturation and protein analysis experiments.

REFERENCES

1. Fissore RA, He CL, Vande Woude GF. Potential role of mitogen-activated protein kinase during meiosis resumption in bovine oocytes. Biol Reprod 1996; 55: 1261-1270.
2. Liu L, Ju JC, Yang X. Differential inactivation of maturation-promoting factor and mitogen- activated protein kinase following parthenogenetic activation of bovine oocytes. Biol Reprod 1998; 59: 537-545.
3. Gordo AC, He CL, Smith S, Fissore RA. Mitogen activated protein kinase plays a significant role in metaphase II arrest, spindle morphology, and maintenance of maturation promoting factor activity in bovine oocytes. Mol Reprod Dev 2001; 59: 106-114.
4. Stephano JL, Gould MC. MAP kinase, a universal suppressor of sperm centrosomes during meiosis? Dev Biol 2000; 222: 420-428.
5. Gautier J, Minshull J, Lohka M, Glotzer M, Hunt T, Maller JL. Cyclin is a component of maturation-promoting factor from Xenopus. Cell 1990; 60: 487-494.
6. Gautier J, Norbury C, Lohka M, Nurse P, Maller J. Purified maturation-promoting factor contains the product of a Xenopus homolog of the fission yeast cell cycle control gene cdc2+. Cell 1988; 54: 433-439.
7. Cuthbertson KS, Cobbold PH. Phorbol ester and sperm activate mouse oocytes by inducing sustained oscillations in cell Ca2+. Nature 1985; 316: 541-542.
8. Deguchi R, Shirakawa H, Oda S, Mohri T, Miyazaki S. Spatiotemporal analysis of Ca(2+) waves in relation to the sperm entry site and animal-vegetal axis during Ca(2+) oscillations in fertilized mouse eggs. Dev Biol 2000; 218: 299-313.
9. Faure JE, Myles DG, Primakoff P. The frequency of calcium oscillations in mouse eggs at fertilization is modulated by the number of fused sperm. Dev Biol 1999; 213: 370-377.
10. Kline D, Kline JT. Repetitive calcium transients and the role of calcium in exocytosis and cell cycle activation in the mouse egg. Dev Biol 1992; 149: 80-89.
11. Kline D. Attributes and dynamics of the endoplasmic reticulum in mammalian eggs. Curr Top Dev Biol 2000; 50: 125-154.

12. Swann K, Ozil JP. Dynamics of the calcium signal that triggers mammalian egg activation. Int Rev Cytol 1994; 152: 183-222.

13. Schultz RM, Kopf GS. Molecular basis of mammalian egg activation. Curr Top Dev Biol 1995; 30: 21-62.

14. Jones KT, Whittingham DG. A comparison of sperm- and IP3-induced Ca2+ release in activated and aging mouse oocytes. Dev Biol 1996; 178: 229-237.

15. Jones KT, Nixon VL. Sperm-induced Ca(2+) oscillations in mouse oocytes and eggs can be mimicked by photolysis of caged inositol 1,4,5-trisphosphate: evidence to support a continuous low level production of inositol 1, 4,5- trisphosphate during mammalian fertilization. Dev Biol 2000; 225: 1-12.

16. Jaffe LF. The path of calcium in cytosolic calcium oscillations: a unifying hypothesis. Proc Natl Acad Sci U S A 1991; 88: 9883-9887.

17. Deng MQ, Shen SS. A specific inhibitor of p34(cdc2)/cyclin B suppresses fertilization-induced calcium oscillations in mouse eggs. Biol Reprod 2000; 62: 873-878.

18. Lawrence Y, Whitaker M, Swann K. Sperm-egg fusion is the prelude to the initial Ca2+ increase at fertilization in the mouse. Development 1997; 124: 233-241.

19. Miyazaki S, Yuzaki M, Nakada K, Shirakawa H, Nakanishi S, Nakade S, Mikoshiba K. Block of Ca2+ wave and Ca2+ oscillation by antibody to the inositol 1,4,5-trisphosphate receptor in fertilized hamster eggs. Science 1992; 257: 251-255.

20. Miyazaki S, Shirakawa H, Nakada K, Honda Y. Essential role of the inositol 1,4,5-trisphosphate receptor/Ca2+ release channel in Ca2+ waves and Ca2+ oscillations at fertilization of mammalian eggs. Dev Biol 1993; 158: 62-78.

21. Williams CJ, Schultz RM, Kopf GS. Role of G proteins in mouse egg activation: stimulatory effects of acetylcholine on the ZP2 to ZP2f conversion and pronuclear formation in eggs expressing a functional m1 muscarinic receptor. Dev Biol 1992; 151: 288-296.

22. Carroll DJ, Ramarao CS, Mehlmann LM, Roche S, Terasaki M, Jaffe LA. Calcium release at fertilization in starfish eggs is mediated by phospholipase Cgamma. J Cell Biol 1997; 138: 1303-1311.

23. Swann K. A cytosolic sperm factor stimulates repetitive calcium increases and mimics fertilization in hamster eggs. Development 1990; 110: 1295-1302.

24. Stice SL, Robl JM. Activation of mammalian oocytes by a factor obtained from rabbit sperm. Mol Reprod Dev 1990; 25: 272-280.

25. Parrington J, Swann K, Shevchenko VI, Sesay AK, Lai FA. Calcium oscillations in mammalian eggs triggered by a soluble sperm protein. Nature 1996; 379: 364-368.

26. Wu H, He CL, Fissore RA. Injection of a porcine sperm factor triggers calcium oscillations in mouse oocytes and bovine eggs. Mol Reprod Dev 1997; 46: 176-189.

27. Yanagimachi R. Intracytoplasmic sperm injection experiments using the mouse as a model. Hum Reprod 1998; 13 Suppl 1: 87-98.

28. Li ST, Huang XY, Sun FZ. Flowering plant sperm contains a cytosolic soluble protein factor which can trigger calcium oscillations in mouse eggs. Biochem Biophys Res Commun 2001; 287: 56-59.

29. Perry AC, Wakayama T, Yanagimachi R. A novel trans-complementation assay suggests full mammalian oocyte activation is coordinately initiated by multiple, submembrane sperm components. Biol Reprod 1999; 60: 747-755.

30. Swann K, Lai FA. A novel signalling mechanism for generating Ca2+ oscillations at fertilization in mammals. Bioessays 1997; 19: 371-378.

31. Wu H, He CL, Jehn B, Black SJ, Fissore RA. Partial characterization of the calcium-releasing activity of porcine sperm cytosolic extracts. Dev Biol 1998; 203: 369-381.

32. Wolny YM, Fissore RA, Wu H, Reis MM, Colombero LT, Ergun B, Rosenwaks Z, Palermo GD. Human glucosamine-6-phosphate isomerase, a homologue of hamster oscillin, does not appear to be involved in Ca2+ release in mammalian oocytes. Mol Reprod Dev 1999; 52: 277-287.

33. Jones KT, Matsuda M, Parrington J, Katan M, Swann K. Different Ca2+-releasing abilities of sperm extracts compared with tissue extracts and phospholipase C isoforms in sea urchin egg homogenate and mouse eggs. Biochem J 2000; 346 Pt 3: 743-749.

34. Sette C, Bevilacqua A, Bianchini A, Mangia F, Geremia R, Rossi P. Parthenogenetic activation of mouse eggs by microinjection of a truncated c-kit tyrosine kinase present in spermatozoa. Development 1997; 124: 2267-2274.

35. Mehlmann LM, Chattopadhyay A, Carpenter G, Jaffe LA. Evidence that phospholipase c from the sperm is not responsible for initiating ca(2+) release at fertilization in mouse eggs. Dev Biol 2001; 236: 492-501.

36. Kono T, Carroll J, Swann K, Whittingham DG. Nuclei from fertilized mouse embryos have calcium-releasing activity. Development 1995; 121: 1123-1128.

37. Mohri T, Shirakawa H, Oda S, Sato MS, Mikoshiba K, Miyazaki S. Analysis of Mn(2+)/Ca(2+) influx and release during Ca(2+) oscillations in mouse eggs injected with sperm extract. Cell Calcium 2001; 29: 311-325.

38. Bos-Mikich A, Swann K, Whittingham DG. Calcium oscillations and protein synthesis inhibition synergistically activate mouse oocytes. Mol Reprod Dev 1995; 41: 84-90.

39. Day ML, McGuinness OM, Berridge MJ, Johnson MH. Regulation of fertilization-induced Ca(2+)spiking in the mouse zygote. Cell Calcium 2000; 28: 47-54.

40. Lorca T, Cruzalegui FH, Fesquet D, Cavadore JC, Mery J, Means A, Doree M. Calmodulin-dependent protein kinase II mediates inactivation of MPF and CSF upon fertilization of Xenopus eggs. Nature 1993; 366: 270-273.

41. Johnson J, Bierle BM, Gallicano GI, Capco DG. Calcium/calmodulin-dependent protein kinase II and calmodulin: regulators of the meiotic spindle in mouse eggs. Dev Biol 1998; 204: 464-477.

42. Kubiak JZ. Mouse oocytes gradually develop the capacity for activation during the metaphase II arrest. Dev Biol 1989; 136: 537-545.

43. Sagata N, Watanabe N, Vande Woude GF, Ikawa Y. The c-mos proto-oncogene product is a cytostatic factor responsible for meiotic arrest in vertebrate eggs. Nature 1989; 342: 512-518.

44. Verlhac MH, Kubiak JZ, Weber M, Geraud G, Colledge WH, Evans MJ, Maro B. Mos is required for MAP kinase activation and is involved in microtubule organization during meiotic maturation in the mouse. Development 1996; 122: 815-822.

45. Verlhac MH, Kubiak JZ, Clarke HJ, Maro B. Microtubule and chromatin behavior follow MAP kinase activity but not MPF activity during meiosis in mouse oocytes. Development 1994; 120: 1017-1025.

46. Liu L, Yang X. Interplay of maturation-promoting factor and mitogen-activated protein kinase inactivation during metaphase-to-interphase transition of activated bovine oocytes. Biol Reprod 1999; 61: 1-7.

47. Hatch KR, Capco DG. Colocalization of CaM KII and MAP kinase on architectural elements of the mouse egg: potentiation of MAP kinase activity by CaM KII. Mol Reprod Dev 2001; 58: 69-77.

48. Gallicano GI, McGaughey RW, Capco DG. Activation of protein kinase C after fertilization is required for remodeling the mouse egg into the zygote. Mol Reprod Dev 1997; 46: 587-601.

49. Ducibella T. The cortical reaction and development of activation competence in mammalian oocytes. Hum Reprod Update 1996; 2: 29-42.

50. Presicce GA, Yang X. Parthenogenetic development of bovine oocytes matured in vitro for 24 hr and activated by ethanol and cycloheximide. Mol Reprod Dev 1994; 38: 380-385.

51. Soloy E, Kanka J, Viuff D, Smith SD, Callesen H, Greve T. Time course of pronuclear deoxyribonucleic acid synthesis in parthenogenetically activated bovine oocytes. Biol Reprod 1997; 57: 27-35.

52. Wang WH, Abeydeera LR, Prather RS, Day BN. Functional analysis of activation of porcine oocytes by spermatozoa, calcium ionophore, and electrical pulse. Mol Reprod Dev 1998; 51: 346-353.

53. Liu L, Ju JC, Yang X. Parthenogenetic development and protein patterns of newly matured bovine oocytes after chemical activation. Mol Reprod Dev 1998; 49: 298-307.

54. Bogliolo L, Ledda S, Leoni G, Naitana S, Moor RM. Activity of maturation promoting factor (MPF) and mitogen-activated protein kinase (MAPK) after parthenogenetic activation of ovine oocytes. Cloning 2000; 2: 185-196.

55. Tatemoto H, Muto N. Mitogen-activated protein kinase regulates normal transition from metaphase to interphase following parthenogenetic activation in porcine oocytes. Zygote 2001; 9: 15-23.

56. Sun FZ, Hoyland J, Huang X, Mason W, Moor RM. A comparison of intracellular changes in porcine eggs after fertilization and electroactivation. Development 1992; 115: 947-956.

57. Nussbaum DJ, Prather RS. Differential effects of protein synthesis inhibitors on porcine oocyte activation. Mol Reprod Dev 1995; 41: 70-75.

58. Cuthbertson KS, Whittingham DG, Cobbold PH. Free Ca2+ increases in exponential phases during mouse oocyte activation. Nature 1981; 294: 754-757.

59. O'Neill GT, McDougall RD, Kaufman MH. Ultrastructural analysis of abnormalities in the morphology of the second meiotic spindle in ethanol-induced parthenogenones. Gamete Res 1989; 22: 285-299.

60. Vincent C, Cheek TR, Johnson MH. Cell cycle progression of parthenogenetically activated mouse oocytes to interphase is dependent on the level of internal calcium. J Cell Sci 1992; 103: 389-396.

61. Collas P, Fissore R, Robl JM, Sullivan EJ, Barnes FL. Electrically induced calcium elevation, activation, and parthenogenetic development of bovine oocytes. Mol Reprod Dev 1993; 34: 212-223.

62. Ozil JP. The parthenogenetic development of rabbit oocytes after repetitive pulsatile electrical stimulation. Development 1990; 109: 117-127.

63. Whittingham DG, Siracusa G. The involvement of calcium in the activation of mammalian oocytes. Exp Cell Res 1978; 113: 311-317.

64. Marcus GJ. Activation of cumulus-free mouse oocytes. Mol Reprod Dev 1990; 26: 159-162.

65. O'Neill GT, Rolfe LR, Kaufman MH. Developmental potential and chromosome constitution of strontium- induced mouse parthenogenones. Mol Reprod Dev 1991; 30: 214-219.

66. Liu L, Trimarchi JR, Keefe DL. Haploidy but not parthenogenetic activation leads to increased incidence of apoptosis in mouse embryos. Biol Reprod 2002; 66: 204-210.

67. Wakayama T, Perry AC, Zuccotti M, Johnson KR, Yanagimachi R. Full-term development of mice from enucleated oocytes injected with cumulus cell nuclei. Nature 1998; 394: 369-374.

68. Swann K. Thimerosal causes calcium oscillations and sensitizes calcium-induced calcium release in unfertilized hamster eggs. FEBS Lett 1991; 278: 175-178.

69. Fissore RA, Robl JM. Sperm, inositol trisphosphate, and thimerosal-induced intracellular Ca2+ elevations in rabbit eggs. Dev Biol 1993; 159: 122-130.

70. Machaty Z, Wang WH, Day BN, Prather RS. Complete activation of porcine oocytes induced by the sulfhydryl reagent, thimerosal. Biol Reprod 1997; 57: 1123-1127.

71. Cheek TR, McGuinness OM, Vincent C, Moreton RB, Berridge MJ, Johnson MH. Fertilisation and thimerosal stimulate similar calcium spiking patterns in mouse oocytes but by separate mechanisms. Development 1993; 119: 179-189.

72. Machaty Z, Wang WH, Day BN, Prather RS. Calcium release and subsequent development induced by modification of sulfhydryl groups in porcine oocytes. Biol Reprod 1999; 60: 1384-1391.

73. Susko-Parrish JL, Leibfried-Rutledge ML, Northey DL, Schutzkus V, First NL. Inhibition of protein kinases after an induced calcium transient causes transition of bovine oocytes to embryonic cycles without meiotic completion. Dev Biol 1994; 166: 729-739.

74. Szollosi MS, Kubiak JZ, Debey P, de Pennart H, Szollosi D, Maro B. Inhibition of protein kinases by 6-dimethylaminopurine accelerates the transition to interphase in activated mouse oocytes. J Cell Sci 1993; 104: 861-872.

75. Moses RM, Kline D, Masui Y. Maintenance of metaphase in colcemid-treated mouse eggs by distinct calcium- and 6-dimethylaminopurine (6-DMAP)-sensitive mechanisms. Dev Biol 1995; 167: 329-337.

76. Moos J, Kopf GS, Schultz RM. Cycloheximide-induced activation of mouse eggs: effects on cdc2/cyclin B and MAP kinase activities. J Cell Sci 1996; 109: 739-748.

77. Alberio R, Kubelka M, Zakhartchenko V, Hajduch M, Wolf E, Motlik J. Activation of bovine oocytes by specific inhibition of cyclin-dependent kinases. Mol Reprod Dev 2000; 55: 422-432.

78. Choi YH, Love CC, Varner DD, Thompson JA, Hinrichs K. Activation of cumulus-free equine oocytes: effect of maturation medium, calcium ionophore concentration and duration of cycloheximide exposure. Reproduction 2001; 122: 177-183.

79. Wang WH, Machaty Z, Abeydeera LR, Prather RS, Day BN. Parthenogenetic activation of pig oocytes with calcium ionophore and the block to sperm penetration after activation. Biol Reprod 1998; 58: 1357-1366.

80. Polejaeva IA, Chen SH, Vaught TD, Page RL, Mullins J, Ball S, Dai Y, Boone J, Walker S, Ayares DL, Colman A, Campbell KH. Cloned pigs produced by nuclear transfer from adult somatic cells. Nature 2000; 407: 86-90.

81. Onishi A, Iwamoto M, Akita T, Mikawa S, Takeda K, Awata T, Hanada H, Perry AC. Pig cloning by microinjection of fetal fibroblast nuclei. Science 2000; 289: 1188-1190.

82. Betthauser J, Forsberg E, Augenstein M, Childs L, Eilertsen K, Enos J, Forsythe T, Golueke P, Jurgella G, Koppang R, Lesmeister T, Mallon K, Mell G, Misica P, Pace M, Pfister-Genskow M, Strelchenko N, Voelker G, Watt S, Thompson S, Bishop M. Production of cloned pigs from in vitro systems. Nat Biotechnol 2000; 18: 1055-1059.

83. Rinaudo P, Pepperell JR, Buradgunta S, Massobrio M, Keefe DL. Dissociation between intracellular calcium elevation and development of human oocytes treated with calcium ionophore. Fertil Steril 1997; 68: 1086-1092.

84. Tombes RM, Simerly C, Borisy GG, Schatten G. Meiosis, egg activation, and nuclear envelope breakdown are differentially reliant on Ca2+, whereas germinal vesicle breakdown is Ca2+ independent in the mouse oocyte. J Cell Biol 1992; 117: 799-811.

85. Eppig JJ, Schultz RM, O'Brien M, Chesnel F. Relationship between the developmental programs controlling nuclear and cytoplasmic maturation of mouse oocytes. Dev Biol 1994; 164: 1-9.

86. Yang X, Kubota C, Suzuki H, Taneja M, Bols PE, Presicce GA. Control of oocyte maturation in cows--biological factors. Theriogenology 1998; 49: 471-482.

87. Salamone DF, Damiani P, Fissore RA, Robl JM, Duby RT. Biochemical and developmental evidence that ooplasmic maturation of prepubertal bovine oocytes is compromised. Biol Reprod 2001; 64: 1761-1768.

88. Capco DG. Molecular and biochemical regulation of early mammalian development. Int Rev Cytol 2001; 207: 195-235.

89. Liu L, Oldenbourg R, Trimarchi JR, Keefe DL. A reliable, noninvasive technique for spindle imaging and enucleation of mammalian oocytes. Nat Biotechnol 2000; 18: 223-225.

Chapter 18

IMPLANTATION: LESSONS FROM A PRIMATE MODEL

Asgerally T. Fazleabas
University of Illinois, Chicago, IL, USA

INTRODUCTION

Implantation is a complex spatio-temporal interaction between the genotypically different embryo and mother. This is a highly coordinated process that is activated when the trophoblast cells of the embryo establish contact with the maternal endometrium. The initiation of pregnancy requires a precisely timed synchrony between endometrial development and the implanting blastocyst. Under the influence of ovarian steroids, the uterine endometrium undergoes profound modifications in cellular differentiation. In primates, at the appropriate phase of the menstrual cycle, the uterus becomes "receptive" and enables the blastocyst to attach. This "receptive window" is initially dependent on estrogen and progesterone. However, further morphological and biochemical changes are induced within the uterus by signals from the developing embryo and following trophoblast invasion.

Understanding the factors involved in embryo-maternal cross talk is crucial for reproductive medicine. However, since there are ethical limitations to studying this process in women we have developed the baboon as a non-human primate model to obtain physiologically relevant data.

UTERINE RECEPTIVITY

The window of receptivity is the period within the uterus that facilitates interactions with the developing embryo for successful implantation. Initially the embryo will appose, attach to, and intrude into the luminal epithelium. After the epithelial basement membrane is penetrated by the embryo, the invading trophoblasts move down into the stromal compartment. On the maternal side, stromal invasion of the embryo is associated with stromal extracellular matrix remodeling, decidualization, angiogenesis, and immunomodulation. Effective biochemical communication between the embryo and the uterus is essential for this highly coordinated, controlled and temporary association (1).

The concept of endometrial receptivity was first established by Psychoyos in the rat and then extended to other species (2-4). Although estrogen and progesterone have long been believed to be essential for developing an appropriate endometrial environment for blastocyst implantation, it is now evident that these effects are further modulated by peptide hormones and peptide growth factors secreted by a variety of cell types within the uterine endometrium. In the baboon, we have divided uterine receptivity into three distinct phases. Phase I is regulated by estrogen and progesterone and is evident between days 8 and 10 post-ovulation of the normal menstrual cycle. The morphological, biochemical and cell biological changes associated with this phase have been reviewed previously (5).

The second phase (Phase II) of uterine receptivity is induced by blastocyst "signals" superimposed on the estrogen/progesterone primed receptive endometrium. In the baboon, and perhaps also in the human, the primary embryonic factor that induces these responses is chorionic gonadotrophin (18). This phase is associated with functional and morphological changes in the endometrium that are distinct from those observed at a comparable time of a non-pregnant cycle (i.e., Phase I of uterine receptivity).

The final phase (Phase III) of uterine receptivity is initiated following attachment and implantation. A universal response is the significant increase in the permeability of the subepithelial capillaries surrounding the blastocyst (2,4). In primates the morphological changes associated with implantation have been extensively studied and elegantly reviewed (6,7). In general together with glandular hypertrophy, stromal cell decidualization is initiated and is accompanied by increased secretion of extracellular matrix proteins.

EMBRYONIC SIGNALS

The blastocyst develops from the morula-stage embryo, and the process requires active gene transcription and translation from the embryonic genome (8). This occurs when the embryo is approximately 4 days of age and composed of 16-32 cells. The developing blastocyst is composed of three different structures within the surrounding of the zona pellucida: 1) the peripherally situated, trophectoderm or trophoblast cells; 2) the embryonic cell proper, inner cell mass, and 3) the fluid-filled blastocoele cavity (9). On day 6, the blastocyst hatches from the zona pellucida. Polar trophectoderm near the site that apposes and attaches to the uterine luminal epithelium differentiates into two different cell types, the cytotrophoblast and syncytiotrophoblast, during penetration between luminal epithelial cells (10). Cytotrophoblasts surround the blastocystic cavity and lie close to the inner cell mass and the syncytiotrophoblast. The syncytiotrophoblast, a multinucleated non-dividing cell that develops from the cytotrophoblast, synthesizes and secretes chorionic gonadotrophin and other placental hormones.

In a cycle in which conception occurs, adequate progesterone secretion from the corpus luteum must be maintained to support the pregnancy (11). This critical process of maternal recognition of pregnancy requires the signal from the developing conceptus that will act as a luteotropic and/or a luteolytic inhibitor prior to the time that normal luteal regression would occur. This biochemical communication is essential for the uninterrupted synthesis and release of progesterone from the corpus luteum. In primates, chorionic gonadotrophin secreted from the trophoblast acts as a LH superagonist (12), extending the life span of the corpus luteum and its ability to continue to secrete progesterone. In addition to rescuing the corpus luteum from regression, the embryonic signals also directly or indirectly modulate the receptive endometrium that will facilitate the implantation process (13).

The idea that embryo-derived factors may directly or indirectly influence endometrial receptivity and implantation in primates is substantiated by several lines of evidence. Studies in the rhesus monkey indicate that endometrial physiology during the midluteal phase in the presence of the conceptus is discernibly different from that in nonfecund midluteal phase (14). Our studies in the baboon demonstrate that chorionic gonadotrophin, when infused in a manner that mimics blastocyst transit, has physiological effects on the three major cell types in the uterine endometrium (i.e., luminal and glandular epithelium and stromal fibroblasts).

MODULATION OF THE PRIMATE ENDOMETRIUM BY CHORIONIC GONADOTROPHIN

Luminal Epithelium

An early maternal response to pregnancy in the luminal epithelium of primate is the formation of the epithelial plaque (15). This response is characterized by hypertrophy of the surface epithelium and cells in the neck glands that round up and form acinar clusters (16,17). The induction of the plaque response requires a synergism between the ovary and the endometrium in response to chorionic gonadotrophin (18). In the baboon, the plaque is restricted to the epithelium immediately adjacent to the implantation site (17,19) whereas in the macaque it is much more extensive (17,20). The large nuclei found in the plaque have been described (20). They were probably polyploid, with very large nuclei as well as examples of bi-, tri- and multinucleated cells being observed (21). The nucleolar channels have also been described by Enders (17), who compared them to those seen in human uterine glandular cells on day 18 of the menstrual cycle. Rossman (22), in his detailed histological study of deciduomata, illustrated examples of multinucleated giant cells as well as cell degeneration. Evidence of phagocytosis, in the form of large vacuoles and the engulfment of whole cells suggests that cell fusion may, on occasions, become lethal to one participant in the process (21).

The function of plaque formation is not clear; it is not involved in the actual process of implantation, that is, the penetration of the luminal epithelium by the blastocyst, but may provide nutrition by means of the intracellular glycogen (20,22). The importance of histiotrophic nutrition with reference to the human embryo has recently been highlighted (23) and it was suggested that glandular secretions may be taken up by the syncytiotrophoblast and yolk sac epithelia prior to the establishment of haemiotrophic nutrition. It has also been suggested that the plaque response might stimulate vascular enlargement over a broad area extending beyond the developing placenta (20). This would bring about a precocious development of the maternal vasculature and in doing so accelerate the development of the placenta as the lacunae communicate with the enlarged vascular bed.

Glandular Epithelium

Chorionic gonadotrophin induces morphologic changes in endometrial glandular structures that resemble those observed in the pregnant baboon of the same gestational age (24). Glands in the functionalis and basalis layers are distended and convoluted. The glandular response to chorionic gonadotrophin infusion is characterized by a marked increase in trascriptional and posttranslational modulation of glycodelin (18). Synthesis of glycodelin by the glandular epithelium parallels the rise and decline of chorionic gonadotrophin in the peripheral circulation (25). Glycodelin is associated with immunosuppression, inhibition of sperm-egg binding, and induction of epithelial cell differentiation (26). However, the true function of glycodelin in the implantation process remains to be elucidated.

Stromal Fibroblasts

In addition to the characteristic stromal edema due to increased vascular permeability (24), the primary effect of chorionic gonadotrophin on stromal fibroblasts is the induction of α-smooth muscle actin (18,27).

In the baboon, the expression of α-smooth muscle actin appears to be hormonally regulated; previous studies have shown it to be absent in the smooth muscle cells of the myometrium and blood vessels in ovariectomized animals, but appears following estrogen treatment (27). During the menstrual cycle, it is not found in the stromal fibroblasts but by day 14 of pregnancy it is apparent in stromal cells beneath the luminal epithelium, and can be demonstrated by immunocytochemistry (21). There is, coincident with this, a change in the morphology of the cells which develop the features of decidualization. They become larger in size with a more rounded profile and at the ultrastructural level, many extensions and processes, which contain actin filaments. There is also evidence of increased biosynthetic activity, with many strands of endoplasmic reticulum and Golgi vacuoles. The decidual cells observed in our studies also produced increased amounts of extracellular matrix as seen by the dense material surrounding the cells under the luminal epithelium (21). This may coincide with an increase in collagen/laminin receptors together with their specific extracellular matrix molecules as has been observed in early pregnancy in the baboon (28), with an increase in α_1, α_3, α_6, β_1 and $\alpha_v\beta_3$ as in the human (29). At this stage in pregnancy, the stromal cells are packed closely together, later, however, they become more spaced out and rounder, and a distinct pericellular basement membrane can be seen around each one from about day 40 of pregnancy (21). Thus, the increased expression of actin occurs in concert with changes in integrin expression and extracellular matrix secretion suggesting alterations in signal transduction pathways (18,28,30). Similar changes in the actin cytoskeleton have previously been described in the differentiation of granulosa cells (31) and has been associated with cellular remodeling by actin-rich myofibroblasts in breast tumors (32). Such myofibroblasts have been shown to synthesize extracellular matrix components including collagen types I, III, V, fibronectin, vimentin and oncofetal fibronectin (32).

Cytoskeletal proteins play a critical role in mitosis, cell growth, cell motility and inhibition of apoptosis (33,34). We have hypothesized that the induction of α-smooth muscle actin by chorionic gonadotrophin may be essential to decrease the progesterone regulated proliferation process in these cells and initiate the differentiation process which is a prerequisite for decidualization (30).

In conclusion, our data have demonstrated that chorionic gonadotrophin has physiologic effects on the primate endometrium during the window of uterine receptivity. These responses to chorionic gonadotrophin occurs in synergism with progesterone. Antagonism of progesterone action by progesterone receptor antagonists during the infusion of chorionic gonadotrophin, markedly affects epithelial and stromal responses induced by chorionic gonadotrophin during the window of uterine receptivity (35).

ENDOMETRIAL RESPONSES TO IMPLANTATION

Primate blastocysts implant in a simplex uterus with the inner cell mass oriented adjacent to the endometrium. In non-human primate species that have been studied to date, syncytial trophoblast forms near the inner cell mass and penetrates the uterine epithelium by intruding between uterine epithelial cells (36,37).

Implantation in the baboon (7,21) is similar to that described for the rhesus macaque (6,38). Specialized villi; known as anchoring villi, facilitate the attachment of the placenta to the uterine wall and provide the source of the migratory cell population that invade into the maternal endometrium.

Coincident with this invasive process, the stromal fibroblasts are enlarged compared to the non-pregnant precursors. Decidualization, which involves the transformation of stromal fibroblasts to decidual cells, is the major change that occurs in the primate endometrium after conception. In primates and rodents, the uterine endometrial stromal cells differentiate to decidual cells following the establishment of pregnancy. Decidual cells play an important role in implantation and provide nutritional support for embryo. Decidual cells are also believed to produce factors that control trophoblast invasion and protect the embryo from maternal immune rejection. During the process of decidualization in the primate, fibroblast-like stromal cells change morphologically into polygonal cells and begin to express specific decidual proteins (39,40). This is manifested by the down-regulation of α-smooth muscle actin expression and the induction of insulin-like growth factor binding protein-1 (41,42).

Previous studies in the baboon have clearly demonstrated that insulin-like growth factor binding protein-1 gene expression in the endometrium is a conceptus mediated response (43). Subsequent studies *in vitro* established that insulin-like growth factor binding protein-1 gene expression in decidualizing stromal fibroblasts requires the presence of both hormones and cAMP (42). This induction is associated with a concomitant decrease of α-smooth muscle actin expression *in vivo* (27) and *in vitro* (42). Since interleukin-1β (IL-1β) is expressed both in the progestational endometrium and in trophoblast cells (44-46) we evaluated IL-1β as one possible factor that could influence differentiation of stromal cells into decidual cells. IL-1β has been reported to be actively involved in fetal-maternal interactions (47-

49), but its role in decidualization has not been clarified. In addition, IL-1 can modulate changes in the cytoskeleton (50,51) and induce cyclooxygenase-2 gene expression (52-54). Our data would suggest that IL-1β activates a signaling pathway that induces cyclooxygenase-2 expression followed by an increase in insulin-like growth factor binding protein-1 expression (55). Coincident with the induction of cyclooxygenase-2, IL-1β also induces metalloproteinase-3 (MMP-3) expression in stromal fibroblasts (56). We hypothesize that the local action of MMP-3 dissociates the surrounding extracellular matrix resulting in the loss of focal adhesion complexes and the alteration in the actin cytoskeleton. This dissociation is the necessary pre-requisite for decidualization and insulin-like growth factor binding protein-1 induction. Thus, the complex signaling pathways activated during implantation may play a critical role in maintaining the appropriate homeostatics required for decidualization and trophoblast invasion.

CONCLUSIONS

Implantation is a complex process that appears to have unique manifestations in each of the species studied to date. Much of our understanding of this intricate process comes from elegant studies in rodent models. In the case of primates and humans, limited data are available but constraints in non-human primates with regards to cost and low fecundity rates does not permit extensive investigations. The obvious moral and ethical limitations in humans preclude obtaining any *in vivo* data. Yet, infertility and pregnancy wastage affects one in every nine couples in the Western world. Thus, we have chosen to study embryo-maternal cross talk in a relevant non-human primate model, the baboon, with the hope that these basic studies will provide crucial information that could be useful in reproductive medicine to improve pregnancy rates in infertile couples.

Our data would suggest that uterine receptivity in the primate can be divided into three phases. Phase I is modulated by estrogen and progesterone secreted by the ovary following ovulation. Further modulation of this critical phase is dependent on the presence of an embryo. Thus, in a conception cycle, the embryonic signal is superimposed on the estrogen-progesterone-primed endometrium to induce functional changes that facilitate the implantation process. Once implantation is initiated, Phase III is activated and this is associated with profound modifications of the stromal compartment and the decidualization process.

354

REFERENCES

1. Lessey B.A. Endometrial receptivity and the window of implantation. Baillieres Best Pract Res Clin Obstet Gynaecol 2000; 14:775-788.
2. Psychoyos A. "Endocrine Control of Egg Implantation," In Handbook of Physiology Female Reproductive System Endocrinology Section 7, Volume 2, 1973, American Physiological Society. pp 187-215 Part 2.
3. Psychoyos A. Uterine receptivity for nidation. Ann New York Acad Sci, 1986; 476:36-42.
4. Psychoyos A. "The Implantation Window: Basic and Clinical Aspects." In Perspectives in Assisted Reproduction, T. Mori, T. Aono, T. Tominaga, M. Hiroi, M. eds. Ares Serono Symposia, Volume IV, 1993, pp 57-62.
5. Fazleabas A.T., Hild-Petito S., Verhage H.G. The primate endometrium: morphological and secretory changes during early pregnancy. Semin Reprod Endocrinol 1995; 13:120-132.
6. Enders A.C. "Overview of the Morphology of Implantation in Primates." In In Vitro Fertilization and Embryo Transfer in Primates R.L. Wolf, R.L. Stouffer, R.M. Brenner, eds. Springer-Verlag, 1993. pp 145-157.
7. Enders A.C., Blakenship T.N., Fazleabas A.T., Jones C.J.P. Structure of anchoring villi and the trophoblastic shell in the human, baboon and macaque. Placenta 2001; 22:284-303.
8. Schultz R.M. "Blastocyst." In Encyclopedia of Reproduction, E. Knobil, J.D. Neill, eds. Academic Press, 1998. p 370.
9. O'Rahilly R., Muller F. Week 1 (stages 2-4): morula: blastocyst. Human Embryology and Teratology. New York: John Wiley & Sons 1996. p 33.
10. Bentin-Ley U., Hort T., Sjogren A., Sorensen S., Falck Larsen J., Hamberger L. Ultrastructure of human blastocyst-nedometrial interactions in vitro. J Reprod Fertil 2000; 120:120-137.
11. Niswender G.D., Juengel J.L., Silva P.J., Rollyson M.K., McIntush E.W. Mechanisms controlling the function and life span of the corpus luteum. Physiol Rev 2000; 80:1-29.
12. Rao C.V. An overview of the past, present, and future of nongonadal LH/hCG actions in reproductive biology and medicine. Semin Reprod Med 2001; 19:7-17.
13. Carson D.D., Bagchi I., Dey S.K., Enders A.C., Fazleabas A.T., Lessey B.A., Yoshinaga K. Embryo Implantation and Uterine Receptivity. Develop Biol 2000; 223:217-237.
14. Ghosh D., Sengupta J. Recent developments in endocrinology and paracrinology of blastocyst implantation in the primate. Hum Reprod Update 1998; 4:153-168.
15. Enders A.C., Lantz K.C., Petersen P.E., Hendrickx A.G. From blastocyst to placenta: the morphology of implantation in the baboon. Hum Reprod Update 1997; 3:561-573.
16. Tarara R., Enders A.C., Hendrickx A.G., Gulamhusein N., Hodges J.K., Hearn J.P., Eley R.B., Else J.G. Early implantation and embryonic development of the baboon stages 5, 6 and 7. Anat Embryol 1987; 176:267-275.
17. Enders A.C. Structural responses of the primate endometrium to implantation. Placenta 1991; 12:309-325.
18. Fazleabas A.T., Donnelly K.M., Srinivasan S., Fortman J.D., Miller J.B. Modulation of the baboon (Papio anubis) uterine endometrium by chorionic gonadotrophin during the period of uterine receptivity. Proc Natl Acad Sci (USA) 1999; 96:2453-2458.
19. Enders A.C., Schlafke S. "Implantation in Nonhuman Primates and in the Human." In Comparative Primate Biology. A.R. Liss ed. Reproduction and Development, 1986. pp 291-310.
20. Enders A.C., Welsh, A.O., Schlafke S. Implantation in the rhesus monkey: endometrial responses. Am J Anat 1985; 173:147-169.

21. Jones C.J.P., Fazleabas A.T. Ultrastructural observations on the formation of epithelial plaque and stromal cell transformation by post-ovulatory chorionic gonadotrophin treatment in the baboon. Hum Reprod 2001; 16:2680-2690.

22. Rossman I. The deciduomal reaction in the rhesus monkey (Macaca mulatta). 1. The epithelial proliferation. Am J Anat 1940; 66:277-365.

23. Burton G.J., Hempstock J, Jauniaux E. Nutrition of the human fetus during the first trimester – a review. Placenta 22, Supplement A, Trophoblast Research 2001; 15:S70-S76.

24. Hild-Petito S., Donnelly K.M., Miller J.B., Verhage H.G., Fazleabas A.T. A baboon (Papio anubis) simulated-pregnant model: cell specific expression of insulin-like growth factor binding protein-1 (IGFBP-1), type I IGF receptor (IGF-IR) and retinol binding protein (RBP) in the uterus. Endocrine 1995; 3:639-651.

25. Hausermann, H.M., Donnelly K.M., Bell S.C., Verhage H.G., Fazleabas A.T. Regulation of the Glycosylated β-Lactoglobulin Homologue, Glycodelin [Placental Protein 14 (PP$_{14}$)] in the Baboon Uterus. J Clin Endocrinol Metab 1998; 83:1226-1233.

26. Seppala M., Kostinen H., Koistinen R. Glycodelins. Trends Endocrinol Metab 2001; 12:111-117.

27. Christensen S., Verhage H.G., Nowak G., de Lanerolle P., Fleming S., Bell S.C., Fazleabas A.T., Hild-Petito S. Smooth muscle myosin II and α smooth muscle actin expression in the baboon (Papio anubis) uterus is associated with glandular secretory activity and stromal cell transformation. Biol Reprod 1995; 53:596-606.

28. Fazleabas A.T., Bell S.C., Fleming, S., Sun, J., Lessey, B.A. Distribution of integrins and the extracellular matrix proteins in the baboon endometrium during the menstrual cycle and early pregnancy. Biol Reprod 1997; 56:348-356.

29. Lessey BA. The role of the endometrium during embryo implantation. Hum Reprod 2000; Suppl 6:39-50.

30. Kim J.J., Jaffe R.C., Fazleabas A.T. Blastocyst invasion and the stromal response in primates. Hum Reprod 1999; 14(Suppl. 2):45-55.

31. Kranen R.W., Overes H.W.T.M., Kloosterboer H.G., Poels L.G. The expression of cytoskeleton proteins during the differentiation of rat granulosa cells. Hum Reprod 1993; 8:24-29.

32. Brouty-Boye D., Raux H., Azzarone B., Tamboise A., Tamboise E., Beranger S., Magnien V., Pihan I., Zardi L., Israel L. Fetal myofibroblast-like cells isolated from post-radiation fibroblasts in human breast cancer. Int J Cancer 1991; 47:697-702.

33. Mashima T., Naito M., Tsuruo T. Caspase-mediated cleavage of cytoskeleton actin plays a positive role in the process of morphological apoptosis. Oncogene 1999; 18:2423-2430.

34. Suarez-Huerta N., Lecocq R., Mosselman R., Galand P., Dumont J.E., Robaye B. Myosin heavy chain degradation during apoptosis in endothelial cells. Cell Prolif 2000; 33:101-114.

35. Banaszak S., Brudney A., Donnelly K., Chai D., Chwalisz K., Fazleabas A.T. Modulation of the action of chorionic gonadotropin in the baboon (Papio anubis) uterus by a progesterone receptor antagonist (ZK 137.316). Biol Reprod 2000; 63:820-825.

36. Enders A.C., Hendrickx A.G., Schlafke S. Implantation in the rhesus monkey: Initial penetration of endometrium. Am J Anat 1983; 167:275-2981.

37. Smith C.A., Moore H.D.M., Hearn J.P. The ultrastructure of early implantation in the marmoset monkey (Callithrix jaccus). Anat Embryol 1987; 175:399-410.

38. Enders A.C., Lantz K.C., Peterson P.E., Hendrickx A.G. From blastocyst to placenta: The morphology of implantation in the baboon. Hum Reprod Update 1997; 3:561-573.

39. Fazleabas A.T., Hild-Petito S., Verhage H.G. The primate endometrium: morphological and secretory changes during early pregnancy. Sem Reprod Endocrinol 1995; 13:120-132.

40. Fazleabas A.T., Kim J.J., Donnelly K.M., Verhage H.G. "Embryo-Maternal Dialogue in the Baboon (Papio anubis)." In Proceedings of the Serono Symposium on Embryo

356

Implantation: Cellular, Molecular and Clinical Aspects D.D. Carson, ed. Springer-Verlag, 1999. pp. 202-209.

41. Tarantino S., Verhage H.G., Fazleabas A.T. regulation of insulin-like growth factor binding proteins (IGFBPs) in the baboon (Papio anubis) uterus during early pregnancy. Endocrinology 1992; 130:2354-2362.

42. Kim J.J., Jaffe R.C., Fazleabas A.T. Comparative studies on the in vitro decidualization process in baboons (Papio anubis) and humans. Biol Reprod 1998; 59:160-168.

43. Fazleabas A.T., Jaffe R.C., Verhage H.G.,Waites G., Bell S.C. An insulin-like growth factor binding protein (IGF-BP) in the baboon (Papio anubis) endometrium: synthesis, immunocytochemical localization and hormonal regulation. Endocrinology 1989; 124:2321-2329.

44. Kauma S., Matt D., Strom S., Eierman D., Tuner T. Interleukin 1-β (IL-1β), human leukocyte antigen HLA-DR α, and transforming growth factor -β (TGFβ) expression in endometrium, placenta and placental membranes. Am J Obstet 1990; 163:1430-1437.

45. O'Neill L.A., Greene C. Signal transduction pathways activated by the IL-1 receptor family: ancient signaling machiner in mammals, insects, and plants. J Eukocyte Biol 1998; 63:650-657.

46. Hu H.L., Yang Y., Hunt J.S. Differential distribution of interleukin-1α and interleukin-1β proteins in human placentas. J Reprod Immunol 1992; 22:257-268.

47. Simon C., Frances A., Piquette G.N., Hendrickson M., Milki A., Polan M.L. Interleukin-1 system in the materno-trophoblast unit in human implantation: immunohistochemical evidence for autocrine/paracrine function. J Clin Endocrinol Metab 1994; 78:847-854.

48. De los Santos M., Mercader A., Frances A., Portoles E., Remohi J., Pellicer A., Simon C. Role of endometrial factors in regulating secretion of components of the immunoreactive human embryonic interleuking-1 during embryonic development. Biol Reprod 1996; 54:563-574.

49. Qwarnstrom E.E., Page R.C., Gillis S., Dower S.K., 1988. Binding, internalization and intracellular localization of interleukin-1 β in human diploid fibroblasts. J Biol Chem 1988; 263:8261-8269.

50. Qwarnstrom E.E., MacFarlane S.A., Page R.C., Dower S.K. Interleukin-1 β induces rapid phosphorylation of talin: a possible mechanism for modulation of fibroblast focal adhesion. Proc. Natl. Acad. Sci. USA 1991; 88:1232-1236.

51. Singh R., Wang B., Shirvaikar A., Khan S., Schelling J.R., Konieczkowski M., Sedor J.R. The IL-1 receptor and Rho directly associate to drive cell activation in inflammation. J Clin Invest 1999; 103:1561-1570.

52. Kniss D.A., Zimmerman P.D., Garver C.L., Fertel R.H. Interleukin-1 receptor antagonist blocks interleukin-1-induced expression of cyclooxygenase-2 in endometrium. Am J Obstet Gynecol 1997; 177:559-567.

53. Huang J.C., Liu D.Y., Yadollahi S., Wu K.K., Dawood M.Y. Interleukin-1β induces cyclooxygenase-2 gene expression in cultured endometrial stromal cells. J Clin Endocrinol Metab 1998; 83:538-541.

54. Guan Z., Buckman S.Y., Miller B.W., Springer L.D., Morrison A.R. Interleukin -1β-induced cyclooxygenase-2 expression requires activation of both c-Jun NH_2 terminal kinase and p38 MAPK signal pathways in rat mesengial cells. J Biol Chem 1998; 273:28670-28676.

55. Strakova Z., Srisuparp S., Fazleabas A.T. Interleukin-1β (IL-1β) induces the expression of insulin-like growth factor binding protein (IGFBP-1) during decidualization in the primate, Endocrinology 2000; 141:4664-4670.

56. Strakova Z, Srisuparp S., Fazleabas A.T. A role of IL-1β during in vitro decidualization in the primate. J Reprod Immunol 2002 ; 55:35-47.

Chapter 19

ASSISTED REPRODUCTION: TECHNIQUES AND PARTICIPANTS

Christoph R. Loeser[1], Thomas Stalf[2], Hans-Christian Schuppe[1], and Wolf-Bernhard Schill[1]
[1]*Center for Dermatology and Andrology, University of Giessen, Giessen, Germany*
[2]*Institute for Reproductive Medicine, Giessen, Germany*

INTRODUCTION

Oocytes of rabbits and mice, successfully cultured *in vitro* in the first half of the twentieth century, laid the foundation for assisted reproductive techniques decades later. *In vitro* fertilization (IVF) studies in rabbits resulted in the first birth of a normal offspring in 1959. Since then, advances in experimental reproductive technology, leading to new insights into sperm capacitation, oocyte maturation and sperm-egg interaction, have tremendously increased our knowledge about the fertilization process in humans.

Infertility is a problem affecting about 10% of all couples in the western world. In 1978, the British gynecologist Patrick Steptoe and the physiologist Robert Edwards reported about the first baby being born from culture of a fertilized oocyte that had developed to an 8-cell stage embryo *in vitro* (1). Although embryonic transfer in rabbits was successfully performed as early as 1890, the work of Steptoe and Edwards was the beginning of an explosive development in the treatment of human infertility. Assisted reproductive techniques (ART) have since emerged as successful treatment. Today, IVF is a standard procedure in many centers all over the world, with pregnancy rates ranging from 10-40% per embryo transfer (ET). A few hundred thousand children have been born after treatment with ART. At present, the take-home baby rate is 15-30% per treatment cycle (2).

PARTICIPANTS IN ART

Concerning the participants in ART, females account for 30%, males for 30%, both women and men for 30%, and 10% of the couples suffer from idiopathic infertility (3). The majority of female participants are with tubal factor infertility. The fallopian tubes in this condition may be blocked on one side or bilaterally, preventing or inhibiting passage of the oocyte. In most cases, these disturbances occur secondary to an infection; less than 5 % of females have blocked tubes from birth. Infertility in women can also be caused by other gynecological diseases, such as polycystic ovary syndrome or endometriosis. The latter condition, referred to as retrograde menstruation (4), is characterized by the occurrence of ectopic endometrial tissue, i.e., invasion of the myometrium by the mucous membrane lining the uterus. The ectopic endometrial tissue is subjected to the same endocrinological, metabolic, and catabolic processes, as the orthotopic uterine cells. As endometriosis may be accompanied by adnexal adhesions, fertility is often directly impaired by this disease. Moreover, poor oocyte quality and low implantation capacity have been discussed in this group of patients (5). Compared with these common causes of infertility, presence of sperm antibodies in the female patients is rare.

Pathological sperm findings play an important role as a cause of infertility, accounting for 50% of the total male infertile population. A wide range of etiologic factors and pathogenetic mechanisms may cause the mere symptom of male infertility. Different anatomical levels as well as specific periods of life have to be considered with regard to disturbances of the complex development and regulation of male reproductive functions. Diagnostic categories of male infertility include disorders of the hypothalamic-pituitary-gonadal axis such as hypogonadotropic hypogonadism, congenital abnormalities affecting the testes, i.g., testicular maldescent or chromosomal abnormalities such as Klinefelter's syndrome (47, XXY karyotype), acquired damage of spermatogenesis due to drugs, life-style or environmental factors, congenital agenesia of the vasa deferentia (and/or seminal vesicles), acquired blockage of sperm transport as a cause of obstructive azoospermia, and sexual and/or ejaculatory dysfunction (6, 7). Moreover, varicocele and infectious or inflammatory disease of the genital tract should not be overlooked as most common and treatable causes of male infertility. In some men, impaired spermatogenic function reflects carcinoma-in-situ or even testicular cancer.

In spite of modern diagnostic methods, underlying causes of pathological semen profiles (Figure 1b-d) remain unclear in up to one third of men undergoing fertility check-up (6). Infertility has to be referred to as idiopathic in approximately 10% of barren couples. As some of these patients can be successfully treated with assisted reproductive techniques, idiopathic infertility is considered an indication for IVF. ART are contraindicated in cases of missing maturable oocytes, i.e., in women beyond the menopause or those who have undergone ovariectomy. A prerequisite is the presence of spermatozoa, at least spermatids, from testicular biopsies. Restrictions may be imposed by the legal situation in

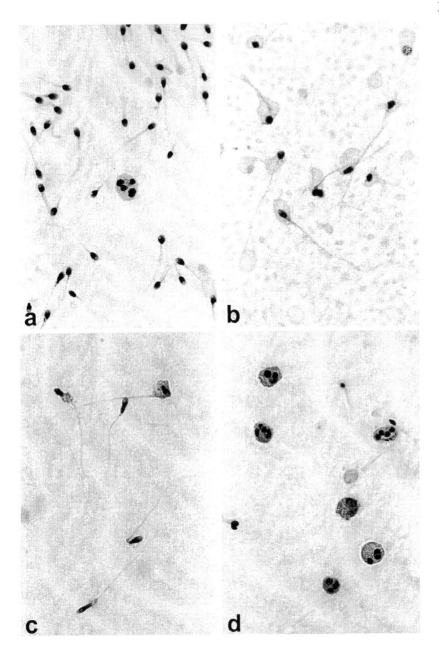

Figure 1. Morphological semen analysis: a, Spermatozoa with morphologically normal heads and some with minor deviations; note a single neutrophilic granulocyte; b, spermatozoa with high degree of acrosomal deficiencies; note double heads and additional defects of midpiece and flagellum; c, hyperelongated sperm heads, some with concomitant acrosomal deficiency; note excessive cytoplasm and bent tails; d, round cells including multinuclear immature germ cells (spermatids) as well as neutrophils and a macrophage.

many countries, although at present, the IVF procedure is prohibited only in Libya.

TECHNIQUES IN REPRODUCTIVE MEDICINE

Collection Of Gametes

ART are generally based on the availability of gametes, both oocytes and spermatozoa. Women from barren couples are usually subjected to hormonal stimulation, aimed at maturation of one or more oocytes. As stimulation therapy is often crucial to later oocyte quality, this step is of utmost importance.

Sperm Preparation

Male gametes for ART are usually obtained by masturbation. Surgical procedures are only necessary for retrieval of testicular or epididymal spermatozoa (see below). Some disturbances of semen transport, such as retrograde ejaculation, can be overcome by administration of an alpha-sympathomimetic midodrin or imipramin (8, 9). In cases of persistent retrograde ejaculation, systemic alkalization of urine or instillation of the bladder with appropriate medium is recommended (7). In certain cases of anejaculation, i.g., those associated with paraplegia, vibro-massage or electro-ejaculation may be successful.

Table 1. Reference values of semen quality according to WHO (10)

Parameter	Value
Volume	$\geq$ 2 ml
pH	$\geq$ 7.2
Sperm concentration	$\geq$ 20 x 10^6 spermatozoa per ml
Total sperm number	$\geq$ 40 x 10^6 spermatozoa per ejaculate
Motility	$\geq$ 50% motile (grades a + b) or
	$\geq$ 25% with progressive motility (grade a) within 60 minutes of ejaculation
Morphology	*
Vitality	$\geq$ 50% live, i.e. excluding dye
White blood cells	< 1 x 10^6 / ml
Sperm autoantibodies	< 50% motile spermatozoa with adherent particles or beads (MAR test, immunobead test)

* Data from ART programs using strict criteria of sperm morphology assessment suggest a threshold value of 15% normal forms

Human semen normally contains more than 40 million spermatozoa (for standard values see Table 1). However, these cannot be directly used for ART, but require further processing. Sperm preparation is firstly aimed at elimination of free seminal plasma which contains substances that may inhibit fertilization. For example, prostaglandins from the seminal plasma may induce contraction when directly applied to the uterus during intrauterine insemination (see below). Secondly, sperm preparation helps increase the number of motile and morphologically intact spermatozoa and reduces immotile or dead gametes as well as leukocytes and cellular debris.

In principle, three different techniques are available for sperm preparation (11). The migration procedure allows immotile spermatozoa to actively migrate from the ejaculate to a new medium. Most widely used is the swim-up technique (Figure 2.1). First, semen is transferred to a

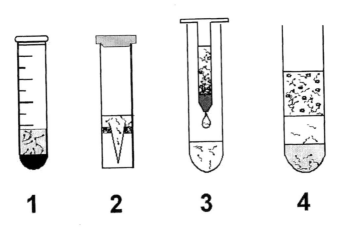

Figure 2. Sperm preparation methods. 1, Swim-up method. Sperm cells are sedimented at the bottom of the centrifugation tube and overlayered with medium. Motile sperm cells are able to swim up into the medium; 2, migration-sedimentation method. The semen is placed in a special tube around an inner conus and overlayered with medium. Motile sperm cells migrate into the medium and are concentrated by sedimentation in the inner conus; 3, glasswool filtration method. The semen is loaded to a column filled with glass wool. Motile sperm pass through the glass wool, whereas immotile spermatozoa and detritus are retained on the column; 4, density gradient centrifugation method. The semen is layered over a two-step gradient. Motile sperm, which change their orientation, are able to penetrate the gradient faster and sediment at the bottom of the centrifugation tube.

centrifuge tube; solids (mainly spermatozoa) are sedimented and the supernatant is discarded. The sedimented spermatozoa are usually washed by suspending in an appropriate medium and centrifugation. Finally, they are carefully overlayered with medium and incubated at 37°C. During the incubation time, motile sperm swim upwards and can be collected. The advantage of this method is maximum quality (90-95% motile spermatozoa), but the yield is only about 10% of the starting number of spermatozoa. The migration-sedimentation technique is a modification of the swim-up procedure, allowing spermatozoa from semen to migrate into the surrounding medium without previous centrifugation, the major part of them accumulating in a funnel (Figure 2.2). Although the sperm quality and yield are not higher than the swim-up protocol, this method avoids several centrifugation steps.

The second method is based on gradient centrifugation (Figure 2.3). The liquefied semen is applied to a single- or multi-layered gradient of a dense (viscous) fluid and is then centrifuged. Sucrose solutions or commercial products, i.e., silan- or polyvinylpyrrolidone-coated beads (Percoll, Ficoll) are frequently used for the gradient separation. The efficacy of the method is based on the fact that spermatozoa, which are located in the direction of the centrifugal force, show a lower resistance against the fluid and, therefore, sediment and pass through the gradient solution more rapidly. Motile and morphologically normal spermatozoa are able to assume this condition, at least partly. Therefore, mainly motile spermatozoa are found on the bottom of the centrifuge tube. The advantage of this method is a higher yield than after swim-up (approximately 20-30%), but the quality is somewhat lower. An additional problem is contamination with the viscous solution, which has to be removed by washing.

The third technique is based on a filter mechanism. In this method, the original or washed semen is passed through a filter that retains immotile and morphologically aberrant spermatozoa, debris and round cells (Figure 1d). The most widely used filtration medium is a column filled with glass wool.

A defined amount of glass wool is inserted into a column, i.e., a tuberculin syringe (Figure 2.4). The semen when applied to the column passes through the glass wool and the filtrate contains a significantly higher number of motile and morphologically normal spermatozoa. In addition to the high yield (50%), this method is advantageous in that no centrifugation step is needed. However, the quality of spermatozoa present in the filtrate is significantly lower than that obtained after swim-up procedure. Whether the potential damage to the male gamete from reactive oxygen species (ROS) can be prevented during this procedure is a matter of debate (12).

ASSISTED REPRODUCTION TECHNIQUES

Intrauterine Insemination (IUI)

IUI can initially be performed in couples who present with sufficient semen quality, open fallopian tubes, and successful ovarian stimulation or ovulation induction (3). By means of a thin catheter, prepared spermatozoa are injected through the cervix into the uterine cavity (Figure 3). Prior to the procedure, a sonographic examination is performed to exclude the existence of too many follicles and thus reduce the risk of multiple pregnancy.

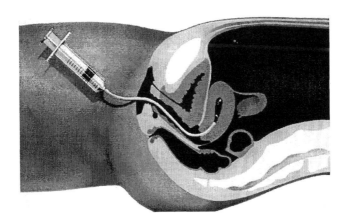

Figure 3. Intrauterine insemination (IUI). A catheter filled with prepared sperm penetrates the cervix, injecting sperm cells into the cavum uteri

In Vitro Fertilization (IVF) Techniques

Extrauterine fertilization is indicated in patients with functional or anatomical uterine disorders, tubal obstruction or dysfunction and in cases of very poor sperm characteristics or repeated failure of inseminations. Depending on semen quality, conventional IVF or intracytoplasmic sperm injection (ICSI) can be performed. The difference between the two methods is the procedure of insemination; all other treatment components are similar. Several oocytes are required for successful extracorporal fertilization, since not every oocyte is able to be fertilized and pregnancy rates are low in cases of only one embryo (see below). Therefore, stimulated ovulation therapy is used to produce an average of 10-15 oocytes.

Initially, oocytes were collected laparoscopically by aspirating fluid of the individual follicles. Since the late eighties, a far less traumatic procedure of ultrasonographically controlled vaginal follicle puncture has been considered the method of choice (13; Figure 4). Under local anesthesia or sedation, an ultrasonographic needle is inserted into the vagina, which allows visualization of the ovary. The vaginal wall is then penetrated by means of a puncture needle, and each available follicle is punctured. Negative pressure produced by a pumping system aspirates follicular fluid together with the oocytes and granulosa cells, which are then collected in a sterile tube. In the laboratory, the oocytes and granulosa cell complexes

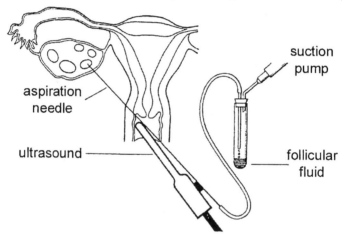

Figure 4. Scheme of an ultrasound-guided transvaginally follicle puncture. The needle penetrates the vagina and punctures the follicle in the ovary. The follicular fluid is aspirated into a tube and then examined under a microscope.

are selected under a binocular microscope and transferred to a sterile culture medium where they usually remain 2-4 hours until insemination.

At the time of oocyte retrieval, semen is usually collected by means of masturbation. If necessary, cryopreserved samples can be used. In selected cases cryopreservation of semen is performed for medical reasons (i.e., prior to chemotherapy). On the other hand, cryopreservation of mature metaphase II-oocytes is still associated with low success rates and the technique is not yet recommended as a routine procedure (14). The same applies to freezing or *in vitro* maturation of ovarian tissue (15).

After semen analysis and sperm preparation (see above), oocytes are inseminated under a binocular microscope. In most cases, 4-well dishes are used. These dishes have four cavities, each with approximately 1 cm in diameter. Usually, several oocytes are incubated together in a culture dish; negative effects have not been observed. Some authors have even reported a synergistic effect which, however, has not been confirmed. Improvement by co-culture with other cells could be observed only in some patients; these conditions, however, do not increase the outcome of IVF in general (16).

In most cases, a carbonate-buffered medium is used which is kept in an incubator at 37°C under 5% CO_2 in air. The oocytes are then mixed with 1 x 10^6 motile and sufficient percentage of morphologically intact spermatozoa, together with the surrounding cumulus mass (cumulus cells) which, at this stage, serves for selection of morphologically intact spermatozoa, but has also secretory properties (17,18). Fertilization usually does not occur in cases of low sperm concentration or insufficient number of progressively motile and/or morphologically normal spermatozoa. Fertilization may also fail because of suboptimal condition of the oocytes that are either immature (germinal vesicle stage or metaphase I) or morphologically aberrant. Both conditions are correlated with the age of the patient and disorders such as polycystic ovary syndrome or endometriosis. Even under ideal conditions, fertilization may not necessarily occur; fertilization rates normally range between 70 to 80%.

Fertilization is a complex process which remains to be fully elucidated. Ca^{2+} oscillation in the cytoplasm immediately after fertilization is characteristic (19). Cytoplasmic factors from both the oocyte and the spermatozoon are probably involved in inducing decondensation of the sperm head (20). This leads to the formation of pronuclei (PN) indicating fertilization. Pronuclei are the haploid, but already decondensed, nuclei of oocyte and spermatozoon before they fuse to form a zygote. Fusion of the male and female pronuclei completes the process of fertilization, creating a new individual, at least at the genetic level.

Under a microscope, fertilized oocytes are easy to recognize by the presence ot two nuclei (Figure 5). They are checked 18 hours after insemination to achieve an identification rate of > 90%. However, it cannot be ruled out that a single oocyte may develop pronuclei at a later stage, or that PN have previously fused and are no longer identifiable. The laboratory personnel should carefully check that anomalies such as polyploidy with >2 pronuclei (Figure 6) are excluded from the culture. The occurrence of triploidy may not necessarily result from the penetration of two spermatozoa into the oocyte; an additional pronucleus may also be produced by diploid spermatozoa, diploid oocytes or unextruded polar bodies. Oocytes with 1 pronucleus are considered irregularly fertilized. However, in conventional IVF, approximately 50% of 1PN stages have a diploid chromosome complement resulting from a pathogenetic activation of the oocyte (20).

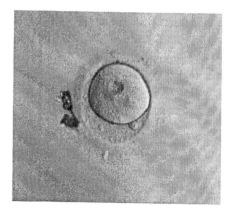

Figure 5. A normally fertilized oocyte with two clearly visible pronuclei.

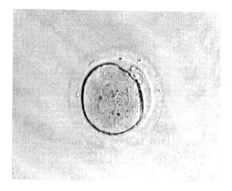

Figure 6. An abnormally fertilized oocyte with three pronuclei.

Studies in the late nineties reported for the first time how to differentiate oocyte quality by means of morphological criteria of the substructures of pronuclei, which might influence the pregnancy rate if properly selected (21, 22). These findings remain to be validated, and standardized criteria for evaluation of oocytes are still lacking. However, it appears that apart from the number of nucleoli in the pronuclei it is also their orientation and pattern, which is important. These small corpuscles, which consist of t-RNA, tend to condense during the pronuclear development and to concentrate on the fusion site of the pronuclei (Figure 7). Thus, good-quality oocytes are considered those which are well developed and synchronous in both pronuclei. Furthermore, morphological criteria of the cell body have to be considered for evaluation of the PN stage. Anomalies such as vacuoles, cell inclusions or darkly granulated cytoplasm are known to reduce the implantation capacity (23).

Correctly fertilized 2PN-oocytes are further cultivated in medium. The two pronuclei fuse within approximately 20-24 hours after insemination (20). The first cell division occurs within 25-26 hours and a 2-cell or 4-cell embryo is usually formed on the second day (Figure 8). Even early embryos show quality features that are indicative of the implantation capacity. Human embryos tend to tie off small anuclear vesicles, so-called fragments (24). This may occur to a small extent, but may also result in disintegration of one or more blastomeres. The consequence is a direct loss of cellular mass or even a loss of cytoplasmic polarity (25). The degree of fragmentation of transferred embryos is known to be significantly correlated with the pregnancy rate. In addition, the rapidity of development plays an important role. For example, cell division is a significant marker indicating good results (26). If selection is possible, embryos with the lowest fragmentation rate and highest number of blastomeres should always be used.

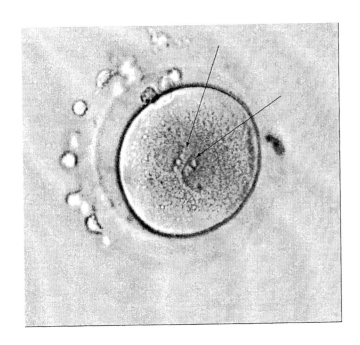

Figure 7. 2PN-stage with an excellent PN-score: the nucleoli in the pronuclei (arrows) are polarized on the fusion front.

In the late nineties, blastocysts were routinely used for embryo transfer (27). The pregnancy rate has, therefore, significantly increased in countries where embryos can be selected (see below). This technique was possible by the development of new sequential medium that has a more complex composition than the traditional salt solution of the standard medium. Under these conditions, a normal fertilized egg develops to an 8-cell embryo on day 3 after insemination, morula (> 32 cells) on day 4 or 5, and, finally, blastocyst (> 64 cells) on day 5 or 6. A normal blastocyst has a larger diameter than the preceding oocyte and early embryonic stages. The trophoblast lying around the inner site of the zona pellucida can be differentiated from compact inner cell mass (ICM). The trophoblast cells later generate the placenta, whereas the ICM forms the embryo. The cavity between ICM and surrounding trophoblast, the so-called blastocoel, is responsible for the name blastocyst. It is known that about 40% of human embryos achieve blastocyst stage *in vitro* (28). Before implantation can occur, the blastocyst must hatch from the zona pellucida. Since this occurs only in a small number of human embryos *in vitro*, the blastocyst is usually transferred prior to hatching. Embryo transfer is normally done by aspirating the cells into a small catheter, which is able to penetrate the cervix uteri. Together with a small amount of medium, the embryos are deposited in the cavum uteri near to the fundus region.

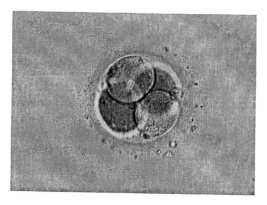

Figure 8. A 4-cell-embryo without fragments or inequal blastomeres (classification A) 42h after insemination.

Recently, a new technique referred to as assisted hatching has been established. In this technique, the zona is opened to facilitate hatching. Opening of the zona may be performed mechanically by means of a micropipette, chemically with acid Tyrode or, more recently and to an increasing extent, by laser (29, 30). The diameter of the opening should be 1.5 times that of the zona; a smaller diameter would interfere with implantation. According to the available data, assisted hatching is performed in a small number of older patients (> 40 years), those with cryopreserved embryos, patients who repeatedly experience implantation failure despite good embryo quality, and embryos with thickened zonae.

ICSI

Severely impaired semen quality minimizes the chance of successful fertilization within conventional IVF treatment. Specific cut-off values cannot be established, because (a) several parameters are involved in fertilization (see above), and (b) assessment of sperm parameters is encumbered with a subjective factor, especially in terms of morphology. The number of couples with infertility caused by severely impaired semen characteristics is very high worldwide, accounting for one third of patients in IVF centers (3). Therefore, methods were developed in the late eighties to compensate for the male factor. A first step was partial zona dissection (PZD): in analogy to assisted hatching, the zona pellucida was opened to allow motile spermatozoa to penetrate the oolemma (oocyte) and achieve fertilization. In this procedure, the oocyte was first freed from surrounding granulosa cells by incubation with the enzyme hyaluronidase. Thereafter, the zona was opened mechanically by means of a fine glass cannula, or chemically by means of acid Tyrode, as the zona pellucida is susceptible to acidic medium. More recently, diode laser has been used. However, the fertility rate has remained unacceptably low, while the polyploidy rate has dramatically increased (31).

In the early nineties (32), the technique of subzonal sperm injection (SUZI) was first applied to humans. In this approach, a limited number of motile spermatozoa were inserted through the zona into the perivitelline space with the aid of a glass cannula. However, this method too, resulted in low fertilization rates and a high incidence of polyploidy. Despite variations in the number of spermatozoa used, the problem of polyploidy remained. It is known that in the human system only acrosome-reacted spermatozoa are able to fuse with the oolemma. While different methods are available for induction of the acrosome reaction in human spermatozoa, a complete reaction could not be induced under physiological conditions. Moreover, it is not possible to determine whether the injected spermatozoa are acrosome-intact or acrosome-reacted.

In parallel with SUZI, the first intracytoplasmic sperm injection was performed by Prof. A. van Steirteghem at Brussels Free University in 1990 that revolutionized reproductive medicine in the following years (33). In this procedure, a single motile spermatozoon was directly injected into the oocyte cytoplasm (Figure 9). With this method, it was possible to compensate the andrological factors, provided that a few motile spermatozoa were available. Birth resulting from ICSI was first reported in 1992 (33). Fertilization rates of 70% per injected oocyte were comparable to those achieved in conventional IVF. Triploidy, which also occurred, was

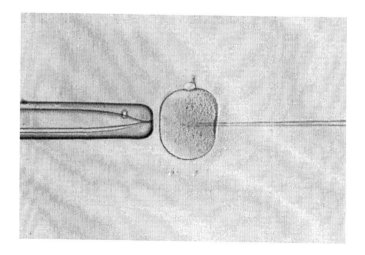

Figure 9. Intracytoplasmic sperm injection (ICSI): A holding pipette is fixing the oocyte by aspiration (left side). A single sperm is injected with a fine glass cannula (right side) while the polar body is in 12 or 6 o'clock position.

explained by diploid gametes or unextruded polar bodies. However, the rate of polyploidy in this method was lower than 5%. Embryonic development and pregnancy rates were comparable to or better than those of conventional IVF, probably due to the more frequent indication because of the male factor. To date, there is no convincing evidence to indicate that microinjection results in a higher malformation rate. However, accessible data is controversially discussed.

Microsurgical Procedures

The ICSI technique is now a standard procedure that exceeds conventional IVF protocols. The fact that only a single spermatozoon is required, the procedure has allowed even patients with azoospermia to be treated, provided that a few spermatozoa were present in the testis or the epididymis.

Today, in many andrological centers specimens of testicular tissue are routinely obtained by biopsy for testicular sperm extraction (TESE). Testicular tissue specimen or extracted sperm can be cryopreserved for future use. Spermatozoa can also be obtained from the epididymis by microsurgical aspiration (MESA) (34). In particular, MESA is suitable for men with obstructive azoospermia and normal spermatogenesis. In contrast, TESE can also be performed in cases of non-obstructive azoospermia, provided that spermatogenesis up to mature elongated spermatids is maintained at least focally in some seminiferous tubules. In most centers, the fertilization rates after TESE are reported to be slightly reduced (approximately 50%), and the pregnancy rates also appear to be lower than after use of mature ejaculated spermatozoa. Nevertheless, since the late nineties, TESE has become part of the routine program in most IVF centers.

The success rate of fertilizing an oocyte is very low when immature spermatids are used; ICSI protocol, when early elongated spermatids are used, has resulted in acceptable fertilization rates but very low pregnancy rate. On the other hand, pregnancy was only achieved in exceptional cases after injection of round spermatids (35). These published data seem to suggest that the degree of sperm maturation plays an important role in embryonic development.

Cryopreservation of Gametes

Cryopreservation of gametes is a routine procedure in most IVF centers. While freezing of spermatozoa or testicular tissue is feasible, cryopreservation of mature metaphase II oocytes has produced poor results. Only freezing of fertilized PN stage oocytes or embryos has become a routine procedure (36). However, the implantation rates were found to be significantly lower when cryopreserved, instead of fresh, PN stage oocytes/embryos were used. In many centers, pregnancy rates using such cells or embryos are only 50% compared with cycles using fresh material. Since cryopreservation of spermatozoa has no such detrimental effects, cytoplasmic damage is suspected to be the reason. On the other hand, no increased malformation rates have been observed with cryopreserved PN stage oocytes. However, to avoid repeated ovarian hyperstimulation, this technique is now widely used.

RESULTS OF ASSISTED REPRODUCTIVE METHODS

The pregnancy rates after both IVF and ICSI are reported to range between 20 and 35% per embryo transferred (37), with a take-home baby rate of 15-25%. Since the number of embryos was, in most cases, more than one, the implantation rate per embryo was estimated to be approximately 10%. However, the cumulative pregnancy rate resulted in 50-60%. The relatively low percentage of successful cycles probably corresponds to the rate of natural conception in healthy couples, calculating pregnancies occurring per month. The biochemical events during the implantation process are complex and remain to be fully elucidated. Failed pregnancy may result from suboptimal endometrial environment or impaired development of the embryo. *In vitro* studies with fluorescent in situ hybridization (FISH) have revealed that 20-50% of human embryos had chromosomal aberrations (38) and probably would not have developed further. The only procedure to circumvent this problem is the selection of embryos which, however, is prohibited in some countries, including Germany and Switzerland. Based on the embryo quality (see above) the developmental potency can be estimated, since embryos with chromosomal damage show increased fragmentation.

For this reason, blastocyst culture was a great success since it allowed selection of embryos offering maximum developmental potentials. Under these conditions, the pregnancy rates increased to > 60% per embryo transfer (ET), although only 1-2 blastocysts are usually transferred (39). As the number of embryos is reduced by use of blastocyst transfer, the rate of multiple pregnancies is lower than that after conventional ET. However, it should be noted that blastocyst cultures is a reasonable approach only if a sufficient number of fertilized oocytes are available. In case of fewer oocytes or embryos, it is better to perform ET early on day 2 or 3 to overcome the suboptimal culture conditions *in vitro*.

The malformation rate after IVF or ICSI is similar to that of the general population. However, the usually higher average age of fertilization patients is reflected by an increased abortion rate, which is not caused by ART. When the age factor is eliminated from these data, there are no differences compared with naturally occurring pregnancies (37).

It should be noted that the use of several embryos results in a higher rate of multiple pregnancies. Twin pregnancies occur in more than 20% and triple pregnancies in 5% of all cases. To solve this problem, it has been suggested that the number of transferred embryos generally be limited to two, considering various parameters such as female's age, quality of embryos, and number of oocytes. Nearly equal pregnancy rates have been achieved in patients where the number of embryos had been reduced to two compared with those who had three embryos. However, in cases where only two embryos were available, the rates were markedly lower, even in non-selective systems, as shown by data from Germany. It appears that blastocyst culture is an excellent method to reduce the rate of multiple pregnancies by restricting the number of embryos to 1 or 2.

CONCLUSIONS

The reasons for reproductive failure are complex. In women with diagnoses such as bilateral tubal disease, ART can be a causative therapy. In men, the etiology of impaired sperm quality/function often remains unclear. ART, therefore, bypasses the "male factor". ART are often successful but always expensive and not without risk. Fortunately, this causes a new awareness of the values and chances of andrology, demanding a critical diagnostic work-up leading to appropriate indications. The recognition of causative factors for male infertility might lead to simpler, cost effective therapies. The array of ART represents last resorts when less invasive treatments failed or were considered inappropriate. ART has dramatically changed the options for infertile couples.

ACKNOWLEDGEMENTS

The authors are indebted to the superb editorial assistance of Mrs. Gudrun Scharfe and Mrs. Loreita Little.

REFERENCES

1. Steptoe P.C., Edwards R.G. Birth after reimplantation of a human embryo. Lancet II 1978; 366
2. Tarlatzis B.C., Bili H. Intracytoplasmic sperm injection. Survey of world results. Ann NY Acad Sci 2000; 900:336-344
3. Assisted reproductive technology in the United States: 1997 results generated from the American Society for Reproductive Medicine/Society for Assisted Reproductive Technology Registry. Fertil Steril 2000; 74:653-654
4. Pellicer A., Albert C., Garrido N., Navarro J., Remohi J., Simon C. The pathophysiology of endometriosis-associated infertility, follicular environment and embryo quality. J Reprod Fertil Suppl 2000; 55:109-119.
5. Adashi, E.Y., Rock, J.A., Rosenwaks, Z., eds. Reproductive endocrinology, surgery and technology, Lippincott-Raven Publisher, 1996.
6. de Kretzer DM. Male infertility. Lancet 1997; 349:787-790
7. Rowe, P.J., Comhaire, F.H., Hargreave, T.B., Mahmoud, A.M.A. WHO manual for the standardized investigation, diagnosis and management of the infertile male. Cambridge University Press, Cambridge, 2000.
8. Köhn F.M., Schill W.-B. The alpha-sympathomimetic midodrin as a tool for diagnosis and treatment of sperm transport disturbances. Andrologia 1994; 26:283-287.
9. Kamischke, A., Nieschlag, E.. Treatment of retrograde ejaculation and anejaculation. Hum Reprod Update 1999; 5:448-474.
10. WHO laboratory manual for the examination of human semen and sperm-cervical mucus interaction. 4th edition. Cambridge University Press, 1999.
11. Leung C.K. Recent advances in clinical aspects of assisted reproduction. Hong Kong Med J 2000; 6:169-176.
12. Speroff, L., Glass, R.H., Kas, N.G., eds. Clinical Gynecology and Infertility. Lippincott, Williams and Wilkins, 1999.
13. Murray A., Spears N. Follicular development in vitro. Semin Reprod Med 2000; 18:109-122

374

14. Mandelbaum J. Embryo and oocyte preservation. Hum Reprod (Suppl 15) 2000; 4: 43-47.

15. Kim S.S., Battaglia D.E., Soules M.R. The future of human ovarian cryopreservation and transplantation: fertilty and beyond. Fertil Steril 2001; 75:1049-1056.

16. Wiemer K.E., Cohen J., Tucker M.J, Godke R.A. The application of co-culture in assisted reproduction: 10 years of experience with human embryos. Hum Reprod 1998; 13:226-238.

17. Canipari R. Oocyte-granulosa cell interactions. Hum Reprod Update 2000; 6:279-289.

18. Katz D.F., Drobnis E.Z., Overstreet J.W. Factors regulating mammalian sperm migration through the femal reproductive tract and oocyte vestments. Gamete Res 1989; 22:449-469.

19. Tesarik J. Oocyte activation after intracytoplasmic injection of mature and immature sperm cells. Hum Reprod 1998; 13:117-127.

20. Hewitson L., Simerly C., Dominko T., Schatten G. Cellular and molecular events after in vitro fertilization and intracytoplasmic sperm injection. Theriogenology 2000; 53:95-104.

21. Scott L.A., Smith S. The successful use of pronuclear embryo transfers the day following oocyte retrieval. Hum Reprod 1998; 13:1003-1013.

22. Tesarik J., Greco E. The probability of abnormal preimplantation development can be predicted by a single static observation on pronuclear stage morphology. Hum Reprod 1999; 14:1318-1323.

23. Serhal PF, Ranieri DM, Kinis A, Marchant S, Davies M, Khadum IM. Oocyte morphology predicts outcome of intracytoplasmic sperm injection. Hum Reprod 1997; 12:1267-70.

24. Antczak M., Van Blerkom J. Temporal and spartial aspects of fragmentation in early human embryos: Possible effects on developmental competence and association with the differential elimination of regulatory proteins from polarized domains. Hum Reprod 1999; 14:429-447.

25. Scott L.A. Oocyte and embryo polarity. Sem Reprod Med 2000; 18:171-183.

26. Hunter R.H. Failure of embryonic development. Reprod Nutr. Dev 1988; 28:1781-1790.

27. Plachot M. The blastocyst. Hum Reprod (Suppl. 15) 2000; 4:49-58.

28. Devreker F., Englert Y. In vitro development and metabolism of the human embryo up to the blastocyst stage. Eur J Obstet Gynecol Reprod Biol 2000; 92:51-56.

29. De Vos A., Van Steirteghem A. Zona hardening, zona drilling and assisted hatching: new achievments in assisted reproduction. Cell Tissue Organ 2000; 166:220-227.

30. Manvelbaum J. The effects of assisted hatching process on implantation. Hum Reprod 1996; 11 (Suppl 1):43-50.

31. Cohen J., Alikani M., Malter H.E., Adler A., Talansky B.E., Rosenwaks Z. Partial zona dissection or subzonal sperm insertion: microsurgical fertilization alternatives based on evaluation of sperm and embryology morphology. Fertil Steril 1991; 56:696-706.

32 Tarin J.J. Subzonal inseminaton, partial zona dissection or intracytoplasmic sperm injection? An easy decision? Hum Reprod 1995; 10:165-70.

33. Palermo G., Joris H., Devroey P., Van Steirteghem A.C. Pregnancies after intracytoplasmic injection of a single spermatozoon into an oocyte. Lancet 1992; 340:17-18.

34. Tournaye H. Surgical sperm recovery for intracytoplasmic sperm injection: which method is to be preferred? Hum Reprod 1999; 14:71-81.

35. Levran D., Nahum H., Farhi J., Weissmann A. Poor outcome with round spermatid injection in azoospermic patients with maturation arrest. Fertil Steril 2000; 74:443-449.

36. Porcu, E. Oocyte Cryopreservation. In Towards Reproductive Certainty: Infertility and Genetics Beyond, D.K. Gardner, W.B. Schoolcraft, R. Jansen, D. Mortimer, eds. Carnforth: Parthenon Press, 1999, pp. 234-242.

37. Society for Assisted Reproductive Technology and the American Society for Reproductive Medicine (1999): 1996 results generated from the American Society for Reproductive Medicine/Society for Assisted Reproductive Technology Registry. Fertil Steril 71; 798-807.
38. Muné S., Cohen J. Chromosome abnormalities in human embryos. Hum Reprod Update 1998; 4:842-855.
39. In vitro culture of human blastocyst. In Towards Reproductive Certainty: Infertility and Genetics Beyond, D.K. Gardner, W.B. Schoolcraft, R. Jansen, D. Mortimer, eds. Carnforth: Parthenon Press, 1999; pp. 223-232.

Chapter 20

THE REPRODUCTIVE EFFECTS OF HORMONALLY ACTIVE ENVIRONMENTAL AGENTS

Benjamin J. Danzo
Vanderbilt University School of Medicine, Nashville, Tennessee, USA

INTRODUCTION

Efforts during the past 50 or more years to improve agricultural productivity and manufacturing processes have led to the introduction of numerous man-made chemicals into the environment. In the course of this chapter, these chemicals, generally, will be referred to as environmental toxicants or xenobiotics. It is now known that many of these chemicals have unexpected effects on the animal populations of the planet and one expects that such effects have occurred or will occur in the human population. The untoward effects that are of particular interest to us in the context of reproductive biology are effects that interfere with the normal development and function of the male and female reproductive systems and which lead to reduced fertility, infertility, or sterility. Although the mechanisms by which environmental toxicants cause their disruptive effects on reproduction were at first elusive, it is now clear that many act by interfering with the physiological regulation of reproductive processes by sex-steroid hormones. Other environmental toxicants act through other mechanisms, such as the thyroid hormone receptor and through the aryl hydrocarbon receptor (1, 2). This chapter will concentrate exclusively on xenobiotics that act through the sex-steroid hormone pathway since these hormones are known to be intimately involved in regulating reproductive processes. The realization that xenobiotics can act as hormone mimics or antagonists has given rise to the terms "endocrine disruptors", "hormonally active agents", or "environmental hormones" also being used to describe these compounds.

Although many toxicants in the environment have weak hormonal activity, their lipophilic nature and long half-lives allows them to accumulate and persist in fatty tissues of the body, thus increasing their concentration and bioavailability. Due to its high lipid content, organochlorine xenobiotics accumulate in breast milk (3). Therefore, nursing infants are exposed to higher concentrations of xenobiotics than at any other time. Endocrine disrupting chemicals are also found in drinking water. Environmental toxicants can be disbursed widely by rivers, streams and ground water. The persistence and worldwide distribution of organochlorine pollutants by global distillation have been demonstrated. Xenobiotics can accumulate in fish, fresh water and sea mammals, and in other animal food sources (4, 5). Many plants commonly used as food, for example soybeans, contain natural hormonally active agents (6, 7). Thus, every human and animal inhabitant of the earth is exposed to some level of potentially endocrine-disrupting environmental agents.

In this chapter, a brief background will be given on environmental toxicants and on the mechanism of action of sex-steroid hormones. A summary of the embryology of the male and female reproductive system will be give so that reader may be aware of the target tissues and processes that are susceptible to disruption by environmental toxicants. The chapter will conclude with specific examples of reproductive deficits that arise from the disruptive action of environmental toxicants, with a discussion of the potential danger of these compounds for humankind, and with a discussion of theoretical considerations that bear on the topic of environmental toxicants and reproduction. Although some references are provided in the text, the student is encouraged to consult references in the cited reviews (1, 2, 5, 8-10) for more detailed information.

WHAT ARE ENVIRONMENTAL TOXICANTS?

Evidence from epidemiological and laboratory studies points to a host of diverse man-made chemicals in the environment that have been implicated in causing abnormalities of the male and female reproductive systems (1, 2, 5, 11). These chemicals fall into several categories based on their chemical structure, Figure 1. The broadest category is that of chlorinated hydrocarbons (organochlorine compounds). Many insecticides such as DDT [1,1,1-trichloro-2,2-bis(p-chlorophenyl)ethane] its metabolites and congeners, the fungicides Vinclozolin [3-(3,5-dichlorophenyl)-5-vinyl-oxazoladine-2,4-dione] and pentachlorophenol, and the herbicide Linuron [3-(3,4-dichlorophenyl)-1-methoxy-1-methly-urea] fall into this category. Another group of organochlorine compounds is the polychlorinated biphenyls (PCBs). They are manufactured through the progressive chlorination of biphenyl, theoretically yielding 209 congeners (12). PCBs have been used extensively for a variety of industrial purposes including as constituents in pesticides as heat transfer fluids, and as dielectrics. Bisphenol A [4,4'-(1methylethylidene)bisphenol], like the PCBs, is based on the biphenyl nucleus, but it lacks chlorine substituents. This compound, which is used in the manufacture of polycarbonate plastics, resins, and dental sealants, has also been shown to have deleterious effects on reproduction.

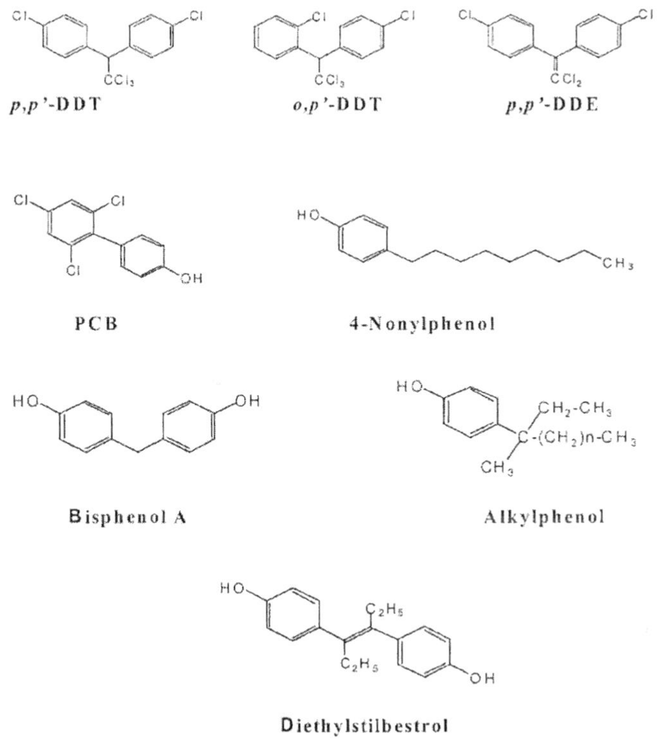

Figure 1. Planar structures of some xenobiotics reported to exhibit endocrine-disrupting activity. It is, however, the three dimensional configuration of the compounds that governs their ability to bind to steroid-hormone receptors.

Another class of hormonally-active environmental agents is the phthalates. Dibutyl benzyl phthalate, used as a plasticizer for cellulose resins, polyvinyl acetates, polyurethanes and polysulfides, has been characterized as a toxicant affecting reproduction and development in several studies (5).

Alkylphenolic compounds are nonionic surfactants that are widely used in detergents, cosmetics, paints, herbicides, pesticides, and other products. Two of these compounds that have been implicated as endocrine-disrupting agents are nonylphenol and octylphenol. Nonylphenol is a technical grade mixture of monoalkyl phenols, predominantly *para* substituted. Its side chains are isomeric branched alkyl radicals. Octylphenol [*p*-(1,1,3,3-tetramethylbutyl)phenol] is available in a highly purified form. Both of these compounds and alkylphenol ethoxlyates and their metabolites may be major contributors to the estrogenic activity that has been detected in sewage treatment plant effluents (13, 14). Some of these compounds, e.g., nonylphenol, have been shown to bind to the estrogen receptor.

In addition to man-made chemicals, other natural chemicals in the environment are mimics of estrogens. These hormonally active agents are the plant-derived, dietary phytoestrogens, a diverse group of nonsteroidal compounds that are present in most plants, fruits and vegetables. The blood concentration of phytoestrogens can exceed that of endogenously produced estrogens in populations where the dietary intake of edibles containing them is high. There are three main types of phytoestrogens—the isoflavones, coumestans, and lignans. There are more than 1000 types of isofalvones; they are the most common phytoestrogens and they are thought to have the highest estrogenic activity. Common sources of isoflavones are soybeans, chickpeas, lentils, beans, and clover. Lignans are abundant in flaxseed, and coumestans are found in sprouting plants. Phytoestrogens bind to the estrogen receptor and, like the man-made xenobiotics, their affinity for the receptor is far less than that of estradiol (6). Some phytoestrogens have been reported to be estrogenic, while others have been reported to be estrogen antagonists. While phytoestrogens may have beneficial uses such as in hormone replacement therapy for postmenopausal women and possibly providing protective effects against breast cancer (6), they have not yet been shown to cause reproductive tract lesions.

Selective estrogen receptor modulators (SERMs) are compounds whose interaction with the estrogen receptor results in different effects in different cells or tissues (15). The classical example of a SERM is the drug tamoxifen. This compound is an antiestrogen in breast tissue (and has been used in the treatment of estrogen-responsive breast cancer), but has estrogenic effects on the uterus and skeleton. Some environmental toxicants and phytoestrogens may be SERMs (16), but this has not been documented extensively. Selective modulators of the androgen receptor action are under development.

In addition to the compounds discussed above, many other chemicals that are widely distributed in the environment have been reported to have effects on reproduction or to have other endocrine-disrupting effects. Any list of such compounds must be considered incomplete since there are literally thousands of man-made chemicals in the environment and more, new, chemicals are being manufactured few of which have been tested for their endocrine disrupting potential. Most of the plant-derived compounds have yet to be evaluated for their ability to interfere with steroid hormone action or to disrupt the development and function of the reproductive tract.

Synthesis And Mechanism Of Action Of Steroid Hormones

Sex steroid hormones are produced from cholesterol by several organs of the body. Primarily the ovary produces estradiol, but the testis also produces it. The skin and adipose tissue contain the enzyme aromatase, which synthesizes estradiol from precursors, primarily testosterone. Leydig

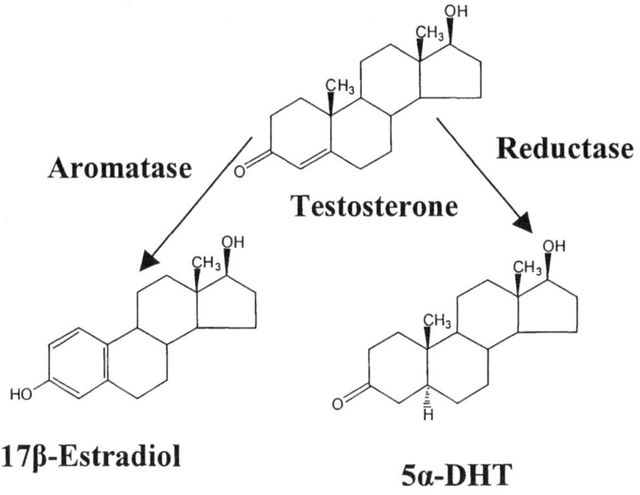

Figure 2. Synthesis of 17β-estradiol and 5α-dihydrotestosterone (5α-DHT) from testosterone.

cells of the testis are the primary source of testosterone. The enzyme 5α-reductace, which is present in target tissues, converts testosterone to dihydrotestosterone. Figure 2 illustrates the synthesis these hormones; many environmental toxicants bring about their effects by mimicking or antagonizing them.

Steroid hormones primarily act by binding to ligand-activated transcription factors termed steroid hormone receptors (17, 18). These receptors belong to the nuclear receptor super family. In common with other members of this family, steroid hormone receptors contain a ligand-binding domain (LBD) and a highly conserved DNA-binding domain (DBD), Figure 3. These domains mediate the recognition and binding of the appropriate steroid ligand (LBD) and the binding of the receptor protein to specific target DNA sequences (DBD), respectively. The DNA sequences to which the DBD binds are referred to as the hormone response element (HRE).

These elements are located in regulatory sequences normally present in the 5'-flanking region of the target gene. The HREs are often found relatively close to the core promoter, however, in some cases they are located in enhancer regions several kilobases upstream from the transcription initiation site. Steroid receptors bind almost exclusively to the HRE as homodimers. The C-terminus of the LBD contains an essential ligand-dependent transactivation function, activation function 2 (AF2). The N-terminus of many nuclear receptors contains activation function 1 (AF1).

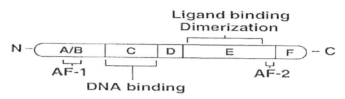

Figure 3. The domain structure of steroid receptors. Taken from Aranda and Pasqual (17) and used with permission.

Members of the nuclear receptor family mediate the physiological effects of steroid, thyroid, and retinoid hormones primarily by regulating the assembly of transcriptional preinitiation complexes in the promoter region of target genes, thus regulating the expression of the genes in response to the hormone. Once the hormone-receptor complex is bound to the hormone response element in the region of the target gene, it recruits members of the SRC family, which is comprised of a group of structurally and functionally related transcription co-activators (17, 19). The receptors also interact with the transcriptional co-integrators p300 and CBP, which are thought to integrate various afferent signals at the promoter. CBP/p300 interacts with a large variety of transcription factors including AP1, Jun, Fos, etc and serves a co-activator role for these proteins potentiating their transcriptional activity. This finding has led to the speculation that CBP/p300 may serve as co-integrators of intracellular and extracellular signaling pathways. SRC appears to be recruited directly by the liganded receptor and then serves as a platform for the recruitment of CBP. CBP/p300 and members of the SRC family have intrinsic histone acetyltransferase (HAT) activity that is thought to disrupt the nucleosomal structure of the promoter. Other co-activators may be also involved in the remodeling of chromatin in the region of the promoter. These proteins and/or others may also serve as adapters between the receptor-hormone complex and the components of the basal transcription apparatus. Subsequent interaction of the complex with polymerase II affects transcription of the regulated gene, Figure 4. In addition to co-activation, systems for co-repression have been demonstrated for the retinoic acid and the thyroid hormone receptors (17-19). While to date co-repressors have not been demonstrated for steroid hormone dependent genes, it is likely that such molecules are involved in inhibiting the transcription of these genes.

Steroid hormone receptors and nuclear receptors in general can also modulate gene expression by mechanisms that are not dependent upon their binding to HREs. They, therefore, can positively and negatively regulate the expression of genes that do not contain HREs. This regulation, which occurs through positive or negative interference with the activity of other transcription factors, is referred to as transcriptional "cross-talk" (17). The estrogen receptors, for example, use protein-protein interactions to enhance the transcription of genes that contain AP-1 sites. The co-integrator role of CBP/p300, discussed above, constitutes another form of cross-talk.

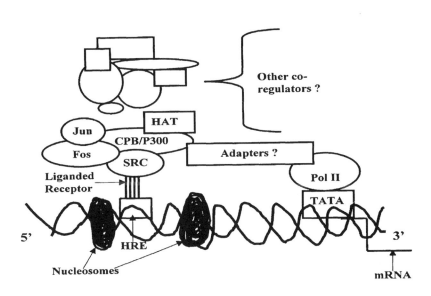

Figure 4. This Figure presents a cartoon of a steroid hormone receptor together with its co-regulators, bound to the HRE on DNA.

The ligand-binding domain, and specifically the structure of the ligand-binding site on steroid hormone receptors, is an important consideration for our discussion on the mechanisms by which environmental toxicants may mimic or interfere with the action of steroid hormones. The crystal structure of the LBD of several nuclear receptors has been elucidated (17, 18). The data indicate that the LBDs of the different nuclear receptors are similar, indicating that a canonical structure is required to form this domain. The LBDs are formed by 12 conserved α-helical regions, which are numbered H1-H12. A conserved β-turn is located between H5 and H6. The LBD is folded into a 3-layered, anti-parallel helical sandwich. A central core layer of three helices is packed between two additional layers to create a cavity, which is the ligand-binding pocket. This pocket is highly hydrophobic and is buried within the bottom half of the LBD (17, 18). Contacts of the binding pocket with the ligand may be extensive and may involve different structural elements within the LBD.

During receptor-ligand interaction, the binding site conforms to the shape of the ligand, and flexible ligands could have their conformation altered upon binding to the receptor. Steroid receptors undergo a conformational change into a more compact form upon binding the ligand. Thus, the receptor can control the shape of the ligand and the ligand can control the shape of the receptor. The interaction of a xenobiotic with a receptor could control its function by inhibiting or stimulating the coupling of the receptor-ligand complex to its effector, that is, to all of the other components (e.g., co-regulators) with which the complex interacts at each regulated gene. It has been demonstrated (20) that some environmental toxicants, even within the same class, can inhibit the binding of radiolabeled physiological ligands to both the estrogen and androgen receptors, whereas other xenobiotics show a greater degree of receptor specificity. Although the environmental toxicants bind to steroid receptors, their affinity for the receptors is 1000-fold or less that of the natural ligands (1). The ability of one xenobiotic ligand and its closely related congeners to compete for binding to more than one receptor implies that they have a greater degree of conformational flexibility and adaptability than do the natural hormones. One might anticipate, then, that environmental toxicants that bind to steroid hormone receptors do not all result in conferring the same shape on their cognate receptor as does the physiological ligand. Those that do are likely to mimic the physiological ligand; those that do not are likely to disrupt normal receptor function. Given the flexibility of most xenobiotics and the ability of the receptor binding-site pocket to conform to the shape of the enclosed ligand, it is possible that a toxicant could act as an androgen/antiandrogen when it binds to the androgen receptor or act as an estrogen/antiestrogen when it binds to the estrogen receptor. Therefore, thinking of the xenobiotic-receptor-complex as the effector may be more meaningful than thinking of a specific xenobiotic as an estrogen/antiestrogen or androgen/antiandrogen.

SEXUAL DIFFERENTIATION AND DEVELOPMENT

To understand how hormonally active agents in the environment (both man-made and natural) could interfere with the development and function of the male and female reproductive systems, it is necessary to know at what points regulation of these systems by steroid hormones occurs. The chromosomal and genetic sex of an embryo are determined at the time of fertilization and are dependent upon the type of spermatozoon that fertilizes the egg. Phenotypic sex, the outward appearance of the individual, is dependent on the appropriate action of sex-steroid hormones. Vertebrate embryos initially exhibit an ambisexual stage in which an indifferent gonad is present. This gonad can differentiate into a testis or an ovary. Both male and female embryos have two pairs of genital ducts, the Müllerian (paramesonephric) and Wolffian (mesonephric) ducts (21). The former structures are the anlage of the oviducts, uterus, and the upper part of the vagina. The Wolffian ducts give rise to the epididymis, vas deferens, and the seminal vesicles. In the early embryos, the external genitalia are in a sexless, undifferentiated, state and consist of a genital tubercle, labioscrotal swellings, and urogenital folds (Figure 5). The genital tubercle gives rise to the penis (male) and clitoris (female). The labioscrotal swellings fuse to form the scrotum in the male. In the female, the labioscrotal swellings fuse anteriorly to form the anterior labial commissure and mons pubis, and posteriorly to form the posterior labial commissure. The unfused portions of the labialscrotal folds form the labia majora. In the male, the urogenital folds fuse on the underside of the penis and form the spongy urethra. The surface ectoderm fuses in the mid plane of the penis, enclosing the spongy urethra and forming the penile raphe. The urogenital folds in the female do not fuse except posteriorly where they form the frenulum of the labia minora; their unfused portions form the labia minora. Sex steroid hormones regulate these developmental events.

Under the influence of the SRY gene on the Y chromosome and of several down-stream effectors and autosomal genes (e.g., SOX9 and SF-1) (22, 23), the indifferent gonad develops into a testis. The Sertoli cells of the testis secrete Müllerian inhibiting substance (MIS), which causes regression of the Müllerian ducts. Testosterone secreted by the Leydig cells of the developing testis induces the differentiation of the Wolffian ducts into the structures of the male reproductive tract mentioned above. All of these structures contain androgen receptors and are, therefore, capable of responding to androgenic stimulation. The masculinization of the external genitalia is thought to be dependent on the conversion of testosterone to 5α-dihydrotestosterone (5α-DHT). Recently, however, mice lacking 5α-reductase 1 and 2, the enzymes catalyzing this process, as a result of gene targeting were shown to have fully formed external genitalia and to be fertile (24). The authors concluded that testosterone is the only androgen required for the differentiation of the male genital tract, including the external genitalia, in mice and that 5α-DHT serves primarily as a signal amplification mechanism. The female reproductive system develops in the absence of the SRY gene and, consequently, in the absence of MIS and androgens. The female pathway is, therefore, considered to be independent of hormonal regulation and is thought to develop in the absence of estrogenic stimulation. However, the fetal rabbit ovary has been shown to produce estradiol at the same age as the testis is able to synthesize testosterone. Estrogen receptors are present in the mouse embryo at the ambisexual stage of development and

386

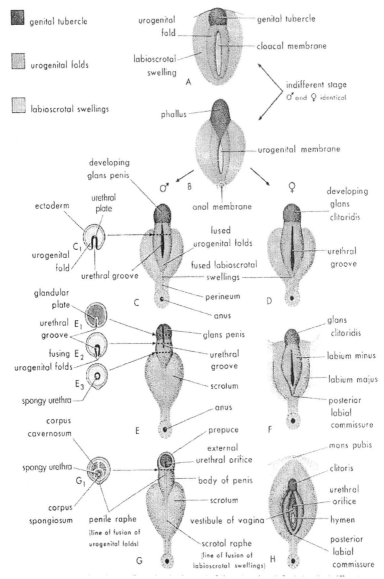

Figure 5. Differentiation of the external genetalia. Figure A and B are diagrams illustrating the development of the external genitalia during the indifferent or undifferentiated stage (4th to 7th weeks of gestation). Figures C, E, and G, represent stages in the development of the male external genitalia at 9, 11, and 12 weeks, respectively. To the left are schematic transverse sections of the developing penis (C1, E1 E2, E3, G1) illustrating formation of the spongy urethra. Figures D, F, and H illustrate stages in the development of the female external genitalia at 9, 11, and 12, weeks of gestation, respectively. This Figure is adapted from Moore and Persaud (21) and is used with permission.

they are present in early embryos of other species. Therefore, it is likely that estrogens are involved in the differentiation of the structures derived from the Müllerian ducts and they may affect the differentiation of Wolffian derivatives. Estrogens regulate the feminization of the indifferent external genitalia, which also contain estrogen receptors.

Sexual differentiation in humans occurs during the 7th though 12th weeks of gestation (21). During the development and differentiation of the male and female reproductive systems, several windows exist during which environmental steroid agonists/antagonists could interfere with physiological development. One would anticipate that the exposure of a chromosomal male to antiandrogenic xenobiotics could interfere with the androgen-dependent differentiation of the structures derived from the Wolffian ducts or with the normal development of the male external genitalia. Estrogen receptors are also present in male reproductive tract tissues. Therefore, exposure of male embryos to estrogenic xenobiotics during critical developmental stages could interfere with androgen-dependent regulation of the tract by curtailing androgen action and/or by inappropriately stimulating estrogen-dependent processes. Since structures in the developing female reproductive tract contain androgen receptors, androgenic environmental toxicants could masculinize the female fetus. Inappropriate exposure of the female fetus to estrogenic toxicants could disrupt normal development the internal and external genitalia.

Anomalies of embryonic reproductive tract development occur spontaneously. Hypospadias is a condition that occurs in about one of every 300 human male infants. In this condition, the external urethral orifice is on the ventral surface of the glans penis (glandular hypospadias) or on the ventral surface of the shaft of the penis (penile hypospadias). Although more severe forms of hypospadias occasionally occur, these two constitute about 80% of the cases. These defects result from the failure of the glandular plate to canalize and/or failure of the urogenital folds to fuse. A rarer type of hypospadias, perineal hypospadias, occurs when the labioscrotal folds fail to fuse. In this case, the external urethral opening is located between the unfused halves of the scrotum. Because the external genitalia in this type of severe hypospadias are ambiguous, persons with perineal hypospadias and cryptorchidism (undescended testes) are sometimes diagnosed as male pseudohermaphrodites or as females. Micropenis is a condition in which the penis is so small that it is almost hidden by the suprapubic fat. Hypospadias and micropenis are both results of inadequate androgen production and/or androgen receptor levels. Failure of the lower parts of the paramesonephric ducts to fuse results in a double uterus. This may be associated with a double or single vagina. A bicornuate (double horned) uterus arises when the lower parts of the paramesonephric ducts fuse, but the upper portions do not. Absence of the vagina and uterus occurs about once in 4000 live human female births. Many of these developmental anomalies have been described in male and female wildlife species exposed to polluted environments. The molecular basis exists for environmental toxicants to cause these abnormalities in humans also.

The genital tracts of males and females are not fully differentiate and functional until the rising levels of steroid hormones that occur at puberty act on them. It is at puberty that the periodic recruitment and maturation of primordial follicles and the shedding of eggs occur in the ovary. The cyclic secretion of estradiol and progesterone by the ovary also begins at this time. The cyclic, steroid hormone-dependent regulation of the female reproductive tract is also initiated at puberty and continues throughout the reproductive phase of its lifecycle. Androgen-dependent initiation of spermatogenesis and regulation of the male reproductive tract organs also occur at puberty. The hormone-dependent functional activation and maintenance of the reproductive systems that occurs at puberty provide other windows of opportunity for agonist/antagonist environmental xenobiotics to disrupt normal development. One would anticipate that the greatest risks to reproductive health posed by environmental toxicants would be during the embryonic, neonatal and pubertal periods when the reproductive systems are undergoing finely tuned modulation by steroid hormones. It cannot be discounted, however, that endocrine disruptors could be effective in adults because the reproductive systems of this age group are far from static. Rather, the hormone-dependent changes that occur in the adult female reproductive tract can be thought of as a cyclic recapitulation of the differentiation process, therefore, its organs are likely targets for endocrine disrupting xenobiotics. Spermatogenesis is a dynamic process that continues throughout the lifetime of the male. It is likely that the steroid-dependent phases of this process always remain susceptible to disruption by environmental xenobiotics. Bear in mind that both female and male reproductive system organs contain androgen and estrogen receptors (25), therefore, endocrine disruption of either of these systems can occur via androgenic/antiandrogenic or estrogenic/antiestrogenic xenobiotics.

EVIDENCE FOR XENOBIOTIC-INDUCED EFFECTS ON REPRODUCTION

The first indication that environmental pollutants might pose a threat to reproduction was derived from field observations by wildlife biologists. They noted that the population of fish eating birds in the Great Lakes basin, an area highly contaminated with organochlorine compounds such as PCBs and DDT, was declining. These workers recorded generalized reproductive problems in the bird populations including thinning of egg shells, abnormal parental behavior, and poor hatchability. Many other investigators obtained further data by observing reproductive dysfunction in numerous species inhabiting areas contaminated with environmental toxicants (4, 5). These data indicated that fish, birds, reptiles, mammals, and other species inhabiting environments polluted with a number of known and unknown synthetic compounds suffered reproductive problems. Male chicks were observed that had oviducts and gonads resembling ovaries and female birds were found in which the oviducts had developed abnormally. It was noted that alligators inhabiting a Florida lake contaminated with the DDT metabolite, *p,p'*-DDE, had reduced penis size and lowered serum levels of testosterone. Field observations continue to be of value in detecting untoward effects of environmental xenobiotics on reproduction. For example, some male fish inhabiting streams that receive sewerage treatment plant effluents express the estrogen-dependent vitellogenin gene in their liver; this yolk protein gene is normally expressed only in female fish. Subsequent laboratory studies have demonstrated that nonylphenol, octylphenol, and some alkylphenol polyethoxlyates, known pollutants in the streams, can stimulate vitellogenin gene expression in trout hepatocytes (26). Thus, making a case for the involvement of these xenobiotics in the effects seen in the wild fish.

Many laboratory studies have determined that exposure of various animal species to environmental toxicants under controlled conditions can replicate some effects reported in wildlife (1, 2). These studies have enabled investigators to attribute effects to specific environmental agents. Unfortunately, relatively few whole animal studies have been conducted and many of these examined a limited number of reproductive parameters. Furthermore, there has been little consistency in the experimental model used. That is, various species have been examined and the animals have been studied at various stages of the lifecycle—*in utero,* prepubertal, adult. These facts complicate the interpretation of the some of the data obtained. In addition, these differences in experimental protocol, especially the lack of experiments in which *in utero* exposure was used, provide little data on the possible effects of environmental toxicants on the differentiation of the male and female reproductive tracts. Below, representative toxicants and their effects will be used to illustrate some of the disruptive effects that have been described by various investigators, for many of the original references see (1, 2, 27). The effects observed by the investigators are presented here in a descriptive manner. However, the toxicant effects are considered to be the result of the disruption of the steroid hormone-mediated physiological regulation of development and differentiation by mechanisms described above. While the terms "androgenic/antiandrogenic" and "estrogenic/antiestrogenic" are used to describe results obtained, they are meant to be descriptive rather than mechanistic.

Effects of Xenobiotics in the Male

The treatment of male rats *in utero* (by administration of the test compounds to the dam), prior to puberty, and as adults, with *p,p'*-DDE, a compound that binds to both the androgen and estrogen receptors (20), yielded the following results. *In utero* treatment resulted in male pups with a reduced anogenital distance and retained thoracic nipples, both characteristics of female pups. Treatment of prepubertal males delayed the onset of puberty, and treatment of adults led to a reduction in the weights of the seminal vesicle and ventral prostate as compared to controls. Neither treatment of prepubertal or adult rats with *p,p'*-DDE had an effect on the serum level of testosterone. While the investigators interpreted their data as being the result of antiandrogenic effects of the test compound (28), estrogenic effects resulting in nipple retention and reduced anogenital distance cannot be discounted.

The fungicide vinclozolin has been shown to interfere with sexual differentiation in male rats. Pregnant rats were treated with the compound beginning on the gestational day when the embryos would be at the ambisexual stage of development. At birth, the males had female-like anogenital distances; at 2-wks of age prominent nipple development was present. Many of the males had undescended testes, a vaginal pouch, and small to absent accessory sex organs. All males had a cleft phallus with hypospadias. Some of the findings of this study resemble results obtained when male animals are treated with the anitandrogen flutamide. Further studies have shown that the compound active in producing the results described was not vinclozolin *per se*, but one of its *in vivo* metabolites

The concept of *in vivo* metabolism of toxicants is important to bear in mind. A xenobiotic that is active *in vitro* may not be active *in vivo* because biotransformation may inactivate it. In contrast, a compound that is inactive *in vitro* may be metabolically transformed into an active compound *in vivo*. These considerations make it clear that one cannot rely solely on *in vitro* screening to determine if a compound has endocrine-disrupting potential. Compounds that are active *in vitro* must be tested *in vivo* to determine if deleterious effects are obtained. The absence of activity *in vitro* does not mean that the xenobiotic will lack *in vivo* activity.

Treatment of male rats with methoxychlor, a DDT congener, reduced the weight of the seminal vesicles, cauda epididymis, and pituitary gland. It also decreased the sperm content of the epididymis, and at higher doses, delayed the onset of puberty. Despite these changes, the animals were fertile.

Linuron treatment of male rats results in a statistically significant reduction in the weight of the epididymis and accessory sex organs in sexually immature animals. It causes a reduction in the weight of the accessory sex organs and the prostate in adults. Increased serum levels of estradiol and luteinizing hormone were observed in sexually mature rats treated with Linuron. The alterations caused by Linuron are similar to those produced by flutamide. Linuron has been shown to compete with testosterone for binding to the androgen receptor, so its effects may be due to inhibiting the action of endogenous androgens.

Alkylphenols also have detrimental effects on male reproductive parameters. Octylphenol administered to adult male rats suppressed testicular function as indicated by reduced testicular and prostate size, suppressed spermatogenesis and resulted in low circulating levels of testosterone. Octylphenol administration also decreased the weight and altered the histological features of the epididymis and increased the proportion of abnormal spermatozoa. These alterations of male reproductive function are presumed to be the result of estrogenic effects of the compound, but antiandrogenic effects cannot be excluded.

PCBs are also known to affect male reproduction. Male rat pups receiving PCBs via the milk of their lactating dams have been shown to have reduced testosterone levels and a reduced number of Leydig cells per testis. Gestational and lactational exposure of male mice to PCBs was shown to have no effect on anogenital distance or testis weight, but at 16 wks of age, the mice were not able to produce sperm capable of fertilizing eggs *in vitro* (29).

Effects of Xenobiotics in the Female

A metabolite of DDT, *o,p*'-DDT, competitively inhibits the binding of estradiol to the estrogen receptor. When administered to immature female rats, it stimulates DNA synthesis and cell division in uterine epithelial and stromal cells in the same manner as estradiol. *O,p*'-DDT produces uterine hyperplasia, a characteristic estrogenic reaction (1). The DDT metabolite *p,p*'-DDE, which has been thought of as an antiandrogen (28), also has estrogenic activity. When administered to castrated adult guinea pigs, *p,p*'-DDE maintained the weight of the female reproductive tract at control, non-castrate levels, Table 1. As can also be noted in Table 1, nonylphenol and diethylstilbestrol (DES) had a similar effect on organ weight. Pentachlorophenol was ineffective at maintaining reproductive tract weight of castrated animals at intact levels. Administration of *p,p*'-DDE to intact or castrated adult guinea pigs resulted in hypertrophy of the vagina, cervix, and uterus. Mucous metaplasia was seen in the uterus, Figure 6, and cyst formation occurred in the rete ovarii (30).

These abnormalities clearly point to disruptive effects of *p,p*'-DDE on the uterus and ovary. The fact that *p,p*'-DDE was able to maintain, or stimulate, the histological features of the reproductive tract of the castrated adult female guinea pig clearly points to estrogenic rather than antiandrogenic effects. The results suggest that when p,p'-DDE binds to the estrogen receptor it acts as an estrogen. Nonylphenol and DES treatment also resulted in alterations of genital tract morphology in the adult female guinea pig (30). The administration of methoxychlor to immature female rats delayed the age of vaginal opening and the first estrus. Anomalies of the reproductive tract were also noted in methoxychlor-treated rats.

Table 1. The Effects of Various Treatments on the Weight of the Reproductive Organs of the Adult Female Guinea Pig

Treatment	Reproductive tract weight (g)*	Ovary weight (mg)
Intact control (10)	[a]1.8 ± 0.14**	[a]50 ± 5
Castrate control (5)	[b]0.9 ± 0.03	-----
Intact p,p'-DDE (4)	[a]2.04 ± 0.09	[a]$60 + 20$
Castrate p,p'-DDE (4)	[a]1.89 ± 0.31	-----
Intact nonylphenol (4)	[c]2.37 ± 0.16	[a]$70 + 20$
Castrate nonylphenol (4)	[c]2.78 ± 0.26	-----
Intact DES (3)	[c]2.90 ± 0.07	[a]45 ± 7
Castrate DES (3)	[d]3.63 ± 0.17	-----
Intact PEN (6)***	[a,c]2.36 ± 0.38	[a]60 ± 20
Castrate PEN (5)	[b]0.99 ± 0.03	-----

*The reproductive tract consists of the vagina, cervix, and uterus, which were removed from the animals as a single unit. The ovarian weight is the weight of both ovaries.
**The values presented are the mean and SEM of (n) animals. In any column, values with the same superscript are statistically identical.
***PEN, pentachlorophenol.

In female rats, subcutaneous administration of octylphenol induced estrogenic effects on the reproductive system. It altered the estrous cycle and increased uterine weight. Oral administration of octylphenol caused a significant increase in uterine weight of immature and ovariectomized rats. Bisphenol A was shown to compete with [^{3}H]estradiol for binding to the estrogen receptor. It has been shown to induce the prolactin gene, but at a 1000-5000-fold lower potency than estradiol. Bisphenol A, octylphenol, and nonlyphenol induce morphological and molecular changes in the rat uterus that are similar to those caused by estradiol (31).

It is likely that all of the reproductive anomalies described above in both the male and female are brought about by environmental toxicants interfering with the physiological regulation of the reproductive system by steroid hormones. Examples of the effects of various environmental

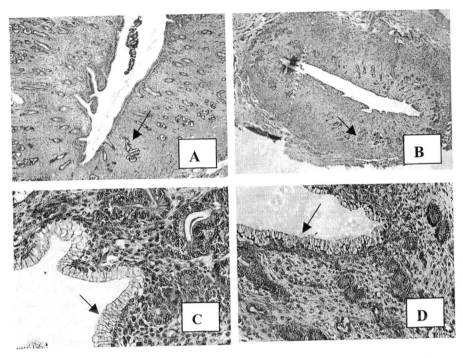

Figure 6. The effects of various treatments on the histological appearance of the adult guinea pig uterus. A, uterus from an intact animal at the estrus stage of the cycle; arrow indicates dilated glands. B, uterus from a 14-day castrated animal; arrow indicates atrophied uterine glands. C, uterus from an intact animal treated with *p,p'*-DDE, note the generalized hypertrophy of the tissue and the mucous metaplasia, arrow, of the epithelium. D, uterus from a 14-day castrated animal treated with *p,p'*-DDE, note the hypertrophy and the mucous metaplasia, arrow.

toxicants on reproductive parameters in several species are summarized in Table 2.

Effects of Environmental Xenobiotics on Humans

The evidence that environmental toxicants have affected human reproductive health is circumstantial. It was reported that there has been a doubling in the frequency of undescended testes in England and Wales between 1962 and 1981. A more recent study showed that the incidence of undecended testes had increased by 65% over the past 2 decades. Hypospadias in boys has more than doubled in the past 40 years. There have been reports that sperm quantity and quality in human males has declined over the past 50 years (8). This reported deterioration in sperm parameters has been controversial. It may be due to regional (country) differences, differences in the way semen parameters have been evaluated, differences in the populations being evaluated, or simply to differences in the statistical methods used to evaluate the parameters, rather than being an actual phenomenon. One of the most adverse trends in male reproductive health over the past few decades has been an increase in the incidence of testicular cancer in several countries (8, 32). There is evidence that the ratio of male to female births is declining in several industrialized countries; the authors speculate that environmental factors may be responsible for the decline. Various investigators have postulated that many disorders in male reproductive function arise from exposure of male embryos to environmental estrogens. This postulate is compelling since the disorders that have been described in the male resemble those one might expect from estrogenic effects on the developing male reproductive system. However, one cannot exclude the possibility that some estrogen-like deficits may actually be the result of antiandrogenic xenobiotics inhibiting the action of physiological androgens on male development (1, 8). Few, if any, abnormalities of the human female reproductive tract have yet been attributed to exposure to environmental hormones/antihormones.

Clinical use of the potent synthetic estrogen DES provides human data that can be compared to those obtained in experimental systems. Clinical studies have demonstrated that DES, which was administered to women with a history of spontaneous abortions with the incorrect expectation that it would prevent these abortions, resulted in a statistically significant number of male and female offspring with reproductive tract abnormalities. Eighteen percent of female patients who were exposed *in utero* to DES had irregular menstrual cycles as compared to 10% in controls. These women also had a lower incidence of pregnancy than in the non-exposed group. Abnormalities of the vagina and cervix were also noted in the exposed women. Reproductive tract abnormalities were also noted in the male offspring of DES-treated women as compared to the male offspring of women that were not treated. These included epididymal cysts, hypotrophic testes, low ejaculate volume and reduced sperm concentration. "Severely pathologic" semen was found in 28% of the DES exposed offspring as compared to 0% in the control group (33). Many of the pathological findings in the reproductive tracts of male and female offspring of DES-treated mothers resemble the defects discussed above in animals that were exposed to environmental xenobiotics under natural and experimental conditions. Had DES been administered consistently to the women during the ambisexual stage of development of their fetuses, greater developmental defects of the genital tracts may have been observed. These findings with DES indicate that the human embryo is a target for exogenous hormonally active agents and imply that the reproductive tract abnormalities in humans can result from exposure to such agents.

Theoretical Considerations

As the reader will have noted, most of the reproductive tract abnormalities attributed to estrogenic environmental toxicants have been reported in male animals and humans. A possible explanation for these observations is that the hormone binding sites on the estrogen receptor of the female reproductive system are occupied by endogenous estradiol and that no, or only a few sites, are available for occupancy by exogenous xenobiotics. Occupancy of the available sites might bring about subtle changes in the reproductive tract that are not distinguishable from normal variation. This could explain why only minor effects of xenobiotic treatment were noted in intact female guinea pigs, while dramatic effects were seen in castrated animals, which are depleted of endogenous estradiol (30). If this is the case, it is possible that during the post menopausal period, when endogenous estradiol is low, the human female may become more susceptible to the disruptive actions of estrogenic environmental toxicants. On the other hand, it is likely that the binding sites of estrogen receptors present in male reproductive tract tissues are not fully occupied by endogenous estradiol. Therefore, these sites can be filled with exogenous estrogenic environmental xenobiotics, resulting in more readily detectable disruption of the reproductive tract. These considerations imply that receptor occupancy may be an important criterion for judging when environmental toxicants may be disruptive. No environmental hormonally active agent has yet been shown to be androgenic. If such agents exist, one would expect them to virilize the female reproductive tract, but to cause only subtle changes in the male reproductive tract. The effects that antiandrogenic toxicants would have on the female reproductive tract are not clear, because androgen-regulated processes in the female are not well described.

The human and animal populations of the planet are exposed simultaneously to all of the toxicants that present in the environment. Many environmental toxicants can act through steroid hormone receptor pathways. The fact that xenobiotics can impinge on these systems provides molecular mechanisms for amplifying their hazardous potential. That is, even if an individual toxicant is weak and is present in low concentration, classes of compounds, acting through common molecular mechanisms, would pose the same risk as exposure to a high concentration of a single toxicant. The question of the environmental relevance of a specific amount of a given toxicant does not seem to be germane in this context, because, as stated above, we are not exposed to one toxicant at a time in the environment.

Thus, even if specific toxicants are weak and relatively rare, their sheer number, their variety, their ability to bioaccumulate, and their ability to act through common molecular mechanisms, reinforces the notion that they may pose a risk to human reproductive health. The manufacture and use of many environmental toxicants such as DDT and PCBs has been banned in industrial countries. The long half-lives of these compounds, however, means that they will be around for a long time. Many compounds that are currently being manufactured are now known to be reproductive toxicants. Furthermore, new compounds are always being produced that may be as harmful, or more harmful, than those that have been banned. Thus, it would appear that the environmental burden of potential endocrine-disrupting chemicals is actually increasing.

To evaluate the potential effects of environmental toxicants on human reproduction, physicians and other health care workers should have a heightened awareness of the possibility that clinical signs, such as precocious puberty, decreased sperm counts, anatomical anomalies of the reproductive tract, decreased fertility, and so on, may be attributable to the endocrine-disrupting effects of environmental chemicals. The establishment of national and state registries to which such data could be reported would help to determine if the incidence of these disorders is increasing among the human population. In-depth surveys of people living in, or who have lived, in areas highly contaminated with environmental chemicals should be taken to determine if long-term effects of these agents on reproductive function are occurring. Since clear-cut effects of environmental toxicants on animals living in such environments have been established, it would be unexpected if such disorders did not also occur in humans exposed to similar conditions.

The dietary intake of man-made estrogenic chemicals is low compared to the dietary intake of the natural hormonally active agents present in plants. This fact coupled with the low estimated hormonal potency of man-made chemicals has led to the argument that estrogenic xenobiotics pose little potential danger to humans (34). While this argument is worthy of consideration, it fails to take into account that some environmental endocrine-disrupting agents are antiestrogens or androgens/antiandrogens. Nor does it take into account the ability of classes of xenobiotics to acting through common molecular mechanisms to effectively increase the body burden of toxicants. Furthermore, the possibility that xenobiotics do not result in the same kind of transactivation as occurs with natural hormonal compounds is not considered. The vertebrate body possesses enzymes that can safely detoxify natural compounds, but they may not be able to do so with some of the xenobiotics, thereby allowing them to accumulate and/or to be converted to harmful metabolic products. The rise in the incidence of reproductive anomalies in human populations that has occurred during the

Table 2. Examples of reproductive deficits attributed to endocrine disruptors

Species	Observation	Contaminant
Mammals		
Humans	Gynecomastia, oligospermia, impotence, hypogonadism, decreased libido, reduced sperm counts and motility, menstrual cycle irregularities	DDT, kepone, oral contraceptive exposure, stilbene derivatives
Cattle	Infertility	Coumestrol
Sheep	Infertility, dystocia	Isoflavonoids, coumestans
Seals	Impaired reproductive functions	PCBs
Mink	Population decline, developmental toxicity, hormonal alterations	PCBs, dioxins
Rabbits	Infertility, failure of ovulation, failure of implantation	Isoflavonoids
Guinea pigs	Infertility	Isoflavonoids, coumestans
Mice	Proliferative lesions, reproductive tract tumors, infertility, inhibition of estrus, inhibition of ovulation	DES, isoflavonoids
Birds		
Japanese quail	Abnormal reproductive behavior, hematology, and feather morphology	o,p'-DDT
Gulls	Abnormal development of ovarian tissue and oviducts in male embryos	o,p'-DDT
Waterbirds	Egg shell thinning, mortality, developmental abnormalities, growth retardation	DDE, PCBs, AhR agonists
Reptiles		
Alligators	Abnormal gonads, decreased phallus size, altered sex hormone levels	o,p'-DDT, p,p'-DDE, dicofol
Red-eared slider turtle	Anomalous reproductive development	*trans*-Nonachlor, *cis*-Nonachlor, arochlor 1242, p,p'DDE, chlordane
Fish		
Mosquito fish	Abnormal expression of secondary sex characters, masculinization	Androstenedione
Roach	Hermaphroditism, vitellogenin in males, altered testes development	Sewage effluent mixture
Lake trout	Early mortality, deformities, blue sac disease	Dioxin, related AhR agonists
White sucker	Reduced sex steroid levels, delayed sexual maturity, reduced gonad size	Bleached kraft plup, mill effluent mixtures
Flatfish	Decreased hormone levels, reduced ovarian development, reduced egg/larvae viability	PAHs
Invertebrates		
Snails	Masculinization, imposex, formation of additional female organs, malformed oviducts, increased oocyte production	Tributyltin, bisphenol A, octylphenol
Marine copepods	Stimulate sexual maturation and egg production	Bisphenol A
Daphnia magna	Delayed molting time	PCB29, arochlor 1242, diethyl phthalate

AhR, arylhydrocarbon receptor; DES, diethylstilbestrol; DDE, dichlorodiphenyldichloroethylene;, DDT, dichlorodiphenyltrichloroethane; PAH, polyaromatic hydrocarbon; PCB, polychlorinated biphenyl. Table, adapted from McLachlan (2), and used with permission.

398

last 50 years does not appear to parallel increases in natural hormone mimics or changes in eating habits, but it does parallel increases in industrial and agricultural xenobiotics. It would be rash to dismiss the possibility that environmental xenobiotics have harmful effects on human reproductive health unless and until evidence proves otherwise.

CONCLUSIONS

Evidence from epidemiological and experimental studies implicates hormonally active compounds in the environment, especially man-made chemicals, as causative agents in the disruption of the reproductive systems of many animal species. There is no direct evidence that human reproductive health has been compromised by environmental xenobiotics. However, circumstantial evidence that such effects have occurred/are occurring is accumulating. Steroid hormones exquisitely regulate the differentiation, development, and function of the reproductive systems of mammalian species. Many hormonally active environmental agents bring about their disruptive effects by acting as steroid hormone mimics or antagonists, thereby interfering with the physiological regulation of reproductive processes, in all species studied. Steroid hormone receptors have an affinity for environmental toxicants that is orders of magnitude less than that for physiological ligands. However, the vast array of toxicants in the environment and the ability of many of them to act through steroid hormone-dependent pathways effectively increases their concentration and consequently their potential danger. It would be unexpected if human reproductive systems, which are regulated by the same mechanisms as in other species, were not adversely affected by environmental toxicants.

ACKNOWLEDGEMENTS

I thank Dr. T. dePaulis for drawing the chemical structures and Drs. D.R.P. Tulsiani and W.J. Kovacs for critically evaluating this manuscript. I am indebted to Ms. Lynne Black for her editorial and technical assistance.

2.   McLachlan JA. Environmental signaling: what embryos and evolution teach us about endocrine disrupting chemicals. Endocr Rev 2001; 22: 319-341.

3.   Craan AG, Haines DA. Twenty-five years of surveillance for contaminants in human breast milk. Arch Environ Contam Toxicol 1998; 35: 702-710.

4.   Colborn T, Clemmens C. Chemically induced alterations in sexual functional development: the wildlife/human connection. Princeton Scientific Publishing. 1992: Princeton, NJ.

5.   National, Research, Council. Hormonally Active Agents in the Environment. Washington, D.C.: National Academy Press; 1999.

6.   Glazier MG, Bowman MA. A review of the evidence for the use of phytoestrogens as a replacement for traditional estrogen replacement therapy. Arch Intern Med 2001; 161: 1161-1172.

7.   Diel P, Schulz T, Smolnikar K, Strunck E, Vollmer G, Michna H. Ability of xeno- and phytoestrogens to modulate expression of estrogen-sensitive genes in rat uterus: estrogenicity profiles and uterotropic activity. J Steroid Biochem Mol Biol 2000; 73: 1-10.

8.   Toppari J, Larsen JC, Christiansen P, Giwercman A, Grandjean P, Guillette LJ, Jr., Jegou B, Jensen TK, Jouannet P, Keiding N, Leffers H, McLachlan JA, Meyer O, Muller J, Rajpert-De Meyts E, Scheike T, Sharpe R, Sumpter J, Skakkebaek NE. Male reproductive health and environmental xenoestrogens. Environ Health Perspect 1996; 104: 741-803.

9.   DeRosa C, Richter P, Pohl H, Jones DE. Environmental exposures that affect the endocrine system: public health implications. J Toxicol Environ Health B Crit Rev 1998; 1: 3-26.

10.   Krimsky S. Hormonal Chaos. Baltimore: The Johns Hopkins Press; 2000.

11.   Colborn T, vom Saal FS, Soto AM. Developmental effects of endocrine-disrupting chemicals in wildlife and humans. Environ Health Perspect 1993; 101: 378-384.

12.   Connor K, Ramamoorthy K, Moore M, Mustain M, Chen I, Safe S, Zacharewski T, Gillesby B, Joyeux A, Balaguer P. Hydroxylated polychlorinated biphenyls (PCBs) as estrogens and antiestrogens: structure-activity relationships. Toxicol Appl Pharmacol 1997; 145: 111-123.

13.   Sonnenschein C, Soto AM. An updated review of environmental estrogen and androgen mimics and antagonists. J Steroid Biochem Mol Biol 1998; 65: 143-150.

14.   Sumpter JP. Reproductive effects from oestrogen activity in polluted water. Arch Toxicol Suppl 1998; 20: 143-150.

15.   Fontana A, Delmas PD. Clinical use of selective estrogen receptor modulators. Curr Opin Rheumatol 2001; 13: 333-339.

16.   Diel P, Olff S, Schmidt S, Michna H. Molecular identification of potential selective estrogen receptor modulator (serm) like properties of phytoestrogens in the human breast cancer cell line mcf-7. Planta Med 2001; 67: 510-514.

17.   Aranda A, Pascual A. Nuclear hormone receptors and gene expression. Physiol Rev 2001; 81: 1269-1304.

18.   Steinmetz AC, Renaud JP, Moras D. Binding of ligands and activation of transcription by nuclear receptors. Annu Rev Biophys Biomol Struct 2001; 30: 329-359.

19.   Lee JW, Lee YC, Na SY, Jung DJ, Lee SK. Transcriptional coregulators of the

21.	Moore KL, Persaud TVN. The developing human: clinically oriented embryology. W.B. Saunders. 1993: Philadelphia, 281-303.22. Foster JW, Dominguez-Steglich MA, Guioli S, Kowk G, Weller PA, Stevanovic M, Weissenbach J, Mansour S, Young ID, Goodfellow PN, et al. Campomelic dysplasia and autosomal sex reversal caused by mutations in an SRY-related gene. Nature 1994; 372: 525-530.

23.	Ikeda Y, Shen WH, Ingraham HA, Parker KL. Developmental expression of mouse steroidogenic factor-1, an essential regulator of the steroid hydroxylases. Mol Endocrinol 1994; 8: 654-662.

24.	Mahendroo MS, Cala KM, Hess DL, Russell DW. Unexpected virilization in male mice lacking steroid 5alpha-reductase enzymes. Endocrinology 2001; 142: 4652-4662.

25.	Williams K, McKinnell C, Saunders PT, Walker M, Fisher JS, Turner KJ, Atanassova N, Sharpe M. Neonatal exposure to potent and environmental oestrogens and abnormalities of the male reproductive system in the rat: evidence for importance of the androgen-oestrogen balance and assessment of the relevance to man. Hum Reprod Update 2001; 7: 236-247.

26.	Andersen HR, Andersson AM, Arnold SF, Autrup H, Barfoed M, Beresford NA, Bjerregaard P, Christiansen LB, Gissel B, Hummel R, Jorgensen EB, Korsgaard B, Le Guevel R, Leffers H, McLachlan J, Moller A, Nielsen JB, Olea N, Oles-Karasko A, Pakdel F, Pedersen KL, Perez P, Skakkeboek NE, Sonnenschein C, Soto AM, Sumpter JP, Thorpe SM, Grandjean P. Comparison of Short-Term Estrogenicity Tests for Identification of Hormone-Disrupting Chemicals. Environ Health Perspect 1999; 107: 89-108.

27.	Kelce WR, Gray LE, Wilson EM. Antiandrogens as environmental endocrine disruptors. Reprod Fertil Dev 1998; 10: 105-111.

28.	Kelce WR, Stone CR, Laws SC, Gray LE, Kemppainen JA, Wilson EM. Persistent DDT metabolite p,p'-DDE is a potent androgen receptor antagonist. Nature 1995; 375: 581-585.

29.	Fielden MR, Halgren RG, Tashiro CH, Yeo BR, Chittim B, Chou K, Zacharewski TR. Effects of gestational and lactational exposure to Aroclor 1242 on sperm quality and in vitro fertility in early adult and middle-aged mice. Reprod Toxicol 2001; 15: 281-292.

30.	Danzo BJ, Shappell, H.W., Banerjee, A., Hachey, D.L. Effects of nonlyphenol, 1,1-dichloro-2,2(bis(p-chlorlphenyl)ethylene (p,p'-DDE), and pentachlorophenol on the adult guinea pig reproductive tract. Reproductive Toxicology 2002; 16: 29-43.

31.	Laws SC, Carey SA, Ferrell JM, Bodman GJ, Cooper RL. Estrogenic activity of octylphenol, nonylphenol, bisphenol A and methoxychlor in rats. Toxicol Sci 2000; 54: 154-167.

32.	Mills PK. Correlation analysis of pesticide use data and cancer incidence rates in California counties. Arch Environ Health 1998; 53: 410-413.

33.	Bibbo M, Gill WB, Azizi F, Blough R, Fang VS, Rosenfield RL, Schumacher GF, Sleeper K, Sonek MG, Wied GL. Follow-up study of male and female offspring of DES-exposed mothers. Obstet Gynecol 1977; 49: 1-8.

34.	Safe SH. Environmental and dietary estrogens and human health: is there a problem? Environ Health Perspect 1995; 103: 346-351.

INDEX